Physics, Logic, and History

Contributors

Hermann Bondi
Julian V. Langmead Casserley
Robert S. Cohen
George Gamow
Jaakko Hintikka
Dmitri D. Ivanenko
David Kaplan
Imre Lakatos
Alfred Landé
Czesław Lejewski
André Mercier
Arne Naess
Richard H. Popkin
Sir Karl Popper
William Van Orman Quine
György Ránki
Herman Tennessen
Håkan Törnebohm
Hans-Jürgen Treder
Jean-Pierre Vigier

Physics, Logic, and History

Based on the First International Colloquium held
at the University of Denver, May 16-20, 1966

Edited by

Wolfgang Yourgrau

and

Allen D. Breck

University of Denver
Denver, Colorado

℗ PLENUM PRESS • NEW YORK–LONDON • 1970

Library of Congress Catalog Card Number 68-32135

ISBN-13: 978-1-4684-1751-7   e-ISBN-13: 978-1-4684-1749-4
DOI: 10.1007/978-1-4684-1749-4

© 1970 Plenum Press, New York
Softcover reprint of the hardcover 1st **edition** 1970
A Division of Plenum Publishing Corporation
227 West 17th Street, New York, N.Y. 10011

United Kingdom edition published by Plenum Press, London
A Division of Plenum Publishing Corporation, Ltd.
Donington House, 30 Norfolk Street, London W.C. 2, England

This volume is dedicated with respect and affection to the memory of George Gamow

FOREWORD

It is a trite and often lamented fact that every academic discipline suffers from the malady of overspecialization and expertise. Who, in his scholarly experience, has not encountered technical gibberish and the jargon of the pundit?

The contributors to this work have attempted to remove the artificial barriers between these respective disciplines. The purpose of this volume is to explore the ever present links between logic, physical reality, and history. Indeed there are not two or three or four cultures: there is only one culture; our generation has lost its awareness of this. Though serious, it is not tragic. All we need is to free ourselves from the fetters of mere "technicalese" and search for a comprehensive interpretation of logical and physical theories.

Historians, logicians, physicists — all are banded in one common enterprise, namely in their desire to weave an enlightened fabric of human knowledge.

It is a current, and perhaps welcome, trend in philosophic inquiry to de-psychologize systems, methods, and theories. However, there is an equally fashionable tendency to minimize or even eschew the historical aspects of logical and physical theories, and analogously, there is a deep-seated mistrust among physicists and cosmologists against the seemingly pure abstractions of logical formalisms.

We are convinced that logic, physical reality, and history form one rational unity. We shall not try to build imaginary bridges between the three topics chosen for this work. We intend to show that real bridges have always existed and that without them the *edifice of human knowledge will remain unfinished.*

The papers printed here differ somewhat in form and in content from the original presentations at the International Colloquium on Logic, Physical Reality, and History held in Denver in 1966. Authors of papers and commentators have availed themselves of the opportunity to revise, i.e. expand or diminish, their remarks. The discussion section following each paper has not been unduly edited, as it was decided that the give-and-take of discussants should be left more or less in the style and form in which the dialogue originally evolved.

The editing of these proceedings proved, in many aspects, an exasperating task. For instance, some scholars highly specialized in their respective domains were compelled to simplify their theories, and it was our duty to see that any vulgarization was minimized as much as possible. Furthermore, all of us were guilty of certain intellectual sins: on occasion, the logicians succeeded in escaping the fine net of logic proper, the physicists fell prey to opacity of articulation, and the historians sometimes flirted with ideas or facts not worth cogitating. Be all this as it may, we hope that the reader will not encounter any truly staggering triviality, but, at worst, here and there some eccentric, bizarre, or bold viewpoints — at best some startling insights.

Best wishes and hopes for great success of this International Colloquium were expressed by the following colleagues who were unable to attend the meeting: Max Black, Marshall Clagett, Henry Steele Commager, A. C. Crombie, Herbert Feigl, E. H. Gombrich, Eric Goldman, F. A. von Hayek, Werner Heisenberg, Carl G. Hempel, Leon Henkin, William Kneale, György Lukacs, Maurice Mandelbaum, Henry Margenau, Ernest Nagel, J. Robert Oppenheimer, Wesley C. Salmon, Wolfgang Stegmüller, James O. Urmson, Sin-Itiro Tomonaga, and Eugene P. Wigner.

The Organizing Committee consisted of Dr. A. D. Breck, Dr. E. A. Lindell and Dr. W. Yourgrau. Dr. Lindell is the Dean of the Faculty of Arts and Sciences. The editors wish to thank him most sincerely for his devotion and tireless aid. Further, we also feel indebted to Dr. Wilbur C. Miller, the Vice Chancellor for Academic Affairs, for all the assistance he gave to the Organizing Committee. The Colloquia were made possible by grants from the Martin-Marietta Corporation Foundation and the University of Denver. Needless to say, without this financial support our experiment could never have been realized. We also wish to express gratitude to our colleague, Dr. A. J. van der Merwe, for his unflagging willingness to aid us in the final stage of scrutiny of the whole MS.

University of Denver   The Editors
Denver, Colorado
May, 1969

PARTICIPANTS

HERMANN BONDI, Mathematics, King's College, University of London. Co-founder with Gold (and Hoyle) of the "Steady-State" theory of the universe. Cosmology, relativity theory, astrophysics, mathematical physics.

ALLEN D. BRECK, History, University of Denver. Co-chairman of the Colloquium. Medieval history, historiography, philosophy of history.

JULIAN V. LANGMEAD CASSERLEY, Philosophy of Religion, Seabury-Western Theological Seminary. Apologetics, philosophy of history, moral dimensions of the social sciences.

ROBERT S. CHASSON, Physics, University of Denver. Cosmic ray research, planetary physics.

ROBERT S. COHEN, Physics and Philosophy, Boston University. Editor, *Boston Studies in the Philosophy of Science*; history and philosophy of science, plasma physics, relations of history and physics.

GEORGE GAMOW, Physics, University of Colorado. Founder of the "Big Squeeze and Explosion" theory of the universe and protagonist of molecular biology. Theoretical and nuclear physics, cosmology, relation between micro- and macro-physics, co-founder with Condon (and Gurney) of the "Tunnel Effect".

YUSUF FADL HASAN, Research Unit in the Sudan, University of Khartoum. Middle East migration into North Africa.

JAAKKO HINTIKKA, Philosophy, Universities of Helsinki and Stanford. Editor-in-chief of *Synthèse*. Semantics, mathematical logic, epistemology, history of philosophy.

DMITRI D. IVANENKO, Physics, The Moscow State University. Classical and quantum field theory, nuclear physics, relation between micro- and macro-physics, co-founder with Tamm of the theory of specific nuclear energies.

DAVID KAPLAN, Philosophy, University of California, Los Angeles. Symbolic logic, semantics, philosophy of science.

DONALD KEYES, Philosophy and theology, Trinity College, University of Toronto. History of philosophy, Christian existentialism, logic.

GEORGE KREISEL, Mathematics and Philosophy, Universities of Paris and Stanford. Mathematical logic, proof theory in general; author of "Kreisel Theorems".

IMRE LAKATOS, Philosophy, Logic, and Scientific Method, at the London School of Economics. History and philosophy of mathematics, logic, philosophy of science.

ALFRED LANDÉ, Theoretical Physics, Emeritus, The Ohio State University. Atomic structure, quantum theory, spectral lines, Zeeman Effect, multiplet theory, "Landé g-factor".

CZESŁAW LEJEWSKI, Philosophy, University of Manchester. Logic (traditional and modern), semantics, ancient philosophy.

ANDRÉ MERCIER, Institute of Theoretical Physics, University of Berne. Secretary General of the International Committee on General Relativity and Gravitation. Philosophical interpretation of physics, mathematical physics, theory of knowledge.

ARNE NAESS, Institute of Philosophy and the History of Ideas, University of Oslo. Editor-in-chief of *Inquiry*. Symbolic logic, history of scientific ideas, theory of knowledge, experimental psychology.

JOHN U. NEF, Center for Human Understanding, University of Chicago. Founder of the Center and of the Committee on Social Thought. Relations between the natural, social, and behavioral sciences and the arts of literature, painting, music, and architecture.

GEORGE POLYA, Mathematics, Stanford University. Induction, inference, proof theories, history of mathematics, problems in pure and applied mathematics.

RICHARD H. POPKIN, Philosophy, University of California, San Diego. Editor of *The Journal of the History of Philosophy*. History of philosophy in the 16th and 17th centuries, history of culture.

SIR KARL POPPER, Philosophy, Logic, and Scientific Method, at the London School of Economics. Philosophy of science, logic, philosophy of history, ethics, foundations of mathematics and physics.

WILLARD V. O. QUINE, Peirce chair in Philosophy, Harvard University. Mathematical logic, theory of knowledge, philosophy of logic, author of "Quine Theorems".

GYÖRGY RÁNKI, Historical Institute, Hungarian Academy of Sciences. History of Europe, economic history in the 19th and 20th centuries, especially East-Central European history.

HERMAN TENNESSEN, Philosophy, Co-Director of the Center for Advanced Studies in Theoretical Psychology, University of Alberta. Logic, philosophy of language, social research, philosophy of science.

HÅKAN TÖRNEBOHM, Methods of Science, University of Göteborg. Logical foundations of relativity theory, explanations in history, history and philosophy of science.

HANS-JÜRGEN TREDER, Mathematics and Theoretical Physics, German Academy of Sciences. Relativity theory, theories of gravitation, mathematical physics, history and foundations of physics.

JEAN-PIERRE VIGIER, Institut Henri Poincaré, Paris. Co-founder with Louis de Broglie and Bohm of a new elementary-particle model; atomic physics, general relativity theory, theory of levels in physical reality.

TOSHITAKA YADA, Law, Hokkaido University, Japan. Habsburg Monarchy in the 19th century, modern European and especially German history, political science.

WOLFGANG YOURGRAU, Philosophy of Science, University of Denver. Co-chairman of the Colloquium. Quantum theory, general relativity, thermophysics, variational principles; co-editor of *Foundations of Physics.*

CONTENTS

FOREWORD
The Editors   v
I. A REALIST VIEW OF LOGIC, PHYSICS, AND
HISTORY Sir Karl Popper, *University of London*   1
II. KNOWLEDGE AND PHYSICAL REALITY
André Mercier, *University of Berne*   39
III. THE SCIENCE OF HISTORY AND NOTIONS OF
PERSONALITY
Herman Tennessen, *University of Alberta*.   59
IV. THE GROWTH OF A THEORETICAL MODEL: A
SIMPLE CASE STUDY
Håkan Törnebohm, *University of Göteborg*   79
V. EXISTENCE
Willard Van Orman Quine, *Harvard University*.   89
VI. THE PROBLEMS OF UNIFYING COSMOLOGY
WITH MICROPHYSICS
Dmitri D. Ivanenko, *The Moscow State University*   105
VII. THE EVOLUTION OF EVOLUTION
Julian Victor Langmead Casserley, *Seabury-Western*
Theological Seminary   115
VIII. PLEA FOR PLURALISM IN PHILOSOPHY AND
PHYSICS
Arne Naess, *University of Oslo*.   129
IX. ON SEMANTIC INFORMATION
Jaakko Hintikka, *University of Helsinki and Stanford*
University   147
X. QUANTIFICATION AND ONTOLOGICAL
COMMITMENT
Czesław Lejewski, *University of Manchester*   173
XI. POSSIBLE INTERNAL SUBQUANTUM MOTIONS
OF ELEMENTARY PARTICLES
Jean-Pierre Vigier, *Institut Henri Poincaré*   191
XII. THE THREE KINGS OF PHYSICS
George Gamow, *University of Colorado*   203

XIII. SCEPTICISM AND THE STUDY OF HISTORY
Richard H. Popkin, *University of California, San Diego* ... 209
XIV. CAUSATION IN HISTORY
Robert S. Cohen, *Boston University* 231
XV. RELATIVITY THEORY AND HISTORICITY OF
PHYSICAL SYSTEMS
Hans-Jürgen Treder, *German Academy of Sciences* 253
XVI. GENERAL RELATIVITY AS AN OPEN THEORY
Hermann Bondi, *University of London* 265
XVII. WHAT IS RUSSELL'S THEORY OF DESCRIPTIONS?
David Kaplan, *University of California, Los Angeles* 277
XVIII. THE NON-QUANTAL FOUNDATIONS OF
QUANTUM MECHANICS
Alfred Landé, *The Ohio State University* 297
XIX. SOME PROBLEMS OF THE CONNECTION
BETWEEN TECHNICAL DEVELOPMENT AND
ECONOMIC HISTORY
György Ránki, *Hungarian Academy of Sciences* 311
XX. IS SCIENCE HUMAN?
Hermann Bondi, *University of London* 321

CONCLUSION
The Editors 325

INDEX... 327

Chapter I

A REALIST VIEW OF LOGIC, PHYSICS, AND HISTORY

Sir Karl Popper
University of London

Man, some modern philosophers tell us, is alienated from his world: he is a stranger and afraid in a world he never made. Maybe he is; yet so are animals, and even plants. They too were born, long ago, into a physico-chemical world, a world they never made. But although they did not make their world, these living things changed it beyond all recognition and, indeed, remade the small corner of the universe into which they were born. Perhaps the greatest of these changes was made by the plants. They radically transformed the chemical composition of the earth's whole atmosphere. Next in magnitude are maybe the achievements of some marine animals which built coral reefs and islands and mountain ranges of limestone. Last came man, who for a long time did not change his environment in any remarkable way, apart from contributing, by deforestation, to the spread of the desert. Of course, he did build a few pyramids; but only during the last century or so did he begin to compete with the reef-building corals. Still more recently he began to undo the work of the plants by slightly, though significantly, raising the carbon dioxide content of the atmosphere.

Thus we have not made our world, and we have not even changed it much, compared with the changes achieved by marine animals and plants. Yet we have created a new kind of product or artifact which promises in time to work changes in our corner of the world as great as those worked by our predecessors, the oxygen-producing plants, or the island-building corals. These new products, which are decidedly of our own making, are our myths, our ideas, and especially our scientific theories: theories about the world we live in.

I suggest that we may look upon these myths, these ideas and theo-

1

ries, as some of the most characteristic products of human activity. Like tools, they are organs evolving outside our skins. They are exosomatic artifacts. Thus we may count among these characteristic products especially what is called "human knowledge"; where we take the word 'knowledge' in the objective or impersonal sense, in which it may be said to be contained in a book; or stored in a library; or taught in a university.

When referring to human knowledge, I shall usually have this objective sense of the word 'knowledge' in mind. This allows us to think of knowledge produced by men as analogous to the honey produced by bees: the honey is made by bees, stored by bees, and consumed by bees; and the individual bee which consumes honey will not, in general, consume only the bit it has produced itself: honey is also consumed by the drones which have not produced any at all (not to mention that stored treasure of honey which the bees may lose to bears or beekeepers). It is also interesting to note that, in order to keep up its powers to produce more honey, each working bee has to consume honey, some of it usually produced by other bees.

All this holds, by and large, with slight differences, for oxygen-producing plants and for theory-producing men: we, too, are not only producers but consumers of theories; and we have to consume other people's theories, and sometimes perhaps our own, if we are to go on producing.

"To consume" means here, first of all, "to digest", as in the case of the bees. But it means more: our consumption of theories, whether those produced by other people or by ourselves, also means criticizing them, changing them, and often even demolishing them, in order to replace them by better ones.

All these are operations which are necessary for the *growth of our knowledge;* and I again mean here, of course, knowledge in the objective sense.

I suggest that it looks at present that it is this growth of human knowledge, the growth of our theories, which turns our *human history* into so radically a new chapter in the history of the universe, and also in the history of life on earth.

All three of these histories — the history of the universe, the history of life on earth, and the history of man and of the growth of his knowledge — are, of course, themselves chapters of our knowledge. Consequently, the last of these chapters — that is, the history of knowledge — will consist of knowledge about knowledge. It will have to contain, at least implicitly, theories about theories, and especially theories about the way in which theories grow.

I shall, therefore, before going any further into my topic, present a general tetradic schema which I have found more and more useful as a description of the growth of theories. It is as follows:

$$P_1 \rightarrow TT \rightarrow EE \rightarrow P_2$$

Here "P" stands for "problem"; "TT" stands for "tentative theory"; and "EE" stands for "(attempted) error-elimination", especially by way of critical discussion. My tetradic schema is an attempt to show that the result of criticism, or of error-elimination, applied to a tentative theory, is as a rule the emergence of a new problem; or, indeed, of several new problems. Problems, after they have been solved and their solutions properly examined, tend to beget problem-children: new problems, often of greater depth and ever greater fertility than the old ones. This can be seen especially in the physical sciences; and I suggest that we can best gauge the progress made in any science by the distance in depth and unexpectedness between P_1 and P_2: the best tentative theories (and all theories are tentative) are those which give rise to the deepest and most unexpected problems.

My tetradic schema can be elaborated in various ways; for example, by writing it as follows:

$$P_1 \begin{array}{l} \nearrow \\ \searrow \end{array} \begin{array}{l} TT_a \rightarrow EE_a \rightarrow P_{2a} \\ TT_b \rightarrow EE_b \rightarrow P_{2b} \\ TT_n \rightarrow EE_n \rightarrow P_{2n} \end{array}$$

In this form the schema would indicate that, if we can, we should propose many theories as attempts to solve some given problem, and that we should critically examine each of our tentative solutions. We then find that each gives rise to new problems; and we may follow up those which promise the most novel and most interesting new problem: if the new problem, P_{2b}, say, turns out to be merely the old P_1 in disguise, then we say that our theory only manages to *shift the problem* a little; and in some cases we may take this as a decisive objection against the tentative theory, TT_b.

This shows that error-elimination is only *part* of our critical discussion: our critical discussion of the competing tentative theories may compare them, and assess them, from many different points of view. The decisive point is, *of course*, always: how well does our theory solve its problems, that is, P_1?

At any rate, one of the things we wish to achieve is to learn something new. According to our schema, progressiveness is one of the things we demand of a good tentative theory; and it is brought out by the critical discussion of it: the theory is progressive if our discussion shows that *it has really made a difference to the problem we wanted to solve*; that is, if the newly emerging problems are different from the old ones.

If the newly emerging problems are different then we can hope to learn a great many new things when we proceed to solve them in turn.

Thus my tetradic schema can be used to describe the emergence of new problems and, consequently, the emergence of new solutions — that is, new theories; and I even want to present it as an attempt to make sense of the admittedly vague idea of emergence — as an attempt to speak of emergence in a rational manner. I should like to mention that it can be applied not only to emergence of new scientific problems and, consequently, new scientific theories, but to the emergence of new forms of behaviour, and even new forms of living organisms.

Let me give you an example. P_1 may be, say, a certain problem concerning the survival of a species, such as the problem of reproduction, of producing offspring. According to Darwin, this survival problem has found a good solution if the species survives; any other tentative solution will be eliminated by the disappearance of both, the solution and the species.

According to my schema, the attempted error-elimination — that is, the struggle for survival — will bring out the inherent weakness of each of the proposed solutions in the form of a *new problem*. For example, the new problem may be that the parent organisms and their offspring are threatening to suffocate one another. This new problem may, in turn, be solved; for example, the organisms may develop a method of scattering or disseminating their offspring; or else the new problem may be solved by the establishment of a common economy, comprising several organisms. Perhaps the transition from unicellular to multicellular organisms proceeded in this way.

However this may be, my schema shows that there may be more than Darwin's alternative, "*survive or perish*", inherent in the process of error-elimination: error-elimination may bring out new emerging problems, specifically related to the old problem *and* to the tentative solution.

In what follows I shall use my schema, sometimes only implicitly; and I shall refer to emergence, assuming that my schema makes this idea sufficiently respectable within what I hope will be a rational discussion. I propose to deal with some aspects of the growth of knowledge under four headings:

1. Realism and Pluralism: Reduction versus Emergence.
2. Pluralism and Emergence in History.
3. Realism and Subjectivism in Physics.
4. Realism in Logic.

1. REALISM AND PLURALISM: REDUCTION VERSUS EMERGENCE

Man produces not only scientific theories but many other ideas—for example, religious or poetical myths or, say, plots for stories.

What is the characteristic difference between a scientific theory and, say, a work of fiction? It is not, I hold, that the theory is true while the descriptions in the story are not true, although truth and falsity have something to do with it. The difference is, I suggest, that the theory and the story are embedded in different critical traditions. They are meant to be judged by quite different traditional standards (even though these standards may have something in common).

What characterizes the theory is that it is offered as a solution to a scientific problem; that is, either a problem that has arisen before, in the critical discussion of earlier tentative theories, or (perhaps) a problem discovered by the author of the theory now offered, but discovered within the realm of the problems and solutions belonging to the scientific tradition.

However, I am not leaving it at that. For the scientific tradition in its turn is, or was until recently, characterized by what may be called *scientific realism*. That is to say, it was inspired by the ideal of finding *true solutions* to its problems: solutions which corresponded to the facts.

This regulative ideal of finding theories which corresponded to the facts is what makes the scientific tradition a realist tradition: it distinguishes between the world of our theories and the world of facts to which these theories belong.

Moreover, the natural sciences with their critical methods of problem solving, and some of the social sciences too, especially history and economics, have represented for quite a long time our best efforts in problem solving and fact finding (by fact finding I mean, of course, the discovery of statements or theories which correspond to facts). Thus these sciences contain, by and large, the best statements and theories from the point of view of truth, that is, those giving the best description of the world of facts, or of what one calls "reality".

Now let us look at certain relations that hold between some of these sciences.

Take physics and chemistry for example; sciences which make assertions about all physical things and physical states, including living organisms.

Physics and chemistry are not very different, and there seems to be no great difference in the kind of things to which they apply, except that chemistry, as it is usually understood, becomes inapplicable at very high temperatures and also, perhaps, at very low ones. It therefore would not be very surprising if the hopes, held for a long time, that chemistry can be reduced to physics, would come true, as indeed they seem to be doing.

Here we have a real paradigm case of a "*reduction*"; by a *reduction* I mean, of course, that all the findings of chemistry can be fully explained by (that is to say, deduced from) the principles of physics.

Although such a reduction would not be very surprising, it would be a very great scientific success. It would not only be an exercise in unification, but a real advance in understanding the world.

Let us assume that this reduction has been carried out completely. This might give us some hope that we may also reduce one day all the biological sciences to physics.

Now this would be a spectacular success, far greater than the reduction of chemistry to physics. Why? Because the kind of things to which physics and chemistry apply are really very similar from the start. Only think how difficult it would be to say whether the atomic theory is a physical or a chemical theory. In fact, for a long time it was both; and it is this common bond which provides the link which may lead, or perhaps has led, to their unification.

With living organisms the situation is different. They are, no doubt, subject to all kinds of physical and biological laws. Yet there appears to be some *prima facie* difference between living organisms and non-living things. Admittedly, we learn from science that there are transitory or intermediate stages, and also intermediate systems; and this gives us hope that a reduction might be achieved one day. Moreover, it seems not at all improbable that recent tentative theories about the origin of life on earth might be successfully put to the test, and that we might be able to create primitive living organisms artificially.

But even this would not necessarily mean a complete reduction. This is shown by the fact that chemists were able to create all sorts of chemicals, inorganic and organic, before understanding even their chemical composition, to say nothing about their physical structure. Thus even the control of chemical processes by purely physical means is not as such equivalent to a reduction of chemistry to physics. Reduction means much more. It means *theoretical* understanding: the *theoretical* penetration of the new field by the old field.

Thus we might find a recipe for creating some primitive forms of life from non-living matter without understanding, theoretically, what we were doing. Admittedly, this would be a tremendous encouragement to all those who seek for a reduction, and rightly so. But the way to a reduction might still be long; and we could not know whether it was not even impassable: there may be no theoretical reduction of biology to physics, just as there seems to be neither a theoretical reduction of mechanics to electrodynamics, nor a theoretical reduction the other way round.

If the situation is such that, on the one hand, living organisms may originate by a natural process from non-living systems, and that, on the other hand, there is no complete theoretical understanding of life possible in physical terms, then we might speak of life as an *emergent* property of physical bodies, or of matter.

Now I want to make it quite clear that as a rationalist I wish and hope to understand the world and that I wish and hope for a reduction. At the same time, I think it quite likely that there may be no reduction possible; it is conceivable that life is an *emergent* property of physical bodies.

My point here is that those believers in reduction who, for some philosophical or other reason, adopt *a priori* the dogmatic position that reduction must be possible, in a way destroy their triumph should reduction ever be achieved. For what will then be achieved ought to have been achieved all the time; so their triumph will be only the uninteresting one of having been proved right by events.

Only those who assert that the question cannot be settled *a priori* can claim that any successful reduction would be a tremendous discovery.

I have dwelt on this point so long because it has some bearing on the position of the next rung of the ladder — the emergence of consciousness.

There are philosophers, called radical behaviourists or physicalists, who think that they have *a priori* reasons, such as *Ockham's razor,* for asserting that our introspection of mental states or events, and our reports about mental states of events, are simply introspections and reports about ourselves *qua* physical systems: they are reports about physical states of these systems.

Two philosophers expected here this morning have defended such a view with brilliant arguments. They are Herbert Feigl and Willard Van Orman Quine. I should like to make a few critical remarks about their views.

Quine says, with a reference to Carnap and Feigl, that if theoretical progress can be "achieved by...positing distinctive mental states...behind physical behavior, surely as much...could be achieved by positing ...certain correlative physiological states and events instead. ... Lack of a detailed physiological explanation of the states is scarecely an objection to acknowledging them as states of human bodies....The bodily states exist anyway; why add the others?" [1].

Let me point out that Quine speaks here as a realist: "The bodily states exist anyway," he says. Nevertheless, from the point of view I am adopting here, he is not what I should call a "scientific realist": he does not wait to see whether science will achieve a reduction here, as perhaps it may one day; instead he applies Ockham's razor [2], pointing out that mental *entities* are not necessary for the theory.

But who knows what Ockham or anybody else might mean here by necessity? If mental entities or, better, mental states should exist — and I myself do not doubt that they do exist — then positing mental states is necessary for any true explanation of them; and should they one day be reduced to physical states, then this will be a tremendous success. But there will be no success at all if we reject their existence by merely noting that we can explain things without them, by the simple method of confining ourselves to physical things and their behaviour.

To sum up my argument in brief: philosophical speculations of a materialistic or physicalistic character are very interesting, and may even be able to point the way to a successful scientific reduction. But they should be frankly tentative theories (as I think Feigl's theories are). Some physicalists do not, however, consider their theories as tentative, but as proposals to express everything in a physicalistic language; and they think these proposals have much in their favour because they are undoubtedly *convenient:* inconvenient problems such as the body-mind problem do indeed, most conveniently, disappear. So these physicalists think that there can be no doubt that these problems should be eliminated as pseudo-problems.

To this I would reply that by the same method we could have eliminated *a priori* all chemical states and problems connected with them: we could have said that they were obviously physical, and that there was no need to specify them in detail: that all we needed to do was to postulate the existence of some physical state correlative to each chemical state.

I think it is clear that the general adoption of such a proposal would have led to the attitude of not looking for the detailed reduction of chemistry to physics. No doubt, it would have dissolved the analogue of the body-mind problem — the problem of the relation of physics to chemistry; but the solution would have been linguistic; and as a consequence we should not have learned anything about the real world.

All this leads me to assert that realism should be at least tentatively pluralistic, and that realists should subscribe to the following pluralistic postulate:

We must beware of solving, or dissolving, factual problems linguistically, that is, by the all too simple method of refusing to talk about them. On the contrary, we must be pluralists, at least to start with: we should first emphasize the difficulties, even if they look insoluble, as the body-mind problem may look to some.

If we can then reduce or eliminate some entities by way of scientific reduction, let us do so by all means, and be proud of the gain in understanding.

So I would say: let us work out in every case the arguments for emergence in detail, *at any rate before attempting reduction.*

To sum up and sharpen the considerations advanced in this section:

The reduction of chemistry to physics, apparently now well on the way, may be described as a paradigm case of a genuine scientific reduction which satisfies all the requirements of a good scientific explanation.

"Good" or "scientific" reduction is a process in which we learn much that is of great importance: we learn to understand and to explain the theories about the field to be reduced (in this case chemistry) and we learn a great deal about the power of the reducing theories (in this case physics).

It is conceivable, although not yet certain, that the reduction of chemistry to physics will be completely successful. It is also conceivable, though less likely, that we may one day have *good reductions* of biology, including physiology, to physics, and of psychology to physiology, and thus to physics.

I call *bad reduction* or *ad hoc* reduction the method of reduction by merely linguistic devices; for example, the method of physicalism which suggests that we postulate *ad hoc* the existence of physiological states to explain behaviour which we previously explained by postulating (though not by postulating *ad hoc*) mental states. Or in other words, by the linguistic device of saying that I report on a *physiological* state of mine when I report that I now feel that I understand the Schrödinger equation.

This second kind of reduction or the use of Ockham's razor is bad, because it prevents us from seeing the problem. In the picturesque as well as hard-hitting terminology of Imre Lakatos, it is a disastrous case of a "*degenerating problem shift*"; and it may prevent either a good reduction, or the study of emergence, or both.

In order to avoid this disastrous method we must in each case try to learn as much as possible about the field which we hope to reduce. It may be that the field resists reduction; and in some cases, we may even possess arguments to show why the field cannot be reduced. In this case we may have an example of genuine emergence.

I may perhaps end my comments on the degenerating problem shift of behaviourism (especially linguistic behaviourism) with the following remark.

Behaviourists and materialists are anti-idealists: and they are, especially, opponents of Berkeley's "*esse = percipi*" or

$$to\ be = to\ be\ observable.$$

According to them, to be is to be material, to behave as a body in space and time. Nevertheless, it may be said that they do adhere, unconsciously, to Berkeley's equation, although they put it in a slightly different verbal form:

$$to\ be = to\ be\ observed$$

or perhaps

$$to\ be = to\ be\ perceived.$$

For they say that only those things exist which can be observed. They do not realize that *all observation involves interpretation in the light of theories,* and that what they call "observable" is what is observable in the light of pretty old-fashioned and primitive theories. Though I am all for common sense, I am also for enlarging the realm of common sense by learning from science. At any rate, *it is not science but dubious philosophy (or outdated science) which leads to idealism, phenomenalism and positivism; or to materialism and behaviourism,* or to any other form of anti-pluralism.

2. PLURALISM AND EMERGENCE IN HISTORY

I shall not speak about the history of the universe, but only say a few words about the history of life on earth.

It seems that a very promising start has recently been made towards reconstructing the conditions under which life *emerged* on earth; and I think we may, perhaps, expect some major success soon. But while sanguine about emergence, even experimental emergence, I feel very sceptically inclined about reduction. This is due to certain thoughts of mine about the evolution of life.

It seems to me that evolutionary processes or major evolutionary changes are as unpredictable as historical processes or major historical changes. I hold this view because I am strongly inclined towards an indeterminist view of the world, somewhat more radical than Heisenberg's: my indeterminism includes the thesis that even classical physics is indeterministic, and is thus more like that of Charles Sanders Peirce, or that of Alfred Landé. And I think that evolution proceeds largely probabilistically, under constantly changing conditions or problem situations, and that every tentative solution, whether more successful or less successful or even completely unsuccessful, creates a new problem situation. This seems to me to prevent a complete reduction as well as a complete understanding of the processes of life, although it does not prevent constant and far reaching progress towards such understanding. (This argument should not be taken to be like Bohr's application of his complementarity ideas to living organisms — an argument which seems to me very weak indeed.)

But I want to speak in this section mainly about human history, about the story of mankind. This, as I have indicated, is very largely the history of our knowledge — of our theories about the world — and, of course, of the repercussions of these products, which are of our own making, upon ourselves and our further productions.

It is obvious that one can adopt a physicalist or materialist attitude towards these theoretical products of ours; and it might be suspected that my emphasis upon the objective sense of knowledge — my emphasis upon theories as contained in books collected in libraries and taught in universities — indicates that I sympathize with the physicalist or materialist interpretation of theories; I mean an interpretation which sees language as consisting of physical objects — noises, or printed letters — and which sees ourselves as conditioned, or dispositioned, to react to these noises or letters with certain characteristic kinds of physical behaviour.

But nothing is further from my intention than to encourage *ad hoc* reductions of this kind. Admittedly, if forced to choose between any subjectivist or personalist view of human knowledge and the materialist or physicalist view I have just tried to sketch, I should choose the latter; but this is emphatically *not* the alternative.

The history of ideas teaches us very clearly that ideas emerge in logical or, if the term is preferred, in dialectical contexts [3]. My various schemata such as

$$P_1 \rightarrow TT \rightarrow EE \rightarrow P_2$$

may indeed be looked upon as improvements and rationalizations of the Hegelian dialectical schema: they are rationalizations because they operate entirely within the classical logical organon of rational criticism, which is based upon the so-called law of contradiction, that is to say, upon the demand that contradictions, whenever we discover them, must be eliminated. Critical error elimination on the scientific level proceeds by way of a conscious search for contradictions.

Thus history, and especially the history of ideas, teaches us that if we want to understand history, we must understand ideas and their objective logical (or dialectical) relationships.

I do not believe that anybody who has ever seriously gone into any chapter of the history of ideas will think that a reduction of these ideas could ever be successful. But I take it as my task here not so much to argue against the possibility of any reduction, as to argue for the recognition of emergent entities, and for the need to recognize and describe these emergent *entia* before one can seriously think about their possible elimination by way of reduction.

One of my main arguments for the emergent character of theories I have given elsewhere [4]. My argument depends upon the conjecture that there is such a thing as a genuine growth of scientific knowledge; or in practical terms, that tomorrow, or a year hence, we may propose and test important theories of which nobody has seriously thought so far. If there is growth of knowledge in this sense, then it cannot be predictable by scientific means. For he who could so predict today by scientific means our discoveries of tomorrow could make them today; which would mean that there would be an end to the growth of knowledge.

On the other hand, unpredictability in principle has always been considered as the salient point of emergence; and it seems to me that my argument shows at any rate that the growth of knowledge must be unpredictable in principle.

But there are other arguments for the emergent character of theories, or of knowledge in the objective sense. I shall only mention an argument or two against the very popular and very naive view that theories can be reduced to the mental states of those who produce them, or of those who understand them. (Whether or not these mental states themselves may then perhaps be reduced to physical states in turn will not be further discussed.)

The idea that a theory in its objective or logical sense may be reduced to the mental states of those who hold the theory takes, as a rule, the form that the theory just *is* a thought. But this is a trivial mistake: it is the failure to distinguish between two senses of the word 'thought'. In its subjective sense, the word "thought" describes a mental experience or a mental process. But two mental experiences or processes, though they may stand in causal relations to each other, cannot stand in logical relations to each other.

Thus, if I say that certain ideas of the Buddha agree with certain ideas of Schopenhauer, or that they contradict certain ideas of Nietzsche, then I am not speaking about the mental thought-processes of these people or about their interrelations. If I say, however, that Nietzsche was influenced by certain ideas of Schopenhauer, then I do mean that certain thought-processes of Nietzsche's were causally influenced by his reading of Schopenhauer. So we have actually these two different worlds, the world of *thought-processes*, and the world of the *products* of thought-processes. While the former may stand in *causal* relationships, the latter stand in *logical* relationships.

The fact that certain theories are incompatible is a logical fact, and holds quite independently of whether or not anybody has noticed or understood this incompatibility. These purely objective logical relationships are characteristic of the entities which I have called theories, or knowledge, in the objective sense.

This may also be seen from the fact that the person who produces a theory may very often not understand it. Thus it might be argued without paradox that Erwin Schrödinger did not fully understand the Schrödinger equation, at any rate not until Max Born gave his statistical interpretation of it; or that Kepler's area law was not properly understood by Kepler, who seems to have disliked it.

In fact, understanding a theory is something like an infinite task, so that we may well say that a theory is never fully understood, even though some people may understand some theories extremely well. Understanding a theory has, indeed, much in common with understanding a human personality. We may know or understand a man's system of dispositions pretty well; that is to say, we may be able to predict how he would act in a number of different situations. But since there are infinitely many possible situations, of infinite variety, a full understanding of a man's dispositions does not seem to be possible. Theories are similar: a full understanding of a theory would mean understanding all its logical consequences. But these are infinite in a non-trivial sense: there are infinitely many situations of infinite variety to which the theory might be applicable; that is to say, upon which some of its logical consequences may bear; and many of these situations have never been thought of; their possibility may not yet have been discovered. But this means that nobody, neither its creator nor anybody who has tried to grasp it, can have a full understanding of all the possibilities inherent in a theory; which shows again that the theory, in its logical sense, is something objective and something objectively existing—an object that we can study, something that we try to grasp. It is no more paradoxical to say that theories or ideas are our products and yet not fully understood by us than to say that our children are our products and yet not fully understood by us, or that honey is a product of the bee, yet not fully understood by any bee.

Thus, the study of the history of our theories or ideas—and a good case could be made for the view that all human history is largely a history of our theories or ideas — should make us all pluralists. For what exist, for the historian, are people in physical, social, mental, and ideological problem situations; people producing ideas by which they try to solve these problems, ideas which they try to grasp, to criticize, to develop.

The student of the history of ideas will find that ideas have a kind of life (this is a metaphor, of course); that they can be misunderstood, rejected, and forgotten; that they can reassert themselves, and come to life again. Without metaphor, however, we can say that they are not identical with any man's thought, or belief; that they can exist even if universally misunderstood, and rejected.

All this may be reminiscent of Plato and Hegel. But there are great differences here. Plato's "ideas" were eternal, unchanging conceptions or notions; Hegel's were dialectically self-changing conceptions or notions. The ideas which I find most important are not conceptions or notions at all. They correspond not to words but to statements or propositions.

In opposition to Plato and Hegel I consider *tentative theories* about the world — that is, hypotheses together with their logical consequences — as the most important citizens of the world of ideas; and I do not think (as Plato did) that their strangely non-temporal character makes them eternal and thereby *more real* than things that are generated, that are subject to change, and to decay. On the contrary, a thing that can change and perish should for this very reason be accepted as *prima facie* real; and even an illusion is, *qua* illusion, a real illusion.

This is important in connection with the problem of time, and of change.

A historian cannot, I think, accept the doctrine that time and change are illusions; a doctrine upheld by some great physicists and philosophers such as Parmenides, Weyl, and Schrödinger. Nothing is more real than an event, an occurrence; and every event involves some change.

That the pluralistic universe in which the historian lives, with its individual men living individual lives, trying to solve their problems, producing children, and ideas about them, hoping and fearing and deceiving themselves and others, but always theorizing, and often seeking not only happiness but also truth — that this pluralistic universe should be successfully "reduced" to one or another kind of monism, this seems to me not only unlikely, but impossible. But this is not my point here. My point is that only after recognizing the plurality of what there is in this world can we seriously begin to apply Ockham's razor. To invert a beautiful formulation of Quine's [2], only if Plato's beard is sufficiently tough, and tangled by many entities, can it be worth our while to use Ockham's razor. That the razor's edge will be dulled in being used for this tough job is only to be expected. The job will no doubt be painful. But it is all in the day's work.

3. Realism and Subjectivism in Physics

There are two important fields in modern physics in which physicists have allowed subjectivism not only to enter, but to play an essential role: Boltzmann's theory of the subjectivity of the direction of time, and Heisenberg's interpretation of the indeterminacy formulae as determining a lower limit to the effect of the observer's interference with the observed object.

There was also another intrusion of the subject, or of the observer, when Einstein brought in the observer in the number of imaginary thought experiments intended to elucidate relativity; but this is a field from which the observer was exorcized, slowly but steadily, by Einstein himself.

I shall not discuss this point further, nor shall I discuss the subjective theory of time which, in trying to tell us that time and change are human illusions, forgets that they are very real illusions which have in no way been reduced to anything else (and which, I conjecture, are not amenable to reduction). I shall not discuss all this because I have done so only recently. I merely want to say a few words about the Heisenberg formulae and their interpretation.

These formulae are usually derived in a fairly complicated manner; there is, for example, an interesting derivation by Weyl [5] and another rather complicated one by Born [6].

Yet in fact the Heisenberg formula for energy depends neither on wave mechanics nor on Heisenberg's matrix mechanics; nor do we need the commutation relations (which according to Hill [7] are insufficient for the derivation of the formulae). It simply does not depend on the revolutionary new quantum mechanics of 1925-26, but follows directly from Planck's old quantum postulate of 1900:

$$E = h\nu. \tag{1}$$

From this we get immediately

$$\Delta E = h \, \Delta\nu. \tag{2}$$

By using the principle of harmonic resolving power,

$$\Delta\nu \sim 1/\Delta t, \tag{3}$$

we obtain from (2) and (3)

$$\Delta E \sim h/\Delta t, \tag{4}$$

which leads at once to

$$\Delta E \Delta t \sim h, \tag{5}$$

that is to say, a form of Heisenberg's so-called *indeterminacy formulae*.

In precisely the same way we obtain the Heisenberg formula for position and momentum from Duane's principle (whose analogy to Planck's principle has recently been stressed by Alfred Landé). It may be written

$$\Delta p_i \sim h/\Delta q_i. \tag{6}$$

According to Landé this may be interpreted as follows: a body (such as a grid or a crystal) endowed with the space-periodicity Δq_i is entitled to change its momentum p_i in multiples of $\Delta p_i \sim h/\Delta q_i$.

From (6) we obtain at once

$$\Delta p_i \sim h/\Delta q_i. \tag{7}$$

which is another form of Heisenberg's indeterminacy formulae.

Considering that Planck's theory is a statistical theory, the Heisenberg formulae can be most naturally interpreted as statistical *scatter relations*, as I proposed more than 30 years ago [8, 9]. That is, they say nothing about the possible precision of measurements, nor anything about limits to our knowledge. But if they are scatter relations, they tell us something about the limits to the homogeneity of quantum-physical states and, therefore, though indirectly, about predictability.

For example, the formula $\Delta p_i \Delta q_i \sim h$ (which can be obtained from Duane's principle just as $\Delta E \Delta t \sim h$ can be obtained from Planck's principle) tells us, simply, that if we determine the coordinate x of a system (say, an electron) then, upon repetition of the experiment, the momentum will scatter.

Now, how can such an assertion be tested? By making a long series of experiments with a fixed shutter opening Δx and by measuring, in every single case, the momentum p_x. If these momenta scatter as predicted, then the formula has survived the test. But this shows that in order to test the scatter relations, we have actually measured, in every case, p_x with a precision far greater than Δp_x; for otherwise we could not speak of Δp_x as the scatter of p_x.

Experiments of the kind described are carried out every day in all physical laboratories. But they refute Heisenberg's indeterminacy interpretation, since measurements (though not the predictions based upon them) are more precise than this interpretation permits.

Heisenberg himself noted that such measurements are possible, but he said that it was "a matter of personal belief" or "personal taste" whether or not we attach any meaning to them; and ever since this remark they have been universally disregarded as meaningless. But they are not meaningless, for they have a definite function: they are tests of the very formulae in question, that is, of the indeterminacy formulae *qua* scatter relations.

There is, therefore, no reason whatever to accept either Heisenberg's or Bohr's subjectivist interpretation of quantum mechanics. Quantum mechanics is a statistical theory because the problems it tries to solve — spectral intensities, for example — are statistical problems. There is, therefore, no need here for any philosophical defence of its non-causal character.

The irreducibility of statistical theories to deterministic theories (rather than the incompatibility of these two kinds of theories) should, however, be established. Arguments to this effect have been offered by Landé, and very different ones by myself.

To sum up, there is no reason whatsoever to doubt the realistic and objectivistic character of all physics. The role played by the observing subject in modern physics is in no way different from the role he played in Newton's dynamics or in Maxwell's theory of the electric field: the observer is, essentially, the man who tests the theory. For this, he needs a lot of other theories, competing theories and auxiliary theories. All this shows that we are not so much observers as thinkers.

4. REALISM IN LOGIC

I am opposed to looking upon logic as a kind of game. I know about so-called alternative systems of logic and I have actually invented one myself, but alternative systems of logic can be discussed from very different points of view. One might think that it is a matter of choice or convention which logic one adopts. I disagree with this view.

My theory is briefly this. I look upon logic as the theory of deduction or of derivability, or whatever one chooses to call it. Derivability or deduction involves, essentially, *the transmission of truth and the re-transmission of falsity*: in a valid inference truth is transmitted from the premisses to the conclusion. This can be used especially in so-called "proofs". But falsity is also re-transmitted from the conclusion to (at least) one of the premises, and this is used in disproofs or refutations, and especially in *critical discussions.*

We have premises and a conclusion; and if we show that the conclusion is false, and assume that the inference is valid, we know that at least one of our premises must be false. This is how logic is constantly used in critical discussion, for in a critical discussion we attempt to show that something is not in order with some assertion. We attempt to show it; and we may not succeed: criticism may be validly answered by counter-criticism.

What I should wish to assert is (1) that criticism is a most important methodological device; and (2) that if you answer criticism by saying, "I do not like your logic: your logic may be all right for you, but I prefer a different logic, and according to my logic this criticism is not valid," then you may undermine the method of critical discussion.

Now I should distinguish between two main uses of logic, namely (1) its use in the demonstrative sciences, that is to say, the mathematical sciences, and (2) its use in the empirical sciences.

In the demonstrative sciences logic is used in the main for proofs — for the transmission of truth — while in the empirical sciences it is almost exclusively used critically — for the re-transmission of falsity. Of course, applied mathematics comes in too, in which we implicitly make use of the proofs of pure mathematics, but the role of mathematics in the empirical sciences is somewhat dubious in several respects. (There exists a wonderful article by Schwartz to this effect [10].)

Thus in the empirical sciences logic is mainly used for criticism; that is, for refutation. (Remember my schema $P_1 \rightarrow TT \rightarrow EE \rightarrow P_2$.)

Now, what I wish to assert is this. If we want to use logic in a critical context, then we should use a very strong logic, the strongest logic, so to speak, which is at our disposal; for we want our criticism to be *severe.* In order that the criticism should be severe we must use the full apparatus; we must use all the guns we have. Every shot is important. It doesn't matter if we are over-critical: if we are, we shall be answered by counter-criticism.

Thus we should (in the empirical sciences) use the full or classical or two-valued logic. If we do not use it but retreat into the use of some weaker logic — say, the intuitionist logic, or some three-valued logic (as Reichenbach suggested in connection with quantum theory) — then, I assert, we are not critical enough; it is a sign that something is rotten in the state of Denmark (which in this case is the quantum theory in its Copenhagen interpretation, as I indicated earlier).

Now let us look, by contrast, at proofs. Every mathematician knows that considerable interest lies in proving a theorem with the help of a *minimum apparatus.* A proof which uses stronger means than necessary is mathematically unsatisfactory, and it is always interesting to find the weakest assumptions or minimum means which have to be used in a proof. In other words, we want the proof not only to be sufficient, that is to say valid, but we want it if possible to be necessary, in the sense that a minimum of assumptions have been used in the proof. This, I admit, is a somewhat sophisticated view. In unsophisticated mathematics we are happy and grateful if we can prove anything, but in more sophisticated mathematics we really want to know what is *necessary* for proving a theorem.

So if one can prove mathematical theorems with methods weaker than the full battery of classical logic, then this is extremely interesting from a mathematical point of view. Thus in proof theory we are interested in weakening if possible our classical logic, and we can, for example, introduce intuitionist logic or some other weaker logic such as positive logic, and investigate how far we can get without using the whole battery.

I think, incidentally, that the term "intuitionist logic" is a misnomer. It is just a name for a very interesting and somewhat weakened form of classical logic invented by Brouwer and formalized by Heyting. I certainly do not want to say anything in favour of the philosophical theory called intuitionism though I should like to say something in favour of the Brouwer-Heyting logic. But I trust it will not be supposed that I am in any sense defending the authority of intuition in philosophy or logic or anywhere else. Leaving aside for the moment Brouwerian logic, one might say that intuitionism is the doctrine that intuitions are not only important but generally *reliable.* As against this I think that intuitions are very important but that in general they do not stand up to criticism. So I am not an intuitionist. However, the Brouwerian or so-called "intuitionist logic" is, from the point of the present discussion, important because it is just a part, a genuine part, and thus a weakened form, of classical logic; that is to say, every inference which is valid from the point of view of intuitionist logic is also valid from the point of view of classical logic, while the opposite is not the case: we have inferences which may be validly drawn in classical logic but which are not valid in intuitionist logic. Thus if I can prove a theorem (so far proved only by classical means) with intuitionist logic, I have made a real mathematical discovery; for mathematical discoveries do not consist only in finding new proofs of new theorems, but they consist also in finding new proofs of old theorems; and a new proof of a theorem will be especially interesting if it uses weaker means than the old proof. A proof using stronger means one can always have for the asking, *a fortiori*; yet finding a weaker proof is a real mathematical discovery.

So intuitionistic logic is a very interesting approach to mathematics because it tries to prove as many mathematical theorems as possible with reduced logical means.

Intuitionistic logic has a further advantage: one can show that in it the so-called "law of excluded middle" is not demonstrable (although it is a well-formed formula of the system). One can also show that if in any system whatsoever some well-formed formula is not demonstrable, then the system must be consistent. Generally speaking, the weaker the logical means we use, the less is the danger of inconsistency—the danger that a contradiction is derivable. So intuitionist logic can also be looked upon as an attempt to make more certain that our arguments are consistent and that we do not get into hidden inconsistencies or paradoxes or antinomies. How safe such a weakened logic is, as such, is a question into which I do not want to enter now; but obviously it is at least a little safer than the full classical logic. I do not suppose it is always safe, but that is not my point. My point is this. If you wish to prove, or to establish something, you should use weak means. But for disestablishing it, that is to say, for criticizing it, we may use strong means. Of course someone might say, "Look here, I can refute you even with weak means; I do not even need to use the whole of intuitionist logic." Still, that is not very important. The main thing is that for the rationalist *any* criticism is welcome—though he may reply to it by criticizing the criticism.

Now this rationalist view is a realist view of logic. First, because it looks upon logic partly in connection with the methodology of the natural sciences which, I have tried to argue, is a realistic affair. Secondly, and this is a very special point, because it looks upon logical inference as truth transmitting or falsity re-transmitting; that is to say, it is concerned with the idea of truth.

I would assert that not the least important of the achievements of Alfred Tarski is that by introducing two ideas into logic, he has actually made logic very much a realistic affair. The first is Tarski's idea (partly anticipated by Bolzano) that logical consequence is truth transmission. The second, I would say, is the rehabilitation of the correspondence theory of truth, the rehabilitation of the idea that truth is simply correspondence with the facts.

I think I may differ here a little from Quine, because I think that this idea of Tarski's ought to be interpreted as destructive of relativism, and

because I think that Tarski's claim that his theory of truth is an "absolutis-
tic" theory of truth is correct. In order to explain this point, I will recount
a very old story with a slightly new point to it. The old story is the story of
the three main theories of truth. The new point is the elimination of the
word "truth" from the story, and with it, of the appearance of dealing
here with words, or verbal definitions. However, for this elimination
some preparatory discussion is needed.

Of the three main theories of truth, the oldest was the correspondence theory, the theory that truth is correspondence with the facts, or
to put it more precisely, that a statement is true if (and only if) it corresponds to the facts, or if it adequately describes the facts. This is the theory
which I think Tarski has rehabilitated. The second theory is the so-called
coherence theory: a statement is regarded as true if (and only if) it coheres
with the rest of our knowledge. The third theory is that truth is pragmatic
utility or pragmatic usefulness.

Now, the coherence theory has all sorts of versions of which I shall
mention just two. According to the first, truth is coherence with our
beliefs, or more precisely, a given statement is true if it coheres with
the rest of our beliefs. This I find a bit disconcerting because I do not
want to put beliefs into logic, for well-known reasons. (If Peter believes
p, and if p and q are interdeducible, we might say that Peter is logically
bound to believe q. Yet he may not know that p and q are interdeducible, and he may in fact disbelieve q.)

According to the second version of the coherence theory a certain
given statement, of which we do not know whether it is true or not, is
to be accepted as true if (and only if) it coheres with the statements we
have previously accepted. This version has the effect of making our
knowledge utterly conservative: our old knowledge cannot be over-
thrown.

The theory of pragmatic utility is especially concerned with the
problem of theories in the natural sciences such as physics. It says that
we should accept a physical theory as true if it turns out in tests, and other
applications, to be pragmatically useful, or successful.

I propose now to use something like a trick. My trick consists in
this. I shall very soon, until very near the end of this paper, stop referring
to *truth*. I shall not any longer ask, "What is truth?" There are several
reasons. My main reason is that I believe that "What is?" or "What are?"
questions or, in other words, all verbal or definitional questions, should
be eliminated. "What is?" or "What are?" questions I regard as pseudo-
questions; they do not all seem to be so pseudo, but I do think they all
are pseudo-questions. Questions such as, "What is life?" or "What is

matter?" or "What is mind?" or "What is logic?" I think should not be
asked. They are typically unfruitful questions.

So I think we should also discard the question, "What is truth?".

My first reason (just mentioned) for discarding the question "What
is truth?" one may call "anti-essentialism". My second reason is even
more important. It is that we should altogether avoid, like the plague,
discussing the meaning of words. Discussing the meaning of words is a
favourite game of philosophy, past and present: philosophers seem to be
addicted to the idea that words and their meaning are important, and are
the special concern of philosophy.

I wish to present here, in order to clarify matters, a formulation which
I have used before:

	IDEAS
	that is
DESIGNATIONS *or* TERMS	STATEMENTS *or* PROPOSITIONS
or CONCEPTS	*or* THEORIES
	may be formulated in
WORDS	ASSERTIONS
	which may be
MEANINGFUL	TRUE
	and their
MEANING	TRUTH
	may be reduced, by way of
DEFINITIONS	DERIVATIONS
	to that of
UNDEFINED CONCEPTS	PRIMITIVE PROPOSITIONS

the attempt to establish (rather than to reduce) by these means their

MEANING	TRUTH

leads to an infinite regress

On the left we have *words or concepts and their meanings*, and on the
right we have *statements or propositions or theories and their truth*.

Now I should like to say, out of the experience of a lifetime in philosophy, that I have always tried to get away from the left side of the table
and to keep to the right side. One should always keep to assertions, to
theories, and the question of their truth. On should never get involved in
verbal questions or questions of meaning, and never get interested in

words. If challenged by the question of whether a word one uses really means this or perhaps that, then one should say, "I don't know, and I am not interested in meanings, and if you like, I will gladly accept *your* terminology." This never does any harm. One should never quarrel about words, and never get involved in questions of terminology. One should always keep away from discussing concepts. What we are really interested in, our real problems, are factual problems, or in other words, problems of theories and their truth. We are interested in theories and how they stand up to critical discussion (and our critical discussion is controlled by our interest in truth).

Having said this, I intend now to stop using the word 'truth'. Our problem is no longer: Is truth correspondence? Is truth coherence? Is truth usefulness? This being so, how can we formulate our real problem?

Our problem can be sharply formulated only by pointing out that the upholders of the anti-correspondence theories all made an *assertion*. They all asserted that there cannot be such a thing as the correspondence between a statement and a fact. This is their central assertion. They say that this concept is meaningless (or that it is undefinable, which, incidentally, in my opinion does not matter, since definitions do not matter). In other words, the whole problem arises because of doubts, or scepticism, concerning correspondence: whether there is such a thing as a correspondence between a state and a fact. It is pretty clear that these doubts are serious (especially in view of the paradox of the liar).

It is also pretty clear that, but for these doubts, the upholders of the coherence theory and of the theory of pragmatic usefulness would really have nothing to argue against. Nobody denies that pragmatic usefulness and such matters as predictive power are important. But should there exist something like the *correspondence of a theory to the facts,* then this would obviously be more important than mere self-consistency, and certainly also much more important than coherence with any earlier "knowledge" (or "belief"); for if a theory corresponds to the facts but does not cohere with some earlier knowledge, then this earlier knowledge should be discarded.

Similarly, if there exists something like the correspondence of a theory to the facts, then it is clear that a theory which corresponds to the facts will be as a rule very useful; more useful, *qua* theory, than a theory which does not correspond to the facts. (On the other hand, it may be very useful for a criminal before a court of justice to cling to a theory which does not correspond to the facts; but as it is not *this* kind of usefulness which the pragmatists have in mind, their views raise a question which is very awkward for them: I mean the question, "Useful for whom?".)

Although I am an opponent of pragmatism as a philosophy of science, I gladly admit that pragmatism has emphasized something very important: the question whether a theory has some application, whether it has, for example, predictive power. *Praxis,* as I have put it somewhere, is invaluable for the theoretician as a spur and at the same time as a bridle: it is a spur because it suggests new problems to us, and it is a bridle because it may bring us down to earth and to reality if we get lost in too abstract theoretical flights of our imagination. All this is to be admitted. And yet, it is clear that the pragmatist position will be superseded by a realist position if we can meaningfully say that a statement, or a theory, may or may not correspond to the facts.

Thus the correspondence theory does not deny the importance of the coherence and pragmatist theories, though it does imply that they are not good enough. On the other hand, the coherence and pragmatist theories assert the impossibility or meaninglessness of the correspondence theory.

So without ever mentioning the word 'truth' or asking, "What does truth mean?" we can see that the central problem of this whole discussion is not the verbal problem of defining 'truth' but the following substantial problem: can there be such a thing as a statement or a theory which corresponds to the facts, or which does not correspond to the facts?

Behind the doubts concerning the possibility of speaking about correspondence, there are various strong arguments.

First of all, there are paradoxes or antinomies which arise out of this correspondence idea. Secondly, there are the countless unsuccessful attempts to say more precisely what the correspondence between a statement and a fact consists of. There is the attempt of Schlick, who said that correspondence is to be explained by a one-one relationship between the linguistic statement and the fact, that is, by uniqueness. A statement, he said, is "true", or corresponds to the facts, if it stands to the facts of the world in a one-one relationship or in a unique relationship: non-correspondence or "falsity" is the same as ambiguity. Of course, this is an unacceptable view, for many vague and ambiguous statements (such as, "there are a few people somewhere in America") may correspond to the facts; and *vice versa*, every general proposition or theory which corresponds to the facts corresponds to many facts, so that there is not a one-one relationship.

Moreover, a statement which does not correspond to the facts may be quite unambiguous. A murderer may say unambiguously, "I have not killed him." There is no ambiguity in this assertion; but it does not correspond to the facts. Clearly, Schlick's attempt to explain correspondence misfires. Another even worse attempt is Wittgenstein's [11]. Wittgenstein suggested that a proposition is a picture of reality and that correspondence is a relationship very much like the one that holds between the groove on a gramophone record and the sounds which it denotes: a kind of projective relationship between facts and statements. The untenability of this view can easily be shown. One is reminded of the famous story of Livingstone being introduced by an interpreter to a Negro king whom he asked, "How are you?". The Negro king answered with one word, and the interpreter began to talk and talk and talk and talk, for ten minutes, translating the word to Livingstone in the form of a long story of the king's sorrows. Then Livingstone asked whether the king was in need of medical assistance, and then the king began to talk and talk and talk and talk and talk. And the interpreter translated it with one word:"No."

No doubt this story is invented. But it is well invented; and it illustrates the weakness of the projection theory of language, especially as a theory of the correspondence between a statement and a fact.

But this is not all. The matter is even more serious; namely, Wittgenstein, after having formulated this theory, said that is impossible to discuss the relationship of language to reality, or to discuss language at all. (Because language cannot be discussed by language.) This is a field in which words fail us. "It shows itself" is his favourite expression to indicate the failure of words. Any attempt to go deeper into the relationship between language and reality or to discuss language more deeply or statements more deeply is, accordingly, bound to be meaningless. And although he says in the Preface of his book, "the *truth* of the thoughts that are here set forth seems to me unassailable and definitive", he ends up by saying, "Anybody who understands me eventually recognizes them [the propositions of the *Tractatus*] as nonsensical." (Because talk about language is meaningless.) No doubt this refers, apart from other things, especially to his theory of projection. His remarks that his readers will see that what he says is meaningless thus confirms what the opponents of the correspondence theory have always said of the correspondence theory, namely, that is is meaningless to speak about the correspondence between a statement and a fact.

So we are back at the real issue. It is this: is there or is there not a tenable correspondence theory? Can we or can we not speak meaningfully of the correspondence between a statement and a fact?

Now my assertion is that Tarski has rehabilitated the correspondence theory. This, I think, is a great achievement, and it is a great philosophical achievement. I say this because it has been denied by many philosophers (for example, by Max Black) that there is something philosophically important in Tarski's achievement.

The key to the rehabilitation of the correspondence theory is a very simple and obvious observation made by Tarski. That is, if I want to speak about correspondence between a statement S and a fact F, then I have to do so in a language in which I can speak about both, statements such as S and facts such as F. This seems to be frightfully trivial; but it is nevertheless decisive. It means that the language in which we speak in explaining correspondence must possess the means needed to *refer* to statements, and to *describe* facts. If I have a language which has both these means at its disposal, so that it can refer to statements *and* describe facts, then in this language — the *meta*language — I can speak about correspondence between statements and facts without any difficulty, as we shall see.

A metalanguage is a language in which we talk about some other language. For example, a grammar of the German language, written in English, uses English as a metalanguage in order to talk about German. The language about which we talk in the *metalanguage* (in this case English) is usually called the "*object language*" (in this case German). The characteristic thing about a metalanguage is that it contains (meta-linguistic) *names* of words and of statements of the object language, and also (metalinguistic) *predicates,* such as "noun (of the object language)" or "verb (of the object language)" or "statement (of the object language)". If a metalanguage is to suffice for our purpose it must also, as Tarski points out, contain the usual means necessary to speak about at least all those *facts* about which the object language can speak.

All this is the case if we use English as our metalanguage in order to speak about German (as the object language under investigation).

For example, we shall be able to say in the English metalanguage such things as:

The German words "*Das Gras ist grün*" form a statement of the German language.

On the other hand, we shall be able to describe in our (English) metalanguage the fact which the German statement "*Das Gras ist grün*" describes. We can describe this fact in English simply by saying that grass is green.

We can now make a statement in the metalanguage about the *correspondence of a statement of the object language to the facts* as follows. We can make the assertion: *The German statement "Das Gras ist grün" corresponds to the facts if, and only if, grass is green.*

This is very trivial. It is, however, important to realize the following: in our assertion, the words, "*Das Gras ist grün*" put within *quotes*, function as a (metalinguistic, that is, an *English*) name of a *German* statement; on the other hand, the English words 'grass is green' occur in our assertion above *without* any quotation marks: they do not function as a name of a statement, but simply as the description of a *fact* (or alleged fact).

This makes it possible for our assertion to express a relationship between a (German) *statement,* and a *fact.* (The *fact* is neither German nor English, although it is, of course, described or spoken about in our metalanguage, which is English: the fact is non-linguistic, it is a fact of the real world, although we need of course a language if we wish to talk about it.) And what our metalinguistic assertion asserts is that a certain (German) statement *corresponds to a certain fact* (a non-linguistic fact, a fact of the real world) under conditions which are precisely stated.

We can, of course, replace the German object language by any other — even by English. Thus we can make the metalinguistic assertion:

The English statement "Grass is green" corresponds to the facts if, and only if, grass is green.

This looks even more trivial. But it can hardly be denied; nor can it be denied that it expresses the conditions under which a statement corresponds to the facts.

Generally speaking, let S be the (metalinguistic) *name* of a statement of the object language, and let f be the *abbreviation* of an expression of the metalanguage that describes the (supposed) fact F which S describes. Then we can make the following metalinguistic assertion:

A statement S of the object language corresponds to the facts if and only if f.

Note that while "S" is here a metalinguistic name of a statement, "f" is not a name, but an abbreviation of an expression of the metalanguage describing a certain fact (the fact which we can name "F").

We can now say that what Tarski did was to discover that in order to speak about the correspondence between a statement S and a fact F, we need a language (a metalanguage) in which we can speak about both the statement S and the fact F. (The former we speak about by using the *name* "S", the latter by using a metalinguistic expression, f, which *states or describes F.*)

The importance of this discovery is that it dispels all doubt about the meaningfulness of talking about the correspondence of a statement to some fact or facts.

Once this is done, we can, of course, replace the words "corresponds to the facts" by the words "is true".

Tarski, apart from this, introduced a method of giving a *definition* of truth (in the sense of the correspondence theory) for any consistent *formalized system.* But this is not, I think, his main achievement. His main achievement is the rehabilitation of talk about correspondence (and truth). Incidentally, he showed under what circumstances such talk may lead to paradoxes, and how we can avoid these paradoxes; and he also showed that in ordinary talk about truth we can and do avoid paradoxes.

Once we have settled that we can use "truth" when we speak about the correspondence of statements to facts, there is really nothing of importance to be added about the word 'truth'. There is no doubt that correspondence to the facts is what we usually call "truth"; that in ordinary language it is correspondence that we call "truth", rather than coherence or pragmatic usefulness. A judge who admonishes a witness to speak the truth and nothing but the truth does not admonish the witness to speak what he thinks is useful either for himself or for anybody else. The judge admonishes a witness to speak the truth and nothing but the truth, but he does not say, "All we require of you is that you do not get involved in contradictions," which he would say were he a believer in the coherence theory. But this is not what he demands of the witness.

In other words, the ordinary sense of "truth" as it is used in courts of law is, no doubt, correspondence. But my main point is that this may be regarded as an after-thought, and as an unimportant after-thought. For if anybody should want to say, "No, in ordinary language, 'truth' is used in a different sense," I should not quarrel with him. I should suggest that we forget all about terminology: I should be prepared to use the terminology of my opponent, pointing out, however, that we have *at least* these three meanings at our disposal: this is the only thing about which I should be prepared to quarrel; but I should refuse to quarrel about words.

I should point out, though, that the correspondence theory of truth is a realistic theory; that is to say, it makes the distinction, which is a realistic distinction, between a theory and the facts which the theory describes; and it makes it possible to say that a theory is true, or false, or that it corresponds to the facts, thus relating the theory to the facts. It allows us to speak of a reality different from the theory. This is the main thing; it is the main point for the realist. The realist wants to have both a theory and the reality or the facts (don't call it "reality" if you don't like it, just call it "the facts") which are different from his theory *about* these facts, and which he can somehow or other compare with the facts, in order to find out whether or not it corresponds to them. Of course, the comparison is always extremely difficult.

One last word about Tarski's theory. Its whole purpose is often misinterpreted: it is wrongly assumed that it is intended to yield a *criterion*

of truth. For coherence was so intended, and likewise pragmatic usefulness; they strengthened the traditional view that any serious theory of
truth should present us with *a method of deciding* whether or not a given
statement is true.

Tarski has proved many things from his definition of truth. Among
other things, he has proved that in every sufficiently powerful language
(and in every language in which we can formulate mathematical or physical theories) there can be no criterion of truth; that is, no criterion of correspondence: the question of whether a proposition is true is not in
general decidable for the languages for which we may form the concept
of truth. Thus the concept of truth plays mainly the role of a regulative
idea. It helps us in our search for truth that we know there is something
like truth or correspondence. It does not give us a means of finding truth,
or of being sure that we have found it even if we have found it. So there is
no criterion of truth, and we must not ask for a criterion of truth. We
must be content with the fact that the idea of truth as correspondence
to the facts has been rehabilitated. This, I think, Tarski has done; and I
think that he has done thereby an immense service to the realistic outlook.

Although we have no criterion of truth, and no means of being
even quite sure of the falsity of a theory, it is easier to find out that a
theory is false than to find out that it is true (as I have explained in
detail elsewhere). We have even good reasons to think that most of our
theories — even our best theories — are, strictly speaking, false; for they
oversimplify or idealize the facts. Yet a false conjecture may be nearer
or less near to the truth. Thus we arrive at the idea of nearness to the truth,
or of a better or less good approximation to the truth; that is, at the idea of
"verisimilitude". I have tried to show that this idea can be rehabilitated in
a way similar to Tarski's rehabilitation of the idea of truth as correspondence to the facts.[12]

In order to do so I have used mainly the two Tarskian ideas mentioned here. One is the idea of truth. The other is the idea of logical consequence; or more precisely, of the set of logical consequences of a conjecture, or the content of a conjecture.

By incorporating into logic the idea of verisimilitude or approximation to truth, we make logic even more "realistic". For it can now be
used to speak about the way in which one theory corresponds better
than another to the facts — the facts of the real world.

To sum up. As a realist I look upon logic as the *organon of criticism* (rather than of proof) in our search for true and highly informative
theories — or at least for new theories that contain more information, and
correspond better to the facts, than our older theories. And I look upon

criticism, in its turn, as our main instrument in promoting the growth of our knowledge about the world of facts.

REFERENCES

1. W. V. Quine, *Word and Object*, 1960, p. 264.
2. W. V. Quine, *From a Logical Point of View*, 2nd rev. ed., 1961, p. 2.
3. K. R. Popper, "What is Dialectic?" in *Conjectures and Refutations*, 1963.
4. K. R. Popper, *The Poverty of Historicism*, 1957, Preface.
5. H. Weyl, *The Theory of Groups and Quantum Mechanics*, 1931, pp. 72 and 393.
6. M. Born, *The Natural Philosophy of Cause and Chance*, 1949, pp. 189-191.
7. E. L. Hill, in *Mind, Matter, and Method, Essays in Philosophy and Science in Honor of Herbert Feigl* (P. Feyerabend and G. Maxwell, eds.), 1966, p. 442.
8. K. R. Popper, *The Logic of Scientific Discovery*, 1959, 1968 (1st German edn. 1934).
9. K. R. Popper, "Quantum Mechanics without 'The Observer'," in *Quantum Mechanics and Reality* (M. Bunge, ed.), 1967.
10. J. Schwartz, "The Pernicious Influence of Mathematics on Science," in *Logic, Methodology and Philosophy of Science* (E. Nagel, P. Suppes, and A. Tarski, eds.), 1962, pp. 356-360.
11. L. Wittgenstein, *Tractatus logico-philosophicus*, 1922.
12. K. R. Popper, *Conjectures and Refutations*, 1963, 1968, Chapter 10 and *Addenda*.

DISCUSSION

QUINE: I agree with what Popper says. For instance, take non-conventionalism of logic. As a matter of fact, I do not know what doctrine I expounded in London that Popper disagrees with, because, of course, non-conventionalism of logic is something I advocated a long time ago. And I also agree about its function of truth transmission and falsity re-transmission. But I would just add a couple of supplementary observations here. Notice that this view about the function of logic draws no clear boundary between logic and other sciences. When we have a case of, say, three premisses and logic being used to transmit truth to a conclusion, and supposing those premisses are true, then we also have a case of two premisses and an inferential step of a material kind in which the third premiss is absorbed, and this is a step that one would not call logic but rather physics or biology, or whatever the subject matter may have been. This, however, is alright, so far as I can see. The function of logic in transmitting truth, and re-transmitting falsity, doesn't necessarily give us a criterion or a definition of logic. In fact, the question, "What is logic?"—and again I agree with Popper—is a rather sterile kind of question.

I would add a further remark on what happens in the case of disagreement in logic between two logicians who claim to have different logics. In fact, this may be a place where I do disagree with Popper after all, in so far as I am not sure that in this case criticism will cease. Practically perhaps, it should, because there are other more fruitful projects to undertake if things have gone so far as disagreement in the logic itself. But, in principle, I think there is something further that that can be done—that if one is to go on from that point, one has to resort to semantic ascent.

I am a logical realist, as Popper is, but I put logic and physics on the same level. They are truths about the world in the broadest sense of the word. But in the case of a disagreement that is so fundamental as a disagreement on logical principles, the only thing we can do is talk about talking. But this, again, isn't peculiar to logic and it doesn't make logic a matter of convention.

The same thing happens in physics. The same thing happens whenever an issue takes the form of disagreement over extremely fundamental matters, matters so fundamental to the conceptual scheme that if you walk at that level you keep begging the question by simply saying, "Well, it is obvious from what you said that this follows." This happens not only in logic. There is something of this kind when we start talking about relativity theory or quantum mechanics. One has to stand off, talk about the system, and appeal to the pragmatic value of this or that system. One says: Here is a simpler formulation that takes in all the data that the old formulations took in, and, moreover, does it more simply.

We talk in those terms and still it doesn't make logical truth any more a matter of convention than it makes truth in quantum mechanics a matter of convention.

MERCIER: I agree with Popper "that one should not ask: What is life? What is matter? What is logic?". But then neither should one ask, "What is reality?". Therefore one can only describe some fictitious reality definable from a formalized frame, for instance, from present physics.

Furthermore, let me recall that examples used in Heisenberg's book *Principles of Quantum Mechanics* are all derived from the relation between p and x and eventually transformed by some formal trick into relations between energy and time. This is connected with the fact that time is not a so-called q-number but is a c-number and has to do with the fact that in physics one never measures time, but time is given. Time is the quantity upon which everything depends. It is the mathematical variable allowing one to make predictions.

If one wants to establish an uncertainty relation between energy and time, one has to be very careful looking for a way to make out of time an observable. This can only be done in analogy with the so-called homogenized classical canonical formalism by which a supplementary degree of freedom is introduced and in which a time conjugate momentum p_t is considered, it being numerically equal to minus the energy.

Conversely, if you want to make out of x and t a vector in special relativity and to elaborate a relativistic quantum theory, the usual coordinates are no longer q-numbers but c-numbers, and the commutation rules, which are the mathematically correct principles from which uncertainties are derived, do not concern the coordinates anymore, but the fields themselves. However neither do the uncertainties concern the coordinates, but e.g. the numbers of particles of the field and their phases.

I just wanted to call attention to these very delicate matters, which I believe render obsolete older considerations about uncertainties.

Popper claims that he can derive the uncertainty relation in a fashion which has not been considered by other authors. However, this very fashion goes back to the use of Fourier transforms (for a Δv assumes a wave-packet), and this was, as far as I am aware, in principle known long ago.

Similar historical remarks can be made in other cases, e. g. about introducing Planck's constant itself: Boltzmann very clearly states that it is impossible to ground statistical mechanics without having finite cells in phase-space. He just did not give the size of these cells, but he might have called it *h* and waited for somebody to state that it is equal to Planck's constant. That was very many years before Planck found it himself.

YOURGRAU: I was delighted to hear that Popper is passionately embracing the correspondence theory of truth, though I cannot agree with his enthusiastic appraisal of Tarski's presentation. He interprets Tarski's original paper without any qualification. I concur with his pleading for that kind of truth theory but I do not share his uncritical interpretation of Tarski's article, so much in vogue today. And now let me express a brief observation in connection with Popper's way of reasoning and that of Quine.

I think it is not true that when two logicians disagree — say, logicians and other formalists — that this is *logical* fact: It is a hard, *empirical* fact. In disagreements concerning methodology or theory we state the premisses $A, B, C, \ldots, N$ and ask step by step: "Do you agree here, with this point?" The reply: "Yes, indeed." "Well, then you *must* accept the conclusion so-and-so." But he may *not* accept it, he may *still* disagree. I think therefore that though we reason in logic from "intersubjective" premisses, agreement in *attitude* is still a necessary prerequisite for a meaningful dialogue or general discussion. In other words, the one with whom we are speaking must give his *consent*, and I believe that this consent is not merely a question of rational ability, of some method or convention: the 'other' mind has to be willing — perhaps psychologically — to accept certain rules, axioms, and norms. How else can we account for the fact that we argue with colleagues about the same topics for years and that they refuse to confirm the conclusion, even though they accepted the premisses?

This is an unsatisfactory state-of-affairs which one encounters in mathematics, in logic, and in physics. It seems to me that some voluntary consent is the *conditio sine qua non*.

Finally, I should like to slightly modify two points raised in Popper's lecture, because without due qualification they might easily be misinterpreted.

First, it is not true that Schrödinger "did not fully understand" his equation, or did so only after Born's statistical interpretation. I discussed this matter with him on many occasions and I can also produce several references where he granted, with resignation, that Born's exegesis appeared to be plausible in the light of present-day quantum theory. None the less, he was convinced that in the future we may be able to arrive at a modification of statistical quantum mechanics. After all, the statistical approach was certainly not foreign to him; to wit, his by now classic *Statistical Thermodynamics* and his last formal paper where he investigated whether or not energy might merely be a statistical conept. (We are reminded of Planck's and Einstein's attitude to acausality and indeterminacy.)

Second, Popper will agree with me that it would be feasible to add a statistical theory of truth to the other three theories he cited in his address. Of course, such a truth theory would apply mainly to scientific events or facts or ideas, and thus be less universal than the other three conceptions of truth.

Since Popper's paper covers such an enormous area of knowledge, I thought it would be helpful to present these comments in order to avoid any possible misunderstanding of certain remarks of his. Indeed, I am sure that there is no disagreement between us in regard to the last three issues mentioned in my reply.

Yet I am quite certain that my views concerning Tarski's paper will arouse his ire.

KEYES: I would like to attempt a thematic clarification and would also like to raise a question. The motivation of my question is from the standpoint of existential phenomenology, and this seems to me to be justified if one is also concerned with the philosophy of history. In addition, there are certain themes which have been suggested which point to phenomenological questions. For example, there is the contrast between reality as self-subsistent and the subjective constructions that we put upon phenomena. And further, there is the question of the relation between objective occurrence in history and subjective involvement within it.

Now, questions of this type, if we are to transpose them to the phenomenological level, imply something similar to Husserl's *Epoche* in which we do not doubt the actual existence of that which is phenomenologically suspended, but rather we refrain from basing our systematic construction upon it. On the transcendental level this points us to the question of the relation of the *noesis* and *noema*. Furthermore, although from a different standpoint, questions of this type point to Heidegger's relation of *Dasein* and *Vorhandensein* and the grounding of reality in the apriority of *Sorge*. And also, as another example, the same question implies Sartre's relation of the "in itself" and "for itself".

If we are confronted with what Heidegger would term the two basic possibilities of being, human existing and objective reality, we can legitimately consider history from the standpoint of man's subjective involvement within history, which — I submit — is a question that cannot be overlooked regardless of what the formal structure of logic may demand. If man's involvement in history is unique because he alone is able to ask the *Seinsfrage*, are we not in danger of a certain inappropriate form of reductionism if we overlook the question of man's subjective involvement in history? And if this is the case, do we not need to admit, also, another criterion of truth, namely, the *Daseins* moment of insight, the moment of truth (*Augenblick*) of its nature as an historical being?

TREDER: Popper has unfortunately entered into a critical discussion of the Copenhagen interpretation of quantum physics. I regret that he did so from a philosophical point of view. Now most physicists are not very interested in the philosophical aspects of quantum theory. For the thinking and practicing physicist, the Copenhagen interpretation is the standard interpretation of the mathematical formalism of quantum theory. In fact, it is extremely valuable for any interpretation of experiments in that domain. The quantum rules of Planck, Sommerfeld, etc., provide a consistent physical theory only if one gives commutation rules for the canonical quantities. These rules reflect the mathematical form of Bohr's complementarity principle.

VIGIER: On the question of the correspondence theory of truth, while I quite agree with Popper, there are some problems still open. First there is a question of the relation of practice and theory and the correspondence of the results of experiments to prove the truth of a given theory.

For example, it is a great mystery why Newton's law, a very good description of the behavior of planets, was built in a certain frame, within a certain mechanical theory, while Einstein's laws are better and are constructed in the frame of a quite different theory.

The question of the relation of practice and theory dominates the philosophy of science and as a physicist I cannot pretend it is solved satisfactorily.

Now, as for the question of quantum theory. The discussion is really about the fourth indeterminacy relation of Heisenberg and the question of the canonical variables. In reality, if one wants to analyze quantum theory, one should carefully distinguish the nonrelativistic level of quantum theory from the relativistic level. In the latter case if one wants to build a covariant canonical formalism, one must treat time and energy on the same footing. This has been shown by Dirac, for example, in 1965 in a very famous paper in *The Physical Review*, where he says that it is false to assume that the fourth indeterminacy relation simply results from a Lorentz transformation applied to the first three relations. What is introduced as a c-number is a proper time which is connected to the rest-mass of the particle. But time and its canonical quantity, which one can take as Hamiltonian, must be treated on the same footing as the others.

I am not going to discuss here in detail the meaning of the relativistic interpretation, but just stress that it is the only correct one. It has given correct results, but only when they happened to coincide with the non-relativistic limits of the relativistic theory. The only correct quantum theory is the relativistic theory. I think this has a very deep meaning. It is connected with the basic fact that only by introducing a light cone and the idea of time and space-like vectors is it possible to give a covariant meaning to the question of time and space ordering.

Now, to come to what is the real issue, in my opinion, in the discussion of Popper. I don't agree with Treder that the Copenhagen interpretation is the only possible interpretation.

Nobody is contesting the fact that statistical quantum theory is the only correct theory and that the commutation relations can be interpreted as relations between measurements of canonical variables. The whole point is whether the current quantum-mechanical description is complete or not. If you accept the idea that quantum theory is in its definitive form and that particles, for example, can only be described by the canonical variables in space and time (meaning position and momentum) then you can argue, as Treder has done, and prove von Neumann's theorem. But the point is that von Neumann's proof is a vicious circle, because nobody knows whether the quantum theory is complete or not. In fact, experiment is beginning to suggest exactly the contrary, unless, of course, one interprets the new quantum numbers with which one describes particles as having nothing to do with the old quantum numbers. Indeed, we know how to connect spin and mass with angular momentum and displacement in space-time. However, the two numbers M and J (mass and the spin) are not enough to describe particles. We also need hypercharge, strangeness, baryon number, etc., and either those new quantum numbers have something to do with new canonical variables (in that case the description given by the Copenhagen school is not complete, and down goes the whole interpretation) or, of course, one might present the argument that one has to somehow drop space and time and the new quantum numbers are not quantum numbers in any known sense of the word.

The question of the completeness of quantum theory is a fundamental problem, and what I am really suggesting is that the question should be finally settled on the level of theoretical physics and on the level of experiment. These will show in the future whether new variables must be introduced in order to get a real understanding of what is going on in very high energy physics and elementary particle theory. The question of the completeness of the description is, I think, the key logical question to the whole business.

I would just like to add that in the past all those who have said that they had reached truth and that the final description of nature had been reached have always been proven wrong. It is a very dangerous statement to make.

POPPER: There seems to be complete — or almost complete — agreement between Quine and myself. Where I thought that we disagreed was on Birkhoff and von Neumann's "Logic of Quantum Mechanics": I thought (perhaps mistakenly) that he had taken this famous paper to show that logic and physics are in the same boat. Up to a point no doubt they are: both are criticizable. But a mere weakening of classical (or Boolean) logic, as suggested by Birkhoff and von Neumann, and also by Reichenbach, does not seem to me an adequate response to a difficulty in one of the empirical sciences. It is a strategy which we ought to exclude if we want to learn from experience, because every sufficient weakening of the underlying logic will make any empirical theory for ever secure. It will make it uncriticizable, irrefutable.

These remarks show that in some cases at least we can critically discuss a proposed change in logical theory.

I do not think that Vigier has caught the significance of my remark that it is demonstrable that *correspondence is not a criterion of truth*, because from this it follows that in general, with the exception of comparatively trivial theorems, the truth of a statement of a theory *cannot* be proved; which means that even if we should ever arrive at a "final description of nature", we could not know it.

Concerning Treder's comment to the effect that the Copenhagen interpretation is true, that the Copenhagen interpretation is *the* true theory, I would refer to what I have written elsewhere about the subject.[1]

Keyes discussed things which I can't discuss. I know that at least 80% of all contemporary philosophers are existentialists or phenomenologists or both. My late friend, Julius Kraft, wrote in 1932 a beautiful book, *Von Husserl zu Heidegger* (republished in 1957), in which he showed very clearly that the early development of this philosophy in Germany, from Husserl to Heidegger, consisted essentially in abandoning the method of rational argument. He showed that rational argumentation was replaced by "verbalization" — by the magical use of words, words, and words.[2] It is, in my opinion, one of the great tragedies of our times that such a great percentage of all philosophers are existentialists; for this means that a young philosopher has no chance; if he wants to learn philosophy he just learns the wrong thing: the use of jargon instead of rational arguments. I can't discuss Heidegger rationally, since he does not argue rationally. Nor does he produce any thesis which is sufficiently clear to be rationally discussed. Apart from this, I cannot condone the important role Heidegger played under the Nazis, and his

[1]K. R. Popper, *The Logic of Scientific Discovery*; also K. R. Popper, "Quantum Mechanics without 'The Observer'," *Studies in the Foundations, Methodology and Philosophy of Science,* Volume 2, Mario Bunge ed., 1967.
[2]See also K. R. Popper, "Julius Kraft, 1898-1960," *Ratio* 4;2-12 (1962)

behavior to his teacher, Husserl. Hitler may be excused; he did not know any better. Heidegger's behavior is inexcusable; he knew what he was doing. Moreover, according to existentialist philosophy, a man is what he does and what he teaches; and *vice versa*. Thus the argument from the person to the doctrine is admissible in this case.

Yourgrau says that he disagrees with my view that it was Tarski's famous paper that rescued the correspondence theory, and he doubts whether my interpretation of Tarski is correct. I can only say that I learned in 1935 from Tarski himself, and from his paper, how to formulate the correspondence theory, how to answer objections to it, and how to get over any of my misgivings about it; and that I found what I had learned from Tarski so important that when I was invited in 1935 by Susan Stebbing to give two lectures on my methodology, I chose instead as my topic Tarski's theory, because I felt it was the more interesting and important subject.

I also disagree with Yourgrau's view that in order to have a rational discussions, we must first agree on certain fundamentals — on a 'common framework'. This view is precisely what I call "The Myth of the Framework", and I have spoken against it for years. My own view is precisely the opposite: the discussion will be the more fruitful and interesting the more profound the difference is between the starting points of its participants, provided always the participants really want to learn from each other (and they have enough time at their disposal). Thus the more different the approach, the more fruitful the discussion. This, I think, is a historical fact, for which many examples can be given. It is at the root of the fruitfulness of what is called a "culture clash" by anthropologists.

I fully agree with Mercier that if we should not ask, "What is life? What is matter? What is logic?", we should not ask either, "What is reality?". I fully agree with this: we should not waste our time on "What is?"-questions and definitions. I don't think that I have asked the question; and if I did, I am ready to withdraw it.

But this does not mean that we should not speak about reality; just as we can speak about life, about the physics and the chemistry of living things, and about logic, and about alternative logical systems, so we can speak about reality. Moreover, such a proposal as Landé's, that we should call a thing "physically real" if it is kickable and able to kick back, may be useful, even though it is not a definition of reality. At any rate, I disagree with Mercier when he draws, from the admitted fact that we should not waste our time on definitions of reality, conclusions of the kind he seems to draw — for example, that we should not speak about physical reality, but only about "a formalized frame", for instance that of the present-day physics. On the contrary, I think that we have a right to say that it seems that physics progresses in the sense that present-day physics is a better description of physical reality than, say, Galileo's or even Newton's physics.

Mercier's criticism of what I said about the indeterminacy formulae has been answered by Vigier much more interestingly and effectively than I could have done myself. I will only add *one* point: the derivation of the second indeterminacy formula, $\Delta p_i \Delta q_i \geq h$, from the old quantum theory is just as easy as that of $\Delta E \Delta t \geq h$. For this reason, the special status which Mercier claims for the time-variable t seems to me irrelevant to my point. Both formulae are derivable from the old optical *principle of harmonic resolving power* (and they have been so derived by Heisenberg and Bohr). Their interpretation must therefore depend on this derivation. But this leads necessarily to their interpretation as scatter relations, as I have tried to show elsewhere.

BRECK: We historians can, I believe, live in Popper's pluralistic world of indeterminacy because we have all too often forgotten the consequences of a neat tidying-up of the intractible matrix of historical events, only to find that we have "neatly" superimposed what he has called "the myth of the framework" on formlessness and confusion. I quite agree that we must recognize "the plurality of what there is in this world" before we can begin to see what elements can be dropped from our analysis of a situation by the application of William of Ockham's *entia non multiplicanda sunt praeter necessitatem.*

But the problem arises from our definition of *necessity.* Every period burgeons with a plethora of events, facts, and problems, and to pare away that rich abundance in the name of order is to deny a rich reality merely because it is intellectually more convenient to make some sort of easy path. "Necessity" may lead the historian to confess himself lost, without a clue to any simple understanding of the complex series of pictures which presents itself to his mind when he tries to survey the "life and times" even of a single man in history.

I am not so sure that all human history is essentially the history of our theories, though I suppose that such terms as "the state" or "the family" are ideas or theories as often as they are physical realities; they are the result of planning by people confronted by problems, and are thus processes that lead to further problems which in their turn lead to further tentative solutions, in the manner of Popper's *schema.* Perhaps I am only mirroring the discontent historians have with those theorists (most frequently psychologists) who have taken us farther from our *res gestae,* our events and things, documents, artifacts, than the evidence will permit. What Freud did to Moses and Leonardo, what Jung did to Job and Norman O. Brown to the founders of Rome—all these exegeses are tantamount to ideas for whose support the evidence is wholly lacking.

Chapter II

KNOWLEDGE AND PHYSICAL REALITY

André Mercier
University of Berne

For the ordinary man, there happens to be practically no problem of having to ask what reality is, for he believes he sees it around him, acting on things and on himself, and standing there to be contemplated and used; he thus thinks he is able to 'prove' reality by the fact that he can see it, touch it, remove it, and so on.

The philosopher sees problems everywhere. What is plain for the non-philosopher is never plain for him. Reality, especially, represents a huge problem, and schools of thought rest upon their respective conception of such a fundamental notion. One may think in that respect of Augustine's paradox concerning time.

For our purpose, a physicist will have to be an individual between an ordinary man and a philosopher. Now, factually and professionally, my experience is threefold, namely, the experience of an ordinary man, of a physicist, and of a philosopher. So let us approach this threefold experience as we would a building with several entrances, and let us enter at the gate marked Physics'.

Immediately, we find ourselves confronted with a very imposing structure. First, two main arrows indicate very different modes of behavior with respect to what we think will be the reality to be revealed by physics: one leads to the experimental division, the other to the theoretical division. On his way to the first of these divisions, one will find recommendations posted on the walls like: Be a good engineer, Test before speaking, Handwork be thy method, etc. On the other hand, along the other division, he will learn that mathematics is the only tool available, that all is speculation, that great conceptual imagination is required, etc. And the distance between the experimental and the theoretical physicist can become so great that they will often seem to deal with completely different matters. At least the realities of the two seem to be distinct, though both claim to look for the same reality. Hence, already in the frame of physics proper, there is reality as it appears to be, different according to the main divisions of that science, and presumably reality as it is behind these appearances.

Now, how does the 'apparent reality' appear to be in the eyes of the experimental physicist? Well, already that depends on further conditions. For he may specialize in high vacuum techniques, in which case the ultimate reality will appear to consist of the tenuous gas which he still cannot get rid of, however powerful his pumps are; whereas, if he works on lasers, he will look for better and better crystals in which the reality appears to reside in the perfection of structures of some geometrical arrangement. Of course, there will be some feature common to both activities of these two specialists, but it is not to be found in the final result. It is located in their common method of approach, which we call experimental. So, for a number of scientists, reality admits of an approach in which, in their opinion, that reality we call physical has to be brought into some special, or extreme, or pure state with regard to a precise experimental device in order to be recognized as real.

The word 'reality' has its etymological root in the Latin *res*, which means a *thing*. The other English word 'realty', also derived from *res*, means real estate, a property of houses and land, very real things to an ordinary man who wants to grow vegetables and establish a farm and earn a living for his family. Yet it is difficult to grow vegetables in a high vacuum or in a perfect crystal. So the real things of the ordinary man are as dirt compared with the apparent realities of the experimental physicist. Indeed, earth and all such plain realities are very dirty and there are some who want to wash their hands as often as possible.

Now, does the experimental physicist looking for perfect crystals or similar pure cases wash this dirt off his hands? Yes and no. Yes, because it is much easier to deal with pure cases, where gases are more or less perfect, surfaces practically smooth, structures visibly undistorted, etc. So we simplify the cases, eliminate impurities, breed, so to say, generations that are as we should like to have them in order to study them at ease. But actual or natural generations are full of distortions, though there is nothing wrong about distortions, for they are natural, they are simply the case, they belong to the facts with which we have to reckon. Therefore all those manipulations invented by experimentalists are a way of washing their hands of the impurities of real matter.

At the same time, even the ordinary man will recognize that nothing can be seen in dirt because it is an awful mixture. So one has somehow to extract things that have a sufficient degree of homogeneity, of purity, of formal structure, if he is to recognize it as a representative of what he is looking for. But then he extracts either a pure gas, or a liquid, or a crystal. Therefore, reality as experimental reality is found to look like a multitude of pure forms, the reality of which is most artificial, in the original sense of the word, i. e. produced by human techniques.

Let us now turn to theoretical physics. First, we should remember that when a textbook on physics is written, all its claims about the nature of the physical world are of a theoretical kind. Even if experiments and observations are described at length, the description of that world is exclusively theoretical. We read, for instance, that in the Newtonian picture of forces producing accelerations, there is one important force, called gravitation, which is inversely proportional to the square of the distance between point masses, etc. However: are there point masses? Are there forces? Are there distances? As *things*, I mean. Is a mass a thing? Well, in a way it is, e. g. if thought of as the quantity of matter I can eat in an apple. But an apple is an apple and not just a mass. I am never eating just a mass, for if I were, I might as well eat paper, or gold, which I never do. One might answer to that, that there are masses attracting each other, whether of gold, or of paper, or of apple-like material, and that even Newton felt an apple fall on his head—which he probably did not; it is a star, or a planet, or an apple. One might then say, knowing about Einstein's special theory of relativity, that the mass is just a form of energy, and that energy is always what we pay for, even when we buy a pound of apples. To that, I can easily answer, that Newton did not know of Einstein's theory, so the case is settled, the more so, as in Einstein's special theory one is by hypothesis not allowed to talk about gravitation, since Newton's law of gravitation is not Lorentz-invariant. Then one has to go over to general relativity, but the notion of mass does not occur there in the old sense.

Are there then distances? Not as a reality, at least if reality is made up of things you can either take with a spoon, or wrap in a piece of paper, or the like. Distances are nothing but relations between things. So they involve things, and that would be the sense in which they are real. All right. But a distance has a precise meaning only between mathematical points, here between point masses, and apart from the fact that mass is not a real thing in itself, there simply are no point masses: for never was it possible to isolate them, nor does it make sense to imagine them without the use of a delta-function, which is not a regular function but what is called a distribution and diverges completely at the *point* of the point mass. One might then argue that there are extended masses of which we can define the center of mass, which is a point from which distance can be precisely defined. But there are no fluid-like extended masses, since bodies are known to consist of particles, and only in the picture of statistical mechanics does it make sense to talk about a fluid made out of innumerable particles, and this presupposes that we may equate differentials and finite though very small magnitudes. This again is at least bold, if not unsound from the classical point of view of philosophy that led in analogous cases to paradoxes like that of Achilles and the tortoise or that of the arrow that does not fly at all through the air. Mathematics has taken care of such paradoxical difficulties. However, there remains very little of the reality of masses and their centers in the sense of a very matter-of-fact reality as thought of by the ordinary man digging gold and eating apples.

Furthermore, what is the reality of a force? The existence of a force is not even as well established as the existence of God. For either God *is* or he *is not*. But this alternative cannot even be proposed about forces: if you teach Newtonian mechanics, you say "there are forces"; if you teach Lagrangian or Hamiltonian mechanics, you do not talk thus, but you rather say: there are Lagrange functions, or Hamilton functions. Now who would believe that there are Lagrange functions as pieces of reality?

If you ask a physicist, "What is an electron?", the answer will depend upon the man who answers. He may be a cosmic-ray scientist, and maybe he will answer that the electron is the thing made responsible for the track seen in an expansion chamber. Now the track is a row of small numerous water drops condensing from the vapor in the expansion chamber. The track is certainly not identical with the electron. Nobody has seen the electron as such in the chamber as we see our friend walking down the street. Our friend does not produce a track like that. A lady might leave a track of perfume, or even, if she is of great beauty, a completely inexistent track consisting of the remembrance of her where she walked; but the track is not her reality.

Sir J. J. Thomson might have said that the electron is the thing of which he measured the ratio e/m of its charge over mass. Apart from the fact that the reality of the mass is disputable — and possibly that of the charge too — J. J. Thomson had to wait for Millikan to measure e before more clarity was at hand about the electron. By the way, Millikan's measure is not the measure of the electron charge *qua* electron charge, but a charge deposited upon a globule of oil which he suspended between the plates of a condenser with the help of an electric field. It needs a whole chain of reasoning in order to get back to something that is the electron's own.

If one should ask a teacher of quantum mechanics, "What is an electron?", he might give several answers. For instance, there is a formula in which the Hamiltonian operator is given by the left-hand side of a Dirac equation, requiring noncommuting algebra; another answer is that the electron is a quantum of a quantizable field susceptible of entering into interaction through a suitable coupling with photons, themselves the quanta of another field, and so on.

Well, let us ask more about quantum field theory, since it is one of the most recent chapters of theoretical physics which should, on that account, best render what is meant by physical reality. Summarizing a textbook which takes the question seriously, we would learn that quantum field theory rests upon the understanding of the following notions. (1) The notion of an abstract space, based on the body of the real or complex numbers, being vectorial, having a finite or infinite number of dimensions, in which linear and bilinear forms have a meaning, admitting of an interior connection which in usual cases allows the definition of a positive definite quadratic form called norm; this norm is then the nucleus of the probabilistic interpretation of quantum theory. (2) Quantum field theory requires the notion of reference systems; such systems are taken to be Lorentzian, which assumes the whole axiomatics of the special theory of relativity. (3) It needs a basic dynamics, which is adapted from canonical formalism of prequantum theory but is not identical with it, since its symbols have completely new interpretation. (4) One, or rather two procedures of quantization, must be agreed upon. It seems from the theory of the abstract Hilbert-spaces that they are the only ones possible. However, they admit of an infinity of representations, hence the pictures we can make of the reality they cover are more than numerous. (5) Free fields have to be defined. Now, to that purpose one considers their possible tensorial or spinorial character, their properties as to being real or complex numbers, one invents reasonable fields by pure mathematical imagination, and lo! Nature seems to have invented the same, for the behavior observed for such and such so-called elementary particles is well described by the theory of such and such fields. Thus the reality of particles is reduced to the invention of fields, and fields are not even artificially made things, but abstractly made concepts which nobody will ever grasp with pincers nor even blow away, though they may be transformed by suitable mathematics into one another, such a transformation being eventually a faithful picture of those explosions told of by experimentalists at CERN, Brookhaven, Dubna, or elsewhere. Such explosion pictures require, by the way, a sixth notion, that of a coupling between different fields. This coupling is the most mysterious feature of all the business, for why should a so-called free field A (if it exists at all) interact with a free field B, losing its freedom and dooming itself to annihilation or creation—which shows that those elements of physical reality we theoretically posit undergo catastrophe upon catastrophe, and never keep their serene autonomy or even their existence as defined by a fixed set of values of the parameters upon which they depend.

Now, to answer this very last question, of why these catastrophes are undergone by the assumed elements of physical reality, we might say that if there were no such microcatastrophes, there would be no changes in general, for physics has up till now shown that reality appears to work mechanistically, i.e. macrophysics is built out of microphysics and not conversely. This should not, however, be interpreted as the reduction of thermodynamics to mechanics. Consequently, we should not know anything in the world if there were not these catastrophic changes at the microscopic level. Catastrophes are just the way changes take place at the smallest level of reality known to us. Every change is a catastrophe for these elements undergoing the change, since they are not the same after as they were before. However, the calamitous connotation of the word 'catastrophe' is to be dropped since only man puts a value on such changes and makes a distinction between disaster and luck. We shall return to that issue and to the necessity of change for reality to enter into our awareness and knowledge.

But let us first discuss quantum field theory and assume its workings understood on the basis of all the notions and postulates introduced. Something in that theory would then bear the reality we should like to ascribe to nature, and I guess this would have to be these quantized fields or the quanta they admit. However, we cannot say that these fields and their quanta simply exist 'in space and time', where space means the space of our visual sensation. Space nearly completely loses its meaningfulness as to the extension to be given to such elements of physical reality. For, as *waves* the fields fill that space completely, and as particles they explode and decay as soon as one tries to localize them (they even decay spontaneously). The coordinates of space and time are nothing more than parameters which help to connect the very rough means of description gained from our gross human sensation with the odd behavior of these elements which we guess within the microcosmos. So from this point of view we should renounce altogether the classical phrase that characterizes physics as the study of that which is in space and time. Indeed, quantum physical reality is not in space and time.

Turning now to another main contemporary theory in physics, that of general relativity, we learn a completely different description of nature. Here we learn that the basic feature of physical reality consists of its property to remain insensible to the most general changes of coordinates possible, that the generality of such changes is yet unlimited, because we do not know for the time being whether they should be restricted to the frame of a particular geometry like Riemannian geometry, since attempts have been made to enlarge it. Neither do we know whether the choice of coordinates should be dictated by our present more or less intuitive notion of a (3+1)-dimensional manifold; however, the tendency at the moment is to keep the idea of a four-dimensional space-time. In that case, it would still make sense to speak of physics as the study of events in space and time, provided we include a complete reinterpretation of space-time structure. In the era of Kant, space and time were forms of our sensible intuition, not reality itself or pieces of reality. In the theory of general relativity, in spite of the great generality of coordinate transformations, space-time is more *real* than it has ever been, for it is that which is formed and curved according to the fundamental quadratic form ds^2; it is a field by itself—and there is nothing else to add. Contrary to what has been maintained by contemporaries of Einstein, it is less a geometrization of physics than a 'physicalization' of geometry. Of course, it is possible to say that space-time is curved by the matter spread into it, but a thing that can be curved is very real, at least in the opinion of an ordinary man who cannot conceive of something curvable if it is unreal. And if gravitation is said (as makes perfectly good sense) to be nothing but the fact that the universe indeed possesses the ds^2, matter might be only a name for a thing that needs not be there, in a way similar to the explanation of charge given by Wheeler in his 'charge-without-charge theory'.

Now, there may be those who will say that physical reality is fundamentally made of things that have a probability to behave in a way well described by quantum theory.

There may be those who think that physical reality is the highly determined family of world-lines of a well-structured space-time.

There may even be some who would accept both and say that reality behaves probabilistically in the microcosmos, deterministically in the macrocosmos.

But if they firmly believe that to be reality itself, I would say that they are all talking nonsense. For, sixty-five years ago, neither quantum nor relativity theory was at hand, and nobody, even among the best scientists, would have been able to make such claims. So why should we today possess the truth that was not possessed by anyone during the thousands of years past? We already begin to have a glimpse of the limits of these twentieth-century theories. Bohr, as well as Einstein, has taught us at least one great lesson in epistemology, namely, that theories are provisional and may be replaced by others, with respect to which the former theories are but approximations—yet not simply approximations, but pieces of abstract thought that admit of a totally different kind of interpretation of the workings of nature, i. e. different kinds of physical realities.

One main error made by many scientists is to claim that what they *say* is identical with what *is*. What we say is a product of our mind. Both Einstein and Bohr knew and explained that when a physicist has succeeded in uttering a statement clad in the language of perfect mathematics, there is nothing in reality that is identical with the statement, and that what is the case in reality cannot be expressed definitely by a perfect formula. Physics makes models; this truth, which goes back to the time of Hertz, is still with us. Hence the reality described by exactly the words printed in the text books universally accepted at any time is a fictional reality, i.e. mind-made. It is a most ingenious reality, but it is not reality as such. Let us avoid the great mistake of identifying the two things. At the same time — and I insist upon it — it is a wonder, or a mystery, if one prefers, that the 'fictive reality' is projectable onto reality as such in so satisfactory a way that we can make sewing machines, motorcars, or electronic tubes, and put them on the market with great success, even though they do not behave identically with the idea behind their blueprint.

This leads me to the subject of *knowledge*. The preceding discussion has shown sufficiently clearly that the idea of a genuine physical reality is, if not a fake, yet a fiction. I am not criticizing the fact that it is a fiction, for it is a very serious one. 'Fiction' is not a bad word at all. Also, in the good literary sense, fiction is not identical with human life, but a novel is a way to present life as it seems to be, even if the reality of life is not actually made of that which is told in the novel.

According to the dictionary, fiction may be defined as an 'imaginary account'. Etymologically, it has the same root as facts; an account of facts that are imaginary — this is exactly what physics intends to be: an account of these imaginary facts that occur in nature.

In Anglo-Saxon philosophical literature, knowledge nowadays is understood as the ready-made collection of information acquired and ordered according to some logical, aesthetic, or other system. On the European continent another conception prevails: the French *connaissance* or the German *Erkenntnis* do not refer to knowledge as something acquired and definitive, but to the act of acquiring more knowledge or cognition. They include the whole process of acquainting oneself with new circumstances, of clearing up the relations involved and abstracting the concepts in order to classify them according to their basic or derived significance; in brief, the process of gaining and ordering all information of, e.g., a physical nature. At first sight, it seems more or less obvious that such information is 'real information', i.e. information about physical reality. But there is no such physical reality outside the fictional reality which the physicist constructs. So what is there 'really', from which we gain and order information?

In order to avoid this paradoxical situation, many people are inclined to believe, or at least to posit, that there is actually a physical reality, and that our information is genuinely real and consequently our knowledge — even if incomplete — is yet the knowledge *of* that genuine physical reality.

I should like to answer to this by distinguishing three things: (1) actual reality that impresses our intellect with marks which stay there as information; (2) the theories we imagine, which create that fictional reality assumed to give as good an account as possible of the actual reality; and (3) the material applications we make of the theories, like motors and electronic tubes and other devices, which compose a realistic synthesis of the originally given reality. This synthesis has again the character of actuality immanent in reality itself, but it is no more identical with the theories than our image of the original reality. Let us describe an example: A living being acts in some respect as what we may call a machine; as such, we can imagine a model implying the principles of thermodynamics, the estimation of efficiency, etc. Then we can fabricate a robot that will do the work which the original did. Think for example of the girl-automaton, in Hoffmann's tales. Well, the poor hero of the tales believes she is a real girl, but when her song falls off and the mechanic quickly winds her up, the audience already 'realizes' that she is not the original reality, not a real girl but only a robot, and finally the hero finds out and his disappointment amounts to genuine distress.

All such manufactured things are robots. Of course, these robots are remarkable artifacts. The better they are, the less 'dirty', i.e. the more 'clean' does actual reality appear to be. Knowledge is that process that purifies actual reality insofar as actual reality still appears unclear. And now I have used 'unclear' instead of 'unclean' not intending to play on words or even on letters, but taking into account the fact that the uncleanness of actual reality is not to be considered as due to ugly and bad forces originated by some enemy opposing the effort of the scientist. It is rather the present impression we have of a situation not yet cleared up, in which we see but smoke and dust and mud. Each cognitive step separates out some element that still was smoke and dust and mud, and gives it its place in an order of our rationalization. However, this separation is at the same time a simplification which neglects many bindings within actual reality; even the word 'neglect' is unsuitable, for a person who says that he neglects something knows what he is neglecting, whereas by simply performing the separation, one does not know what is left ignored. So an unknown amount of ignorance is always connected with each act of cognition. Hence the cognition is tantamount to an artificial ordering.

Indeed, it needs a technique and an art to order anew and put into suitable places the few elements thus isolated. These elements cannot be reality itself; they are but images of pieces which we believe play their proper role in actual reality. The only assurance we have lies (1) in the possibility of creating those artifacts or machines which in their turn are parts of an actual reality that was not there before, and (2) in the circumstance that as robots they behave within reasonable limits like actual reality that was given (or present) already.

This way of explaining what is usually called 'experimental verification' is different from the ordinary criterion invoked by epistemologists, for epistemologists usually base the correctness of scientific findings on the claim that these findings are public, i.e. they can be verified by everybody. The reasons that I am not using that criterion of publicness are, first, the fact that it is not true that everybody can verify the findings of science, since only scientists can, and only those who are highly competent specialists, and provided they have the necessary tools and devices at their disposal, to be found only in laboratories, and if sufficient financial funds are available, etc., etc. To react to these truisms by saying that not everybody *can*, but everybody could perform the verification if he had all the means required, is begging the question, for it supposes the verification in order to make it possible. Second, modern epistemology (Popper) has called our attention to the fact that verification is an impossible undertaking if conceived rigorously; only falsification is possible *in concreto*. And I should like to add that each attempt to verify definitively is doomed to end in a falsification precisely because our cognition ignores a certain residue that will always falsify a little the projection of our images onto actual reality. A third reason is that the idea of publicness is by many authors wrongly restricted to science and excluded from other modalities of judgment.

The scientific judgment consists of the distinction between truth and untruth. It has two fundamental aspects: the theoretical and the practical, or, in other words, the mathematical (or logical) and the positive aspect. A theorem established without mistake from the premises is mathematically true. A law of nature guessed by induction from the observation of natural bodies is positively applicable to physical (or biological, or other) reality within the limits proper to the scientific frame or chapter to which it belongs, and, within certain limits of its projectability, reverts to the actual reality. But the public audience has no other means with which to believe in the truth of the whole structure than by acquiring the gadgets and using them to their satisfaction, by reading and hearing from others that Galileo, Newton, or Einstein were geniuses and by hoping confidently that the professor living next door is not a charlatan.

The same public audience has similar means of verifying the authenticity of other judgments of value. They can buy tickets to the concert and enjoy the music, which is as public as using the 'gadget', and they can hire a lawyer to present their case before the court, and in all cases anybody has the potential possiblility of verifying the beauty of artistic creation and the goodness of moral teachings. But let me come back to knowledge and physical reality.

We have learnt that physical reality is man-made. Obviously the physics of the eighteenth century was not what twentieth-century physics is; there is a change. We must therefore state that physical reality changes with time. It even changes from man to man, because Dr. X in Denver has gained some knowledge which Dr. Y in Paris has not, and conversely. Hence X and Y project different sets of order onto actual reality. Grossly, contemporary physicists of the same educational level conceive of the same gross physical reality. But a few brilliant people anticipate extensions of that gross physical reality. Their anticipations manifest themselves in a specific, preferred (selected) direction, and become with time an additional part of the gross physical reality conceived by everyone. This preferred direction defines progress; not the progress thought of by Auguste Comte in his positivistic doctrine, but something like a multiple pilgrim's progress. I should like to call it the *progress of scientific knowledge*. This progress is made through two channels. One is the channel usually thought of: the accumulation of additional information and its ordering into patterns that make sense for our understanding. The other one consists of unification. Einstein showed us clearly the importance of that second channel. When we look at physics from the point of view of its history, we notice that great achievements have been made each time a unification succeeded, and we notice, too, that on the whole there *is* a tendency to unify physical theories. At the same time, a total and definite unification seems to be impossible. The reason? On the one hand, there will presumably be always more to add to the unity seemingly reached at one time, and on the other hand, there is something like an intrinsic resistance to unification. Think, for instance, of electromagnetism. The unification of electricity and magnetism was certainly a great achievement. Another achievement occurred when optics was included into the electromagnetic frame. But electrodynamics as developed in the nineteenth century was elaborated as a theory of the same type as rational mechanics, though it made use of a notion alien to mechanical conception, viz. the electromagnetic field. We recall that there was always a kind of latent resistance to total conceptual integration into mechanics. This was clearly recognized by Einstein, who undertook a bold analysis before trying a new synthesis that would avoid the difficulties. These difficulties were not exactly logical contradictions, but rather discrepancies on the physical plane. Indeed, we know today that gravitation and electromagnetism are two interactions so fundamental that in all probability they cannot be unified into one single picture, though it is not a *priori* impossible that one day we shall succeed in finding their common heading.

To conclude, I should like to come back to what I said about those microcatastrophes undergone by the minute elements of what we call 'physical reality'. As already said, the word 'catastrophe' should not be taken with its disastrous connotation. What I intended to explain is that changes must take place, and obviously they do take place already at the very lowest level of organized matter. Changes must take place, else we would never know about anything. Indeed, if nature consisted of free fields alone, well described by plane monochromatic de Broglie waves, without interactions producing their modification, the whole cosmos would remain in the same huge stationary state and no single part in it would ever be able to notice anything. This would be a cosmos of dullness, in which not even the sensation of time would manifest itself. Only differences between an 'after' and a 'before' are noticeable. Knowledge therefore implies temporal changes. Hence, also, reality and temporality are, from a strictly philosophical point of view, one and the same. I said once in a private discussion with Karl Popper that matter is only observable in time and through its changes in time, and that conversely time is that property prior to anything else which we induce from the fact that we notice the changes in matter. Matter thereby assumes the most general form it can. Popper summarized the conversation by saying that my doctrine was a kind of materialism. To which I answered that it could as well be called 'temporalism' or 'theory of temporality'.

Anyway, doctrine or not doctrine, knowledge is the process that takes man along through its interplay with that temporality and allows him to replace it by ordered structures produced by the power of his intellect in such a way that they (the ordered structures), in contradistinction to reality itself, are timeless like mathematical truths. This is not Platonism. For if it were, these ordered structures would be Plato's ideas invoked by the power of our reminiscence of the world of forms; whereas—as science suggests—we must content ourselves with a more modest state of affairs, that is, provisional theories which may and will be replaced by ever better theories in the future.

DISCUSSION

LANDÉ: I find it somewhat problematical to attach less reality to more complicated mathematical expressions exemplified by modern field theory. Have bank accounts become unreal because we calculate them with the help of negative numbers? Have waves become less real or abstract because we use imaginary numbers to describe phases? Of course, there is a progress or regress to less and less visualized things, but I cannot see any qualitative revolution in going from ordinary mechanics to quantum mechanics or into field theory. Things and states-of-affairs turn out to be more complex all the time; mathematicians have fortunately developed formal symbolic methods to describe these very complicated things or situations in what, for them, is simple shorthand language which the outsider has to learn during years of study. But I don't think there is any trend away from realism involved.

MERCIER: I do not say that by developing more and more mathematics reality becomes less and less real. I have given no such qualitative connotation to "physical things". I have only asserted that in all cases the physical models we invent are, at best, *models,* and they should not be identified with the reality the man in the street thinks of.

TREDER: We have the notion of force and the resulting concept of dynamics; these are man-made ideas. Of course, the conception of dynamics due to Newton has been extremely fruitful in physics. But it may be the case that in the quantum theory of fields the traditional notion of dynamics is not necessary and not even possible. Phenomenological quantum theory has only kinematical laws and no dynamical equations. Therefore this form of quantum theory starts with pure kinematical axioms: the existence of one vacuum (one Hilbert-space), Lorentzinvariance, and Einstein's causality. In classical physics the kinematical equations are integrals of the dynamical equations. In the S-matrix form of quantum theory we have only integral conditions — and not dynamical differential equations.

I think that this breakdown of dynamics is tantamount to a new revolution in physics.

Now, I agree with Mercier that we have two pictures of the physical world today, namely, quantum theory and general relativity theory. In the Riemannian space-time of general relativity there is no vacuum. Utiyama proves that, in general, at each point of the curved space-time, an independent (autonomous, so to speak) Hilbert-space must be construed. The first axiom of quantum theory is definitely in contradiction to general relativity.

Finally, the physically important properties of a crystal are the properties of the real crystals, not of the ideal crystals. The theory of real crystals is very much more interesting, too, than the theory of ideal crystals.

MERCIER: I agree that it is more important to make a theory of a real rather than of an ideal crystal, of real gases rather than of ideal gases, and so on. But each time one develops such a thoery, one creates the theory of an ideal, and not of a real, crystal.

YOURGRAU: First of all, Mercier worries, like anybody else concerned with these problems, how to define existence, reality, experience. Now, these are questions which, for theologians, historians, logicians, and physicists alike are of equal weight. We all are required to employ explicit definitions — at least working definitions. Mercier quite rightly introduces the argument that we operate with artifacts.

Well, one such artifact is, I agree, a point-mass. But I don't think any sensible theoretical physicist would ever say that the mass-point (or point-mass) exists as a reality in the sense of a billiard ball. But I claim the question is entirely different when one mentions the electron. I do not know a single physicist who works in the laboratory and works with electrons who doesn't say electrons are *not* artifacts, are *not* concoctions of the imagination or imaginary entities, but possess reality — no less than billiard balls and galaxies, and so forth. In that sense I am a devout realist, and I concur with Popper and Treder in this regard, and certainly with Landé.

Mercier is right in noting that many physicists have flaunted the idea that one should eliminate the concept of force altogether. On the other hand, Treder talked about the interaction of forces; but we need not use that particular expression: call it the interaction of zoom-zooms. One still has to use the term, it seems to me, although it is a potentially risky term. But Mercier's insistence on the field concept I can't accept. I am personally supporting Chew. I think it is possible that in a few years' time we may jettison the concept of the field altogether and resort solely to local particle physics.

MERCIER: Yourgrau wanted me to define existence, reality, experience. Let me just quote Pascal, who said that we cannot define our primitive terms, and have to begin with some undefined notions named by words. Thus, I used notions which are more or less familiar and went on talking about them.

Yourgrau maintained that electrons are not artifacts. But what is an electron, since it is never seen as such or registered as such by any apparatus? One only sees a track of drops in a chamber or hears a kick in a counter, or one defines a system by giving the mathematical form of its Hamiltonian. So either the electron *is* an artifact as inferred from the interpretation of the origin of the track or of the kick, or it is an artifact as defined by such and such terms in a Hamiltonian. There may be a vast number of real electrons in the world; as a matter of fact, I *believe* at present that there are a lot of things of that sort, but I do not *know* and shall never know whether they are identical with the representation I make of them. There is of course a theory that certain fundamental particles are made from quarks. As long as such a theory remains unchallenged, the quark model will show the artificiality of such particles; we have a model comparable to that of ideal gases or crystals. Therefore, the only reality we can assign to an electron is that of an ideal electron! That is its (present) physical reality.

So I believe that my distinction between the mere but hidden reality of things and the physical reality of man-made pictures remains fully correct.

As to forces, we certainly can go on talking about them. Before Newton, people used the word 'force' in a loose manner; then a precise notion was put forward by Newton. But Treder called attention to the fact that such words are losing their original meaning, that quantum theory, for instance, completely replaced the original notion of force by that of interaction. On the other hand, along the lines of general relativity, we are trying to eliminate interaction altogether. That may be a silly way of trying to theorize, but such is the case. Summarizing, I should like to say that the very changes in the precise meaning borne by the same words intended to designate general or fundamental ideas or concepts prove that the physical reality they determine does not coincide with the "really" real, which is unchangeable. Anyhow, we shall never prove the existence of the 'really real'.

LAKATOS: The problem of knowledge and reality has been discussed in the last two and a half thousand years by *philosophers*. The attitude toward this problem divided them into dogmatists and sceptics of various kinds. ('Dogmatists' was the sceptics' term for those who thought that certain knowledge was possible – according to sceptics it was not.)[1] It is an interesting question how and why scientists, normally 'naive dogmatists',[2] tend to become 'sceptics'. This problem may be connected with certain *internal* situations of science, like the obvious unreality of Ptolemaic crystal spheres, the early misfortunes of nineteenth-century atomism, or the particle-wave dualism in quantum theory. Or it may be connected with *external* – religious, philosophical, or other – pressures, as in the case of Osiander's preface to *Copernicus*.

The dogmatist's appraisal of scientific theory is clear: it is either true and hence scientific, or false and then not scientific. But for the sceptic the problem of appraisal becomes difficult: what is the value and what are the standards, if any, of scientific theory if it is not knowledge in the sense in which both dogmatists and sceptics interpret it? As Locke says: "What comes short of certainty, cannot be called knowledge." Some of the 'mitigated sceptics' of the seventeenth century solved this problem by *instrumentalism*: by claiming that science is not knowledge, but only a useful tool, an *instrument for providing predictions*. Other opted for *pragmatism:* they stressed the *social usefulness* of science in giving us power over nature and our fellow men, and saw the justification of scientific pursuits in the pragmatic satisfaction or happiness produced by its gadgetry. Still others despaired of the rationality of scientific enterprise and claimed that each age or even each person had 'his own reality' and there was no criterion to decide which is the better; there are only unprovable *beliefs*. Newton's incredible success in verification, however, defeated the sceptics, radical or mitigated, and Kant, the Newtonians' house-philosopher, proclaimed the official victory over Hume and restored the dignity of infallible knowledge. However, Einstein's refutation of the most miraculously confirmed scientific theory of all time restored the sceptic-dogmatic dialectic.

Must one then fall back again on instrumentalism or pragmatism, or reduce science to the level of subjective beliefs? The answer of some modern philosophers was to experiment with 'probabilism', a watered-down version of dogmatism,[3] but their efforts, I think, failed, and some of them are now seen to fall back on pragmatism. The only new answer was provided by Popper, who refused to compromise with some instrumentalist or pragmatist version of "miti-

---

[1] See Richard H. Popkin's book *The History of Scepticism from Erasmus to Descartes* (1960) and his papers covering the century after Descartes.

[2] Usually this position is described as 'naive materialism', or 'naive materialism'. These are misnomers: for those terms are intended to denote a position according to which it is the case not only that the real world (or "matter") *exists* independently of its being conceived, but also that it is *knowable*.

[3] Cf. my "Infinite Regress and the Foundations of Mathematics," *Arist. Soc.* Suppl. **36** 163ff. (1962).

gated scepticism". He discovered that scepticism and dogmatism share one common property: the identification of knowledge with proven, certain knowledge. If this position is abandoned, the struggle of dogmatism with scepticism vanishes and a new epistemology, the study of the growth of fallible knowledge, of knowledge without foundations, is born. (To appreciate the novelty of this new discipline one has to remember that its subject matter, fallible knowledge, has been regarded — both by the dogmatists and by the sceptics — as a contradiction in terms.) Popper thus solved the problem created by the downfall of Newtonian physics by going beyond both dogmatism and scepticism, though he retains a vital grain of dogmatism (he is a staunch realist and a protagonist of the correspondence theory of truth) and a vital grain of scepticism. And yet he insists that scientific knowledge *is* knowledge, victoriously growing knowledge about the world; he maintains that it is not 'justified knowledge', but knowledge without foundations.

Where does Mercier stand? It is difficult to see. Some passages of this paper leave one with the impression that he is a pragmatist. But surely pragmatism and instrumentalism were crushed by Russell's and Popper's arguments. Other passages may create the impression that he is a fallibilist of the Popperian brand. But then he would have to say whether he agrees with Popper's criteria for deciding that one or the other false theory is the better; no fallibilist appraisal of scientific theories can leave Popper's crucial ideas about empirical content and about truth-content of theories unmentioned.[4]

Finally, according to Mercier, science progresses through two channels: (1) accumulation and ordering of information and (2) unification. He does not mention Popper's now classical criticism of the first channel or his theory that science grows through conjectures with maximum explanatory power (the progressive aspect of Mercier's second channel can be subsumed under this point) and through severe criticism.

NAESS: I think Mercier takes a pragmatic and fictionalistic position in the philosophical landscape. By physical reality, he means the reality of which the theoretical physicists are talking about. If he is a philosopher *and* physicist and says that this so-called physical reality is a fake, or a fiction at best, this is an important commitment. It is indicative of a basic attitude towards the picture of fundamental positions roughly named fictionalist or pragmatic. I don't know of anybody who has attempted to formulate such positions in a precise way.

The realism of common sense is, of course, no position at all in so far as it is not precisely articulated. We may, no doubt, all agree that there *are* many electrons indeed.

What is important here for the philosopher is the term "are"; maybe there are several kinds and criteria of "are's". The tremendous discussion about theoretical constructs suggests at least half a dozen possibilities.

From my own point of view I am seriously interested not merely in all the fundamental positions elucidated by Mercier (roughly termed pragmatism, realism, subjectivism), but also in the fact that, as far as they are fundamental positions, they should be worked out as deeply and comprehensively as possible. The physicist with philosophical training must help to bring them up to date, i.e. connect them with the latest findings in the various physical sciences. This is, I

[4]Cf. his *Logic of Scientific Discovery* (1959) and *Conjectures and Refutations* (1963).

think, more than ever a task for philosophy today: to get the fundamental positions clear, make them explicit, comprehensive, state them in accordance with the latest scientific views. This explication must come prior to refutations. To reduce their number to one or zero is an impossible task as long as the fundamental positions are vaguely or ambiguously expressed. Alternatively, their statement cannot depend on one's friendly or hostile attitude. This is, in my view, the present situation in the realm of philosophy of science. The specialists in refutation will have to wait a while and not spoil their sharp instruments on old wrecks.

MERCIER: I do not object to being put by others into these various 'isms', though personally I do not feel committed to any 'ism'. If there are commitments, it is rather physics itself that is committed, for I described physics *as it is*, and I think that is what Naess understood. Then, of course, it is important to analyze these commitments, in order to show in what way science admits of meanings attached to these various philosophical positions.

VIGIER: I am going to try to be as naive as possible. The first point is clear: here is a physical reality outside the reality which the physicists construct. If the physicists were all to vanish, electrons would go on existing and satisfy in the first approximation the laws of quantum mechanics, even if there were no quantum physicists to build up the laws of quantum mechanics.

The real problem is a simple one. I must say that I don't understand certain parts of Mercier's statement. I think all physics is like a map of motion on space and time. Whether the map is complete or whether it should be a probabilistic one, as the Copenhagen school claims, is another question. But all physics reduces to an ordering in space and in time, a description of motions. From that point of view I don't believe there is anything different in quantum mechanics and in field theory from what thinkers like Einstein have been trying to do. We ask, for instance, what a Hilbert space is. It is just an intermediate tool which is constructed, for example, on Fourier components of states. It is an intermediate instrument which we are using in order to describe the distribution of density and the motion of that distribution of density within space and time; and everybody knows that the word 'space' in that Hilbertian connotation has nothing to do with what experiment finally tells us on physical space structure. It is just a space in a mathematical sense. We can describe it like that, but we shouldn't be victims of our own formulas. In my opinion, physics is just ordering physical events in space and time.

This also holds for Einstein's general relativity theory. The difference between general relativity and special relativity is simply that Einstein has reduced the description of general relativity to local transformations and to a connection between the transformations at different points. But physics could not even exist without the local description of motions, and in that sense I do not think that there is any difference between the most sophisticated present forms of quantum theory and what scientists like Einstein were trying to do.

As for the question of the relation of knowledge of objective reality, I am, like Popper, a hopeless realist. My only questions are: How do we construct the map and what is the meaning of the map? By map, I mean a dynamical model. I quite agree that this is a process. I accept Popper's schema as to the way we construct the process. But if we accept a clear and sharp distinction between the objective real world and the maps which we construct and which are precisely what we mean by physical knowledge, then I think most of the problems which have been raised simply vanish, since we know that we construct maps. Of course, those maps are only approximate because reality is complex and infinite.

We try, naturally, to improve our maps and I agree that they transform qualitatively, and this means we can build even better ones. The history of physics shows that better maps are not merely an improvement of the old maps; the new ones often happen to be built in quite a different way. For example, the general theory of relativity is a completely different conception of mechanics from classical mechanics, but it has the same fundamental air: we want to describe the motions of bodies around the sun. If one accepts this simple and naive wording of the problem of knowledge, then one sees two things. First, it is clear that one cannot predict knowledge, because if one could, it would imply that one already had a map of the total reality, which is obviously not true. It is also clear that nature will go on surprising us — and this is a good thing if we physicists want to stay in business.

In addition, there is a fundamental change in the orientation of the epistemological vector, and that brings me back to an issue raised in Popper's paper. What is really happening is that the problems of epistemology are no longer derived from philosophical speculation but from the practice of modern science.

Finally, we should understand that the maps of reality are built according to the various levels of the structure of reality itself. I don't think, for instance, that one can reduce chemistry to physics, because macromolecules acquire properties which are not discussed by physicists. DNA, for example, carries information, and notions of cybernetics enter into chemistry; they cannot be reduced to the usual concepts of physics. In other words, the various levels of structure and organization of reality have their own intrinsic characteristics. What we are really trying to build is a many-dimensional map with different configurations. But one cannot imbed all knowledge into a general picture. I believe that no such general picture will ever be stable or permanent. We are just moving forward, step by step, into more and more complex knowledge and we know we shall never really recover the infinite complexity of nature. The question of time-ordering and space-ordering, however, is fundamental to all such descriptions, and this is what I think physicists are really trying to do.

MERCIER: I don't think there is a basic difference between Vigier's views and my own. He noted that there is surely a reality outside of the construct of physics, and I agree. The opinion that all physics, including quantum theory, reduces to ordering in space and time may fit into an epistemological tradition going back to the influence of Brunschvicg in France.

POPPER: Concerning a remark of Mercier's assertion that in the eighteenth century physical reality was different from that of the twentieth century, I think it is correct, undoubtedly, that eighteenth-century physical reality was different from twentieth-century physical reality. There were fewer people living; there were more horses and fewer motorcars. All these are *physical things,* and so there is a distinction between physical reality in the eighteenth century and that of the twentieth century. But Mercier has something else in mind. He said that because the *views* of the physicists concerning the physical reality of the eighteenth century were different, therefore the eighteenth-century physical reality was different.

Now, this view, I contend, is clearly wrong. I mean, we have to distinguish between my opinion of a person and the person. Of course, I may be quite wrong. I may think, for example, Joan of Arc was a witch. But this doesn't make her a witch: I am just wrong. Thus I may have some eighteenth-century ideas of physical reality which are simply wrong. And certainly a lot of twentieth-century ideas of physical reality are also wrong.

This point is very important in connection with the electron. I think I agree completely with Yourgrau, and I think he agrees completely with me, although I formulate it a little bit differently. He says that all physicists believe that electrons exist. I would say perhaps there are some who don't believe that. Heisenberg, for example, who is a great physicist, tried to work without the existence of electrons. I think he was wrong.

Now, the only important question for physics is not whether physicists believe that electrons exist, but whether the belief is right or wrong. That is to say, if I believe that electrons exist—and I do happen to do so—this is quite unimportant. What is important is that I may be right or wrong. That is to say, electrons may exist or may not exist. What I believe, or what is just at the moment the state of my mind, is quite irrelevant. I may believe today that electrons exist, I may believe tomorrow that witches exist. All this is not to the point. The point is, that the question of whether electrons exist or not has a *Yes-or-No answer*.

I do think that it is a fundamental error to say that reality changes because our opinion changes. If I now think today that someone is tall and tomorrow that he is not tall, this has no bearing on him. It doesn't change his size. If the eighteenth century has one idea of physical reality and the twentieth century another idea of physical reality, it doesn't affect physical reality. But if you invent a motorcar and eliminate horses from the road, this does affect physical reality. I think this is the simple position.

Now, Vigier's map is quite a good metaphor for many purposes, but it has its limits. The most important point where the metaphor breaks down is connected with argumentation. That is to say, *you cannot argue in terms of maps*, but *you can argue in terms of physical theories*; and you *do* argue in terms of physical theories. This is the fundamental difference between a physical theory and a map. A map is quite a good example, but we should not emphasize the map too much, because argument is formulated in words and maps are not formulated in words.

MERCIER: I completely agree that the word 'theory' would have been more appropriate than the word 'map' in the point made by Vigier. But he intends something specific by the use of word 'map', viz. that we are mapping 'in space and time'. This assertion is questionable.

Now, Popper attacks my view that eighteenth-century physical reality is different from the twentieth. My answer is plain: you see, when Popper uses the phrase 'physical reality' he is talking about the reality which I called just 'reality,' whereas what I called physical reality is not what he calls physical reality. As soon as this difference in meanings is understood, no divergence remains between Popper and me, and the only thing Popper added which I did not state explicitly, though it is an important one for history, is that by inventing motorcars, eliminating horses from the road and all that, we do affect 'reality *tout court*'.

Finally, Popper and I agree about the existence of the electron, insofar as we seem to believe in its existence. But Popper says that the important question is one of Yes-or-No existence, independently of our belief. I agree partly; however, he forgets a necessary premiss: Because, when asking a question of Yes-or-No existence, one must first either agree with whom one is discussing the problem or define in a *ne-varietur* manner what one means by an electron. If one succeeds in doing so — and I have my doubts about it — then and only then does it make sense to put the question of Yes-or-No existence. If one has not put and answered the first question, one cannot ask the second.

VIGIER: I agree with Popper's criticism of my map comparison. It would be better to use the notion of a dynamical map. In a sense any theory is a map-drawing machine, and, of course, by map I mean the four-dimensional description of nature.

Chapter III

SCIENCE OF HISTORY AND NOTIONS OF PERSONALITY

Herman Tennessen
University of Alberta

Among the most tedious platitudes that ever haunted a philosopher of history, or a "metahistorian", as I should prefer to say, are those which are suggested as answers to the perennial (pseudo-) problem: *Is* (written) *history* (i.e. historiography in at least one sense of "historiography") *art, or science, or neither, or both*? Whatever may be thought of the rich variety of available "answers" (their name is legion), one thing, I believe, it is safe to say is that "history" has this in common with "art" (almost regardless of how these notions are conceived): When confronted with a presumed (or alleged) work of either art or history, reactions like "that is not art", "that is not history", do not admit of plausible interpretations ("rephrases") in a purely descriptive ("constative", "verative") direction. They are, on the contrary, rather reasonably seen as more or less genuine value judgements ("prescriptives", "optatives", etc.). Similar implications seem suggested by such question as to what history — really, actually, essentially — is. They manifest themselves most patently whenever a metahistorian attempts to rule out some more or less general approaches to history, or certain methods or procedures (employed by historians) as being impossible in history: "It just can't be done!"[1] Then, invariably, another metahistorian will point to some historians who did just that, which allegedly could not be done. Equally predictable are the objections to such "contrary cases", viz.: "That isn't history!" (Or: "That isn't art", as the case may be.) What is it then? It may be religion, metaphysics, Spenglerism, Toynbeeism — or worse: social science, sociology, and even psychology!

---

[1]Needless to say, this and the following remark should not be construed as attempts to support the general position of Spengler, Toynbee and other "great speculators". I fully subscribe to Popper's ridicule of "holistic" approaches.

In short, there cannot be much doubt that the question: "What is history?" is in an important aspect on a par with: "What is art?" Perhaps art ('art', "art") is somewhat more notorious than history ('history', or "history", respectively). Perhaps we more readily resign *vis à vis* any attempts at "finding out" — empirically by conceptual analysis, or by dialectical method — what art is, and, *pari passu*, are more inclined to settle for suggestions as to what art *ought* to be. It will be my contention, however, that it is an equally hopeless task to try to determine what history (or "history") *is* — or how "history" is used — by observing what so-called "historians" do. And to seek orientation in a search for *the* nature, *the* concept, *the* platonic form, *the* idea of "history" (I find it more than safe to say) is just as futile with regard to "history" as it is with "art". We have consequently equally valid reasons for giving up any *descriptive* task of determining "history" and for settling for a slightly different assignment: to decide what we would wish to see history developed into.

Fortunately, we shan't have to bother about the (pseudo-) distinction between "art" and "science", as the extreme impreciseness of "art" and "science" makes for an entirely unenlightening distinction. The impreciseness of "history", moreover, allows us free reins, when we switch from a descriptive ("verative") to an optative mood. No option is precluded by the notion ("nature", "idea", "concept", "use") of "history", as there is nothing sufficiently precise or specific inherent in that notion to lead to anything like contradictions, "logical odditites", etc. Or is this an exaggeration?

It has been maintained that at least the *problem area* pertinent to historical enterprise, may — even if only tentatively and roughly — still be limited, to some extent. To Cicero this area was *res gestae*. To German metahistorians: *"Geschichte ist was geschehen ist"*. "History", claimed James Harvey Robinson, "is all we know about everything man has ever done, or thought, or hoped, or felt." And the "prime duty of the historian" is, according to Collingwood, *"a willingness to bestow infinite pains on discovering what actually happened"* (my italics). However, needless to say (it has been commonplace for centuries): it takes even more pains to discover and sort out, in any tenable or justifiable way, from all that actually happened, that which is *worthwhile* reporting.[2] I shan't deal with

---

[2]The problem rose to some prominence around the turn of the century, during the debate between Heinrich Rickert and Aloys Riehl. Rickert strictly maintained the distinction between the *"nomothetische"* methods of the *"Naturwissenschaften"*, and the *"ideographische"* methods of the *"Kulturwissenschaften"* [18], and saw, within the latter, *value judgments* as serving the same function as *laws* within the natural sciences. Riehl, on the other hand, was satisfied that historically significant facts could be selected by ascertaining their relative importance for a satisfactory explanation of *status quo*. The "disagreement" may well have been a pseudo-disagreement.

this platitude in the present paper, but assume that any event or sequence of events mentioned below, is, for some reason or other, found to be historically remarkable. It is, in any event, not a problem unique to history. All inquiries are obviously selective, as somewhat redundantly emphasized by Dray.[3]

The answer to the question, what history ought to be, is obvious: it should be good. Less obvious are the alleged *specific* difficulties or obstacles which presumably hamper the achievement of this objective. In point of fact: I don't believe there are any difficulties which apply to history, *sui generis.*

Problems of preciseness and objectivity, for instance, are more or less painfully felt within any ramified branch of any "hard" or "soft" science discipline. True enough, there was a time when philosophers and scientists dreamt of "incorrigible sentences". But for one thing: such sentences weren't, at any time, or by anyone, considered incorrigible with regard to level of *preciseness.* They were, however, expected—or, maybe more likely, invented—to fully satisfy any reasonable, or even possible, demand for truth or tenability. They were meant to constitute the *terra firma* of all non-analytical, non-tautological, synthetic, empirical knowledge. Unfortunately, these "confirmations" or "reports" were either *acognitive,* and *eo ipso* totally unequipped to serve as basis for any segment of (cognitive) knowledge. Or, they were to be conceived as cognitive sentences, in which case they would—unknowingly, perhaps—presuppose or imply a conceptual orientation, in fact: a total view, a (coordinate) system, *"eine komplette, vollständige Weltanschauung..."*, the tenability of which were, to say the least, contentious. More generally speaking, it is not at all easy to make any sense of any observer's or narrator's claim to "merely report what actually happened", or "just give the facts and data themselves",[4] However, as I have pointed out, these are all problems which are encountered within any—rigorous or moldable—discipline. They are most certainly not reflecting difficulties specifically faced by historians. It is more than doubtful whether there are any fundamental, methodological problems peculiar to "the craft of historians" as such.

[3]In [16].

[4]Cf. Skinner's example: "She slammed the door and walked off without a word. Our report is a small bit of history. History itself is nothing more than a similar reporting on a broad scale." [12], p. 15. As Arne Naess points out ([9], pp. 56, 57): "other witnesses might have reported the 'same' event" in a number of different ways. What is indeed difficult is to make a report of the event 'itself' which can be used as a common basis for various interpretations. . . ."

The most frequently mentioned "problem" is: *the problem of the autonomy of history.*" There should be no reason for more dreary repetitiousness on this topic (references are here innumerable). Suffice it to realize that the explosion-like expansion and differentiation of *all* disciplines in the last couple of centuries have rendered the very notion of "autonomous discipline" so obsolete that it has only, as it were, historical interest.

In his Cambridge inaugural lecture, 1895, Lord Acton said truly "that historical studies had entered upon a new era . . . It would be an understatement to say that since 1800 history has passed through a Copernican revolution. Looking back from the present day one sees that a much greater revolution has been accomplished than that associated with the name or Copernicus."

As James Harvey Robinson has claimed:

> The 'New History' is escaping from the limitations formerly imposed upon a study of the past. . . . It will avail itself of all those discoveries that are being made about mankind by anthropologists, economists, psychologists and sociologists. . . ,discoveries which during the last fifty years have served to revolutionize our ideas of the origin, progress and prospect of our race . . . . History must not be regarded as a narrow, stationary subject which can only progress by refining its methods and accumulating, criticizing, and assimilating new material, but *it is bound to alter its aims with the general progress of the social sciences, and it will ultimately play an indefinitely more important role in our intellectual life. . . .*

In the final analysis this "super Copernican revolution"—if one wants to go overboard with Lord Acton and Collingwood—does not amount to more than the practical elimination of the notion of "autonomous disciplines" in the traditional sense of "discipline"—which has made itself felt, particularly in the last forty to fifty years, within all fields of human inquiry. It is simply a result of the fact that few clear-cut classifications are apt to survive sudden, spectacular increase of information within their relevant field of knowledge, without rendering the relevant distinctions somewhat arbitrary. The snails on Celebes were not likely to have presented a problem to the zoological systematist, had the cousins Sarasin not taken their task quite so seriously.[5] Their enormous material revealed "an unbroken series of varieties where the extremes were as different as that of different species (*verschiedene Arten*)". Similarly, the venerable distinction "organic"/"inorganic" was initially destroyed

[5]See [17].

when Friedrich Wöhler (1828) converted ammonium cyanate into urea (NH_2CONH_2). The presumably most fundamental borderline intended to divide "living organisms" and "inanimate objects" really became a problem when the electron microscope and the ultra-centrifuge isolated the viruses of the mosaic diseases, and radioscopy revealed their crystalline nature. Well known also is the effect of the kinetic gas theory on the barrier between "mechanics" and "theory of heat". Other examples may be borrowed from "acoustics" and "mechanics", from "optics" and "electrodynamics", and so forth *ad infinitum, ad nauseam usque.* Difficulties have more recently arisen in connection with the increasingly arbitrary discrimination between "physics" and "chemistry", due to peculiar characteristics of the whole explanatory system, originally designed to deal with the discovery of blackbody radiation. Finally, Darwin could still permit himself some optimism with regard to deciding at what particular point in the series of hominidic fossils the application of the term 'man' would be appropriate. He did not foresee skull finds (from the second interglacial period — between Mindel and Riss) showing a rich variations of combined neanderthaloid and sapiens traits, thus rendering indefensible the common claim, presumably aimed to boost the feelings of human dignity, that man forms an entirely new dimension, fundamentally distinct from all other beings.

However exciting such examples may seem, they are all merely illustrating the trite, but true, saying that "modern science is characterized by its ever-increasing specialization, necessitated by the enormous amount of data, the complexity of its techniques and of theoretical structures within every field. . .". On the other hand, this does not seem "to lead to a breakdown of the science as an integrated realm". Quite contrary: it leads to a breakdown of *the borderlines* between the various major disciplines or "sciences", and a realization of "the structural uniformities of the various applied schemes", hitherto considered expressive of *the* "nature", "essence", "concept", "character" of one or another particular discipline, by the representatives of that major discipline. No wonder that Toynbee[6] feels that he has been trespassing across the different so-called "disciplines", which he finds to be "a curious piece of medieval sculpture", utterly obsolete, and solely maintained as a "self-defence against the fear of being overwhelmed by the sheer mass of information" in which the historian would seem to drown, were he "to bestow

---

[6]See [1], p. 34: "C'est le sentiment que j'ai franchi les lignes de démarcation entre les différentes prétendues disciplines telles qu'on les définit pour l'enseignement scholaire. A cet égard je déclarerais volontiers la guerre aux spécialistes et je dis que je ne crois pas du tout à la notion de 'discipline'. Je crois qu'il s'agit là d'une vieille sculpture médiévale, qui ne présente plus qu'une valeur de curiosité mais qui a eu la chance de se conserver jusqu'à nos jours."

infinite pains in discovering" (Collingwood) "*wie es eigentlich gewesen*" (Ranke), viz. "everything man has ever done or thought, or hoped, or felt" (J. H. Robinson). "People feel that the only hope of saving ourselves from being swamped by this, is to maintain water-tight compartments between one branch of knowledge and another—even if these compartments are arbitrary and out of date... ." It seems absurd to Toynbee "that we should say we are now inhibited from asking the questions that are really of vital importance..., just because we have too much knowledge to cope with them". And he points to the necessity "to find a means of studying human affairs as a whole, ... which would not be superficial and at the same time, comprehensive".

I have tried above to demonstrate that this is not a problem unique to history, or even to the more general field of knowledge concerned with human affairs and human personalities. And, although everybody is yearning these days for a "general practitioner" to synthesize smaller or larger chunks of information, there cannot be much doubt that one neither can, nor even ought to, halt (let alone reverse) the accelerating trend to ever subtler specialization. No road leads back towards a *unification* of physics, astronomy, metallurgy, biology, psychology, history, or any other discipline. We may as well face the fact that the name of an institution or department does not any longer offer any decisive clues as to what its researchers are doing. In the United Kingdom and North America, where it is common to lump university faculty members together in departments with labels borrowed from the names of more or less traditional disciplines, we constantly see not only how departments overlap, but how they are floating over into different *faculties* as well. Particularly interesting are the cases where a growing college finds it necessary to divide (so-called) "arts" and (so-called) "science" into two independent faculties. If the college is fairly advanced, tricky cases may well occur which lend themselves only to a most arbitrary decision. Some departments are divided right down the middle. I find these cases especially interesting because they shed some rather revealing light on the naive, antiquated notion of "the two cultures".[7] No such dichotomy is to be found. There may be two thousand "cultures", but certainly not just two. Moreover, one apparent result of the gradual (although intensely accelerated) shift from a macro- to a micro-specialization is the proportional

[7] It may be convincingly argued (cf. e.g. Helen Liebel in [6] pp. 16 and 17) that the division science-art, *Naturwissenschaften-Geisteswissenschaften*, nomothetic-ideographic [14], etc. is rather *more* pronounced the further we go *back* in the history of man. Maybe the real conflict is to be found fifty thousand years ago, viz. between, on the one hand, poets and musicians, and, on the other hand, toolmakers and men of outstanding dexterity!

degree to which it seems to interfere with the researcher's ability to rest comfortably within his narrow intellectual confinement. Another effect is the researcher's discovery of related methodological problems, similar conceptions and approaches appearing in the most diverse fields of scientific enquiry and other types of intellectual endeavor. Thus the long-dreaded "overspecialization" takes an unexpected turn. Dead is the feudal organization of disciplines and disintricate dichotomies; emerging is an imperspicuous integrality of research.

The historians should therefore, as I see it, not only *"avail themselves"*, as J. H. Robinson suggested, "of all those discoveries that are being made", but *make* them, if necessary. I can foresee historians undertaking (historically relevant) pioneer-studies in pollen-analysis, in radiocarbon dating as well as in climatology (or at least in effects of climatic fluctuations). The principal concern, however, should naturally lie with those *behavioral* sciences that J. H. Robinson mentioned: psychology, sociology, etc. This co-agency, however, is not to be compared—as is often done—with the relationship between geology and physics. Geology, so the argument goes, admittedly *avails* itself of what is commonly referred to as "laws of physics", but without any geologist ever attempting to *establish* such laws: that *would* be absurd! However, it is a contingent truth, simply a function of the fact that physics is relatively rigorous—particularly when compared to the "soft sciences" with which history should seek a symbiotic existence. In order to make the partnership mutually profitable, the craft of historians will undoubtedly be transformed beyond recognition. Not only will historians exploit, say, "latent-attribute-" and "reason analyses",[8] *à la* Lazarsfeld, but, by theoretical and experimental studies, attempt to measure the utility of such and similar techniques. Mathematics and (at least certain branches of) physics constitute peculiar buildings, where order, peace and rigor, and an almost complete unanimity, prevail in a number of "the middle floors", as it were; whereas on "the top floors" (i.e. "the frontiers of knowledge") and in "the basement" (viz. *"Grundlagenforschung"*) the situation is rather shaky, with perennial controversies and fierce fighting. As we move on to "softer sciences", we find that the number of "middle floors" is steadily decreasing. The historian who intends to avail himself of the results of "soft-science" research should be prepared to find himself involved in altercations "as to what are matters of established fact, what are the reasonably satisfactory explanations (if any) for the assumed

---

[8]"A research technique for determining why people behave as they do on given occasions."

facts, and what are some of the valid procedures in sound enquiry" ([7], p. 448). One can easily sympathize with the historians who shy away from such turbulent entanglement to keep their hands clean, even at the cost of ignorance in relevant matters.[9] However, I cannot imagine how the historians would plan to escape the involvement, were they to take to heart Collingwood's code, to shun no pain to find out what actually happened, what "man has done, thought, hoped or felt". To take an example: The *Egils Saga* portrays the most versatile and original poet of Icelandic antiquity, Egil Skaldagrimson. To his contemporaries Egil was even more celebrated as one of the most terror-striking, petrifying *berserkir*.[10] The following incident, however, is reported to have taken place after Egil's final settling as a fierce but highly respected autocrat on his paternal estate in Iceland. As Egil is passing one of his men, who is bending over, arranging his footwear, Egil draws his sword and swiftly beheads the man. Egil's excuse (or explanation) is famous: "He posed so conveniently for a blow." If we assume the veracity of the narrative, to what extent are we entitled to say that we have found out what actually happened, what the people involved thought, hoped, or felt? The *Saga* seems to suggest that Egil held no particular grudge against his victim. Perhaps he just "took out his aggression on him" — an aggression, say, caused by some unrelated, recent frustration to which Egil happened to have been exposed...; so historians may hypothesize. But the mere contemplation of this hypothesis plunges the historian headlong into the metapsychological dispute between the proponents of (a) the "learning model" and (b) the "frustration-aggression model". According to the former, a perfectly adequate explanation of the decapitation is offered in Egil's fervent devotion to the *berserkir* profession — he had learned to react aggressively on a minimum of provocation. According to the latter mode, however, Egil's long career as a *berserkir* should have prevented any built-up aggression, because of the frequent "discharge" in numerous battles.

It would clearly be preposterous for any historian to endeavor to give an account of Egil's behavior in terms of either of the indicated models without an experimental *coup d'essai*, designed to assist him in a tentative assessment of the relative utility or tenability of the two models. It is commonly admitted that this area of research does not lend itself very readily to rigorously controlled experiments. But it is equally obvious

[9]It seems most charitable to see in this light Gallie's [3]absurd persistence in conceiving history as "purely narrative".

[10]Wild Norse warriors who were leading Viking attacks with a frenzied fury allegedly caused by the consumption of *amanitine*, the active narcotic of poisonous fungi, particularly of "fly-bane" or *amanita muscaria*.

that there is no solution in writing experiments off as impossible. "Experiments in history" (or in other empirically oriented fields of behavioral research) is not a locution that entails a logical contradiction.

The ominous task of carrying out such experiments is not inherently insuperable. One may initially have to lower (drastically) the level of ambition with regard to rigor and controllability, etc.; but surely the most reliable road to methodological improvement goes through a tireless employment of the available methods in full awareness of all their potential shortcomings and inadequacy. Historians have often wondered[11] *whether* Antony fled from the battle of Actium blinded by an all-devouring adoration (or unsatiable desire) for Cleopatra — who, he was informed, had already withdrawn her support and pulled out; *or* he simply realized that the battle was lost anyway ... etc. Somehow it seems as though historians have resigned: they rest in the riddle, waiting for the earth to erupt with some miraculous "historical" evidence. It is hard to understand why any evidence whatsoever should be inadmissible to the historians, if it has the slightest bearing on a historical conjecture. There is no reason why history should remain a psychology (or whatever it may be) *in the past tense*. It is perfectly possible to do *something* to find out, say, how subjects of Antony's general description are inclined to react under external stress, and when engaged (in the extreme) in some kind of heterosexual relationship.

Results of such "experiments in history" may not be terribly enlightening, at first. The important thing here is to give precedent to a practice, to break the spell of "an autonomous discipline", and open doors of potential insights and knowledge perfection, which, with hard work and a little luck, might one day contribute to establish "*was eigentlich gewesen*".

Needless to say, there is nothing new or exciting in the above argumentation. "*Es entspricht nur der fortschreitenden Scientifizierung, welche die Historie im 19. Jahrhundert erfahren hat, indem sie mehr und mehr aus dem Bereich der schönen Literatur in das der wissenschaftlichen Forschung hinübergezogen wurde*".[12] Qualms about accepting this "scientification" of history may most plausibly be ascribed to a combination of the following two, rather recalcitrant, misconjectures: (1) that "scientific" *must* be used in a way so as to properly designate only certain particularly exact and rigorous branches of physics; (2) that "history" *must* be defined in terms of what historians *pro tempore*, e.g.

---

[11]See [7], pp. 549, 550 and 555.
[12]See [14], p. 24. See also [18].

anno 1969, actually do. One may deplore the increasing dilution of the
venerable notions of "science" and "scientific"—it is nevertheless precisely this wide or diluted sense which currently has come to characterize
"scientification of history". As to the latter point (2), suffice it to note the
utter fatuity of even an attempt to demarcate, at any given time, the ultimate cincture around any branch or field of research. Had, say, genetics
a century ago been delimited according to what Mendel at that time was
doing, most of the work of modern geneticists would have to be subsumed
under some other heading (than "genetics"). There is no reason to believe that future scientists would find it less preposterous if "history"
were restricted to specify "the craft" of our contemporary historians.
Chances are, moreover, that some metahistorians' numerous pretexts for
renouncing "a science of history"—or declining to recognize it as a possibility—will (prospectively) sound equally absurd. The mere idea that all
history, or a substantial part of professional historiography, ought to have
(or actually has) the general character of scientific research (even in the
above wide or diluted sense of "scientific") is undoubtedly the most
valuable feature of the whole historical tradition, from Ashurbanipal,
Hecataios, and Hesiod through Herodotus and Thucydides to von Ranke,
Lamprecht, Rickert, and Riehl. That such an idea is now "in", now "out",
is obviously irrelevant. It is important to realize that any attempts, systematic or not, to avoid a "history" which *resembles* scientific research,
are clearly without foundation.[13]

Finally, and curiously enough, there appears to be a notion of *"personality"* which on occasion seems to afford a last refuge to those who
are most heritant in accepting a "science of history". After a careful
examinations of a great number of definitions of "personality"[14] I think it
is safe to say that the most common notions of "personality" offer no intrinsic or insuperable difficulties for a science dealing with *decisions* or
actions carried out by *personalities*. Certainly nothing hinges upon the
fact that such actions or decisions are commonly said to be understood,
explained or predicted only on grounds of *reasons* or *norms*, etc. (accep-

---

[13]Although distinctions like *nomothetic-ideographic* [14] are long gone, as are *intentiona-
lity-causality, questio juris-questio facti, vérités de raison-vérités de fait*, etc., a contrast
is occasionally sought between the scientific *laws* of certain branches of physics, and the
not-so-scientific laws employed elsewhere, say, in the behavioral sciences. When a statement is pronounced a "law" within an explanatory system, it implies that its (and its en-
tailments') verity is not to be questioned, even in the face of *quite* a bit of apparent evidence
to the contrary. The less rigorous a system, the more susceptible are the laws to evidential
influence. Conversely, laws "on the middle floors" of e.g. physics, take on a near-analytic
character.

ted by the person, or characterizing the personality in question), rather than in terms of *causes*. Let us say: (1) that Naess has given an adequate reconstruction of Gandhi's norm system; (2) that, in conflict situations, Gandhi was always aware of his norms; and (3) that Gandhi was self-consistent and resolute. It is hard to imagine a sense of "understand", "explain", "predict", such that, if the three above conditions were met, we wouldn't be abundantly well equipped to understand, explain and predict any historically significant decision ever made by Gandhi. True enough, Gandhi may be a particularly fortunate case; but again, it is only a matter of degree between Gandhi's case and that of, say, a somewhat bewildered scatter-brain, or even a mentally deranged person. And the interesting thing to note in this connection is that most typically (or maybe only) in the Gandhi case would particularly enthusiastic employers of this "personality" notion be inclined to admit that "the *personality* was active in making the decision". Oddly enough it seems as though, according to this apparently ponderous notion of "personality" — often termed "integrated" — the personality, P, *must make* the decision, D_i, *but in such a way so as to ensure that D_i cannot be said to be caused by P*. The rationale behind this clause is unclear. It seems most likely connected with scruples against accepting roughly the following reasoning: If D_i is *caused* by *P, P* could not have been able to make any *other* decision than D_i e.g. D_j. It is thereby excluded that *P* could be considered "an integrated personality", viz. "a master in his own house, determining which dispositions will be enacted and how and when they will become overt" or words to that effect.

One would certainly be well advised against accepting this peculiar piece of argumentative discourse, but not for any more interesting reasons than the difficulties of making sense of it. To say that Gandhi's decisions were *caused* by Gandhi's *norms* (for instance), may sound a bit exotic, at first. On second thought, however, not half as exotic as the claim that there were no connections whatsoever between Gandhi's decisions and his norm system. Such a claim could only be based on new informations revealing that Gandhi did not consider his norms at all, when making historically significant decisions. If he *did* consider his norms, however, when deciding (which he did), then certainly Gandhi's norms — together with tenable assessments of his self-consistency, resolutesness, etc. — give us a most valuable key to understanding, explaining, or predicting Gand-

---

[14]On more or less definoform statements about "personality", I am indebted to my colleague in the Department of Psychology, Wm. Blanchard, for furnishing quite a few of the examples.

hi's decisions and actions. Whether the connections are to be seen as "causal" or not is inconsequential. Suffice it to realize that in no sense of "understanding" — viz. *"Begreifen"* or *"Verstehen"* — are we to attempt to understand Gandhi's choice of "enacted dispositions" without having acquainted ourselves with his system of ethical norms. And in no sense of "causal", "caused", or "causality" does it make any difference to *P's degree of power* to choose D_j rather than D_i, that there *is* causality, or that D_j or D_i *is* caused, or *has causal connexions* with something inside (or outside of) *P*. In the case of Gandhi, it is certainly true that *if* Naess's reconstruction of Gandhi's norm system is correct, and *if* we could foresee which norms would apply, and *how*, in each choice situation (as conceived by Gandhi), then *Gandhi couldn't have decided otherwise*, i.e. in this sense: the probability was 100% that Gandhi would decide as he either actually decided, or was predicted to decide (say, D_i). We have in other words constructed a case such that there was by definition no probability left for D_j. But we have *neither* impoverished Gandhi's capacity for decision making, nor for moral effort, endurance, consistency, resoluteness, etc. On the contrary: we are indeed counting on the fact that we have — at least in Gandhi's case — subtracted nothing from (or added nothing to) what may be called *"P's predecisional activity"* eventually leading up to the decision, D_i. D_i is here not itself to be seen as an activity (or an action), but as an *achievement* accomplished through the predecisional activity, the deliberation, the more or less laborious efforts to "make up one's mind", etc. In any case of decision-making, *P* then either pronounces D_i, say, or his predilection for D_j. Hence, since *P* always (?) would have it in his power to utter any other constellation of phrases than those intended to convey D_i (e.g. those meant to express D_j), *P* is always, so it might be argued, completely "free" to make any decision whatsoever. It goes without saying that this sense of "*P* deciding (freely) D_i $(D_j, D_k, \ldots)$" should be ignored as anything else but an example of a primitive linguistic pitfall. A pitfall, however, which, under particularly favorable circumstances, may become a fatal trap. Students of decision-making do in most cases, but not in all, use "decision" in a way which renders "decision" cognitively similar to "judgement", "verdict", "assessment", "determination", "estimate", "appraisal" and the like.

The concern is here most typically with cases, say, where subjects are asked to "decide" (less misleadingly: "conjecture" or "bet") whether, in a deck of cards with positive and negative numbers, the mean will be above, or below, zero. It seems rather reasonable to assume that such practice is quite likely to pave the road for the aforementioned fatuous use of "decision". Be that as it may, in everyday use, if P decides (on) D_i, he also commits himself *ipso facto* to a post-decisional activity, A_i. In point of fact, if P does not make some manifest and serious attempts to carry out A_i, we should hardly consider D_i a decision (or we should say that P had reconsidered his decision). If P decides to marry Miss S, it implies that he eventually takes steps to marry her. This is clearly different from P "deciding" ("finding out") whether he is in love with Miss S. (Although again an unlimited number of difficult borderline cases may be found.) Now, insofar as post-decisional activity is conceptually included in the notion of "decision", the question as to what a personality, P, at any time had it in his power to decide, becomes a straightforward empirical problem, a problem—if P is a historically significant person—for an empirical, experimentally oriented, behavioral science: *history*.

REFERENCES

1. R. Aron (ed.), *L' Histoire et ses Interpretations*, Paris, 1961.
2. R. Brandt, "Personality Traits as Causal Explanations in Historiography," in *Philosophy and History: A Symposium*, New York, 1963.
3. W. B. Gallie, *Philosophy and the Historical Understanding*, London, 1964.
4. P. Gardiner, *The Nature of Historical Explanation*, London, 1952.
5. E. T. Gargan, (ed.), *The Intent of Toynbee's History*, Chicago, 1961.
6. H. Liebel, "History and the Limitations of Scientific Method," *University of Toronto Quarterly*, **XXXIV**, Oct. 1964.
7. G. H. Nadel, (ed.) *Studies in the Philosophy of History*, New York, 1965 (selected essays from *History and Theory*).
8. A. Naess, "Systematization of Grandhian Ethics of Conflict Resolution", *Journal of Conflict Resolution*, **II** (June, 1958), 140–155.
9. A. Naess, "Science as Behaviour," in *Scientific Psychology* **B. B.** Wolman, (ed.), New York, 1965.
10. P. Rossi, *Storia e storicismo nella filosofia contemporanea*, Milan, 1960.
11. G. Simmel, *Die Probleme der Geschichtsphilosophie*, München and Leipzig, 1923.
12. B. F. Skinner, *Science and Human Behavior*, New York, 1953.
13. J. Vogt, *Wege zum historischen Universum: Von Ranke bis Toynbee*, Stuttgart, 1961.
14. W. Windelband, *Präludien*. Tübingen, 1884, **II**, pp. 136–160, "Geschichte and Naturwissenschaft."
15. K. Popper, *Of Clouds and Clocks*, Washington University, St. Louis, 1966, p. 3.
16. W. Dray, "The Historian's Problem of Selection," in *Logic, Methodology and Philosophy of Science, Proceedings of the 1960 International Congress*, Stanford, 1962.
17. P. and F. Sarasin, *Materialen zur Naturgeschichte der Insel Celebes*, Berlin, 1898–1906.
18. H. Rickert, "Geschichtsphilosophie," in *Festschrift für Kuno Fisher: Die Philosophie im Beginn des 20. Jahrhunderts*, Berlin, 1904.

DISCUSSION

BONDI: It seems to me that the difficulty concerning experiments in history is, in a sense, quantitative. In physics, if we want to repeat an experiment we must make a judgment concerning which circumstances of the original experiment were relevant and which were irrelevant, and we must then repeat the relevant ones and disregard what happens to the irrelevant ones. This is always a question of judgment. However, often the answer to this question may be obvious; yet there are many examples in the history of physics of this judgment being wrong. One thus reaches the case of an experiment whose result is at first thought to be inexplicable. Later it turns out that some circumstances which were thought to be irrelevant were highly relevant, and the apparently inexplicable becomes plain. So it would be quite wrong to say that in physics it is always obvious what is relevant and what is not relevant. But would I not be right in saying that in physics it is far easier to decide on relevance than in history? Therefore, to do experiments in history (which necessarily involves the idea of repeatability) requires a degree of judgment about what is relevant and what is not, on which it would be so difficult to reach any form of agreement as to make the whole idea of experimentation in history of dubious value.

TENNESSEN: I have never done any experiment either in history or in physics except on a very elementary level. However, I suppose that in physics, as in many other sciences, it is fairly simple to reach agreement in the middle floors, as it were; but when it comes to the attic or the basement, you get into as much the same difficulties in physics as you do in some softer sciences.

BONDI: I cannot think of an experiment in history for which this question of relevance would not be so debatable that there would be almost no point in making the experiment.

TENNESSEN: Any experiment is better than nothing, I should think. Take the particular case of Antony: Here one is throwing in quite a few assumptions which could be tested psychologically. If a historian is one who takes pains to find out what actually happens, he will have better grounds for establishing what really happened if he knew what was going on within Antony. And is this an explanation, is there a relationship between Antony's being in love, Cleopatra's running away and his running away with her? Is this *the* explanation? In other words, if we can get people enormously in love and see that they do all sorts of stupid things, it seems to me we have to some extent supported that notion, at least more than by just sitting around and saying, "Oh, you know, being in love you do all sorts of stupid things". Admittedly, the empirical support may be rather insignificant. But we have done *something*, and the more we do of that sort of thing, the more we know about the battle at Actium.

BRECK: I would agree with Tennessen that the historian make use of the methodology and reasoned conclusions of experimenters and researchers in other fields of the social and biological sciences – all the way from numismatics to psychology. On the other hand, however, the professional historian is worried about the ability of modern man to understand with any degree of ease the remoter past. Is it possible that our contemporary thinkers, with their absorbing desire to understand our ancestors and their actions, have read more into the past than the facts will support? Thomas Carlyle "knew" Oliver Cromwell, Sigmund Freud "understood" both Moses and Leonardo da Vinci, yet in all three cases we see more of the writer than of the man written about. Brecht on Galileo, for instance, is good theater but poor history, even though the playwright was armed with considerable sociological and political theory. Sometimes we believe we can discern the "spirit of the times" and know what a person in a particular segment of the world at a particular time would have thought.

CHASSON: There is a much closer relationship between physicists and historians than one might think, particularly in astronomy and astrophysics, where only recently has there been any degree of possiblility of interfering with the source of information in an observational manner. Still, we cannot decide what we are going to look at; we simply look and try to assess in reasonable terms what might have happened.

TREDER: The essential point of a physical experiment is that each physicist is in principle able to repeat it at each point and at each time. If one physicist asserts a property of a physical system, any other physicist is able to prove this proposition in his institute and with his instruments. But we can't repeat a historical situation. Marc Antony and Cleopatra were unique people who participated in a unique, unrepeatable situation.

NAESS: Experiments in historical narrative are nonsense; experiments within historical research are common sense.

YOURGRAU: In physics, biology, and psychology we know today a lot about the methods of experiment. We call this methodology; in fact, we deal quite thoroughly with methods of experiments. I do not see how one would develop such a method in history because the question raised is almost a vacuous one: "Is history an autonomous discipline?" I think we all agree that every discipline of human beings would today have recourse to auxiliary disciplines like psychology and sociology.

One speaks about the stupid things people do when they are in love. I made some imaginary experiments. You see, I cannot make any other experiments because I have not been given a method. Hence, I had to make, out of sheer necessity, imaginary ones. I think if one now chooses, let us say, a couple, where the man resembles Antony and the lady Cleopatra, even if one then reproduces that situation, one still is not allowed, according to Tennessen's premises, to say "This is an analogy or faithful picture of Cleopatra and Antony between that date and now." So, at best, one might entertain the notion that something similar has happened. After all, one might say that the biological urges of men haven't changed since cave men — I would question this, by the way — but one could have a very precise, concrete, and scientific viewpoint.

Now the only situations where repetition leads to results are in logic, mathematics, and theology. In mathematics I can always have the same theory, the same proof, and that is the same theory, the same proof as in theology, which latter fact mathematicians do not like to hear. I would say they are aprioristic, and here I can always have a reversible situation. I don't have this, in my opinion, in history. In that respect, history is very close, strangely enough, to astronomy and cosmology, because galaxies too have a history and one cannot reproduce it.

TENNESSEN: There are very many things that are difficult to do. One would like to be able to produce, for instance, something in support of certain theories in geology. There is nothing in principle to prevent it. It is just enormously difficult to establish the circumstances under which one could do so.

VIGIER: Concerning the question of experiments in history and physics, it is true that we can't always experiment in physics in the active sense (for example, in cosmology); but it is also true that there are various levels of experiments. In a sense, the problem in history is something like the distinction one has to make between collective variables and individual behavior. There are various structural levels. One tries to study the behavior of great human masses and to abstract causal laws from that behavior, neglecting fluctuations. Then when one wants to study fluctuations down to the use of a sword, for instance, in this special case one has to utilize another type of knowledge and science. This implies that history is a science but a science with various structural levels and various structural disciplines.

These are not just my ideas; there are comments made by Marx on the relation of his philosophy to historical material, to the effect that history is an autonomous science, dealing with these various structural levels, and that one should have a very complex approach to such a science. However, history has a fundamental new specific aspect which should always be taken into account—namely, that we live in history and we are conscious of history. This means that experiments in history have a specific aspect. We are imbedded in history. Each time we are acting as citizens we are really experimenting in history, starting from a certain background.

So we have a real problem there: a structural problem. What I would like to see analyzed, in a specific, scientific way, is precisely the distinction between these various structural levels and the way in which the specific laws act at each level and the way in which they are interconnected. It is clear that the notion of experiment and the notion of predictability have a very specific meaning in the domain of history. With these presuppositions, I agree with Tennessen's statement that history is a science.

MERCIER: In Tennessen's paper the notion of freedom has been ignored. It seems to me that freedom is necessarily bound to both what personality may be and what history is, freedom not merely interpreted as free will but as it is conceived in philosophy. I don't think that any particular science would ever be able to explain to us what is to be understood by the term "freedom". And philosophy has never given an explanation satisfactory for everybody; but still, philosophers consider freedom as one of the biggest problems.

TENNESSEN: I did intend to say something about freedom, and in fact under the heading of "personality" I managed almost to avoid "personality", and, in my opinion, talked about freedom instead. What I intended to do in the latter part was to show that I can't think of any concept or any notion of "freedom" that would be such that it would preclude the possibility of a science of history.

NEF: The subject matter of history is enormously complicated and varied, and the subject matter treated has been increasing immensely during the last centuries. Now, it appears to me it would be much more helpful if instead of talking about experiments we talked about comparisons. Material of all sorts is available for comparisons in history, and this is an especially fruitful subject for exploration. I would also like to suggest that in citing what is worth comparing, one should decide what is important.

TENNESSEN: I suggest that the reason some are hesitant about accepting "experiment" as well as "predictability" and some of the other words I have used, is that historians, as well as philosophers, orient themselves much more towards what I call the "hard" sciences rather than the "soft" sciences. Greater acquaintance with what is going on under the name of "psychological experiments" would help reduce the hesitancy in accepting the word "experiments". I will give an illustration.

One of the greatest psychologists working with rats in Norway started out trying to test the hypothesis that homosexuality has nothing to do with low sexual potency. The homosexual is, on the contrary, more sexually potent; but he is frustrated by the influence of Christian morality. Now, as all decent psychologists do, our researcher tries to test this theory in rats. It wasn't so difficult to test the sexual potency; he just invented an electric grill and counted the times the rat went over the electric grill in order to copulate. It was somewhat more difficult to introduce Christian morality into the rats. But he finally decided that an electric shock of about 200 volts would be roughly equivalent to Christian morality. So after having gone through that absolutely fabulous and painstaking work of finding out that the rats — 1200 of them — could copulate, the experimentor put them into a laboratory situation, and when the male made the initial step, he turned on Christian morality, and the rats jumped three feet in the air. They all became homosexuals, of course ... .

In this watered-down sense of "experiment" we surely could do much better in history, it seems to me.

BRECK: Granted, historians must learn to use at least some of the techniques of the various social sciences. But I cannot see the work of the scientist in the laboratory as anything directly usable in any attempt to reconstruct the past or find out much about human behavior from rat behavior. We deal so often with the nonlogical elements of human behavior, with the unique event and the single leader or thinker. Exceptions outnumber rules in our historical trade. In sum, the historian cannot experiment. Again, he *can* make fruitful comparisons, indicate trends, expropriate methods in his search. But all this is far from reduplicating the past at will. Such a possibility, it seems to me, is ruled out by the infinite variety of human motives and actions, by our own inability to live completely in the past, even by the changing aspects of what we choose to call "human nature".

BONDI: I agree with Breck. These are all experiments in psychology leading to psychological knowledge that may be of value to science, but are they "historical"?

TENNESSEN: I have tried to argue that in order to live up to Collingwood's commandment: "one should take infinite pains in finding out really what is going on", one cannot take infinite pains in finding out everything that's going on if one excludes certain sources of information. And I happen to stress psychology and sociology, since these are my fields. But I am in this context solely concerned with experiments pertinent to historical theory, so as to ascertain *"wie es eigentlich gewesen"*. I think historians themselves should conduct such experiments; they shouldn't wait for psychologists to do it.

COHEN: Is it really true that an experiment should be repeatable? We may agree that it is perhaps not true, and one can give examples from cosmology to support this, although even there I suspect some methodologists might simply say it is in principle repeatable; thus, if such-and-such happens it could repeat, i.e. something cyclical could occur. But if we accept an experiment as not necessarily repeatable, then what remains of the objection to history on the part of the others? What are we fighting against any longer? It seems we are fighting against an objection which doesn't exist. I think in the United States probably 98% of professional historians hold that history is not a science, in the American sense of the word 'science'. I take it that what the historians have meant is that there are no predictions possible; and we natural scientist come back then and say that this is not a firm objection, and we make our conclusion deeper by saying no experiment is possible in some fields of study. We say, "Well, you are thinking of the naive view that Boyle's law is repeated in every freshman laboratory and we are above that now in physics, and we always were, in fact, and it is good for you to learn this truth from us." And then Mercier responds, if I understand him correctly, that, even if all my arguments are correct, nevertheless, something else remains on the individual level, namely, that the fluctuations, or the individual behavior, not the collective behavior, are in some sense not predictable by historians.

If I understand Tennessen correctly, without discussing freedom explicitly, he would then say, "Ah, that is a branch of science other than history". I suppose he would agree with Vigier and say history, by and large, is on a different level of explanation, call it collective for want of a more precise word, but that psychology will deal with the fluctuations. Would this mean, then, that personal or individual (atomic) behavior, for which psychology is the proper science, deals with question as to whether there is any science to deal with freedom? Or does this mean there is still an open question as to whether there is any science to deal with freedom?

MERCIER: One should never quote oneself but I have written a book on love, and Tennessen's paper mentioned love, and the conclusion in my book is that you can't talk about love (not just with reference to sex but love in a metaphysical sense) without talking about freedom, and you can't talk about freedom without talking about love.

TENNESSEN: I think the notion of 'freedom' is not so clear to me that I could say if there is a science to deal with it and, if so, what we should do about it. I think the first step in "the question of freedom" should be dealt with semantically, and when we have dealt with it in semantics, then we can do something. What I am aiming at is something like that. What one would discover, I suppose, is that there isn't *one* problem, but an enormous number of problems.

Let's say somebody wonders whether man is fundamentally evil. Someone may be interested in that. Well, it appears to me one could do some experiments to find out whether aggression is inherited. And if you find out it is *not*, that a baby is never aggressive, it seems that this *could* indicate that man is not basically evil. The level of aspirations with regard to any inquiry into freedom would have to be on that level, to begin with. One cannot take 'love' or 'freedom' and simply go out and do an experiment like Millikan's oil-drop experiment, which can be repeated any time and is of enormous significance.

VIGIER: Now, geology is a science which is similar to history, although there are significant differences. History is much more complex, but in both cases we have a doman where you cannot make any direct experiments. However, you can build theories. You can utilize knowledge obtained in other domains such as chemistry, and you can make predictions when you are looking for oil fields, for example. The practice of experiment is something completely different from the naive idea that people customarily have.

NAESS: I think I can formulate a main point made by Tennessen in terms of the theory of argumentation. Let F be an explanation of the behavior of a historical person (e.g. Egil's beheading of his comrade Tor), and let us talk about historical research rather than so-called historical science. Historical research must include attempts to evaluate (by arguments *pro et con*) adequate historical explanations. If the correctness of a learning model of aggression is involved in assessing the strength of the (higher order) *pro*-arguments, the strength of rival frustration-aggression models must be accounted for in the *con*-arguments. Thus, historically relevant arguments of higher order lead us to consider the status of experimental evidence concerning rival theories of aggression. If the historian is "task-minded" rather than "discipline-minded", and he needs more data in his evaluation of arguments, he must himself experiment on aggression or use the experiments of others. The main point is that if F definitely is a historical assertion, it is part of the specific task of the historian to evaluate any argument relevant *pro* or *con* F. Thus, if a psychological experiment is required, *the psychological experiment is a genuine part of historical research*. The outcome of the experiment is expressed in a psychological assertion, but the *argumentational pattern* including that assertion is historical, centering on F.

There is a strong argument for historians doing or administering experimental research themselves: they are anyhow heavily influenced by scientific or quasi-scientific theories like those of Marx or Freud. But they often pick them up too late, or make use of uncritical generalizations. If historical research teams more often went ahead themselves, they could get more pertinent data more quickly, and thus be better able to evaluate historical argumentations.

It is sufficient to agree that if an assertion expressing the outcome of an experiment is also an assertion with non-zero relevance within the pyramid of *pro et con* arguments related to the historical assertion F, the experiment is part of historical research.

Chapter IV

THE GROWTH OF A THEORETICAL MODEL: A SIMPLE CASE STUDY

Håkan Törnebohm
University of Göteborg

I will present a case study of the development of knowledge about the solar system in Newton's time and then attempt to draw some general conclusions from it.

Historical accidents which affected the growth of this knowledge are not of interest in this case study. I will therefore take the liberty of making a reconstruction of the historical development. The cast consists of two people whom I have invented, Wenton, a theoretician, and Pekler, an observational astronomer.

TRANSITION FROM PRESYSTEMATIC TO SYSTEMATIC KNOWLEDGE

Wenton began his work with two bodies of hypotheses, (1) a set K_0 of tested and accepted hypotheses, and (2) a set S_0 of speculative hypotheses.

K_0 consists of two hypotheses:

1. The planets move in circular orbits with the sun in the center.
2. The velocity of a planet in its orbit is constant.

$K_0 1$ and $K_0 2$ will be referred to as Copernicus' "laws" of planetary motions.

S_0 consists of four hypotheses:

1. The sun is permanently at rest in the center of the universe.
2. The sun is not affected by the planets.
3. The planetary motions are caused by the sun.
4. The causal link from the sun to a planet is weaker the more distant the planet is.

Wenton transformed these hypotheses into model conditions imposed on a two-particle model consisting of s representing the sun and p

representing a planet. The model sp without model conditions is a bare model. The model conditions dress the bare model.

The model conditions are of two kinds, kinematical and causal model conditions. The former are expressed in the language of the kinematics of mass-points. The causal model conditions are expressed in a special symbolism for causality invented by Wenton. Figure 1 shows the dressed model.

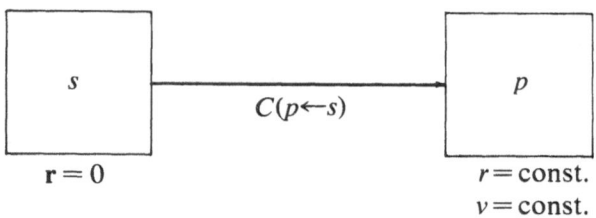

Fig. 1.

We have:

C1: $E(s|p)=0$
C2: $E(p|s)=C(p\leftarrow s)$
C3: $E(p|s)=\mathbf{a}$,

where $E(s|p)$ represents the effect of p on s, $E(p|s)$ the effect of s on p, $C(p\leftarrow s)$ a causal link (coupling) from s to p, and $\mathbf{a}$ the acceleration of p. C1 and C2 are transcriptions of S_02 and S_03. C3 links the causal and kinematical formalisms. Galileo's work suggested C3 to Wenton.

The obvious thing for Wenton to do next was to find an expression for the causal link $C(p \leftarrow s)$. How is p accelerated, if it moves with constant velocity in a circular orbit? The answer is given by the formula

$$\mathbf{a} = C(p \leftarrow s) = -\frac{4\pi^2 r}{T^2}\mathbf{r}^0.$$

According to S_04, $C(p\leftarrow s)$ should decrease when r increases. Wenton was therefore led to set forth the following model condition:

$$T^2 - \frac{4\pi^2 r^\alpha}{\gamma}.$$

This model condition is readily transformed into a testable hypothesis. The quantity T is operationally defined as the time of a complete revolution and r is the distance from the sun to the planet as measured by contemporary astronomers.

The astronomer Pekler confirmed the hypothesis and found that $\alpha{=}3$; he also obtained the value of the constant γ.

Wenton used this result, which was a new piece of knowledge, and obtained the two new model conditions:

(g) $$\mathbf{a} = C(p \leftarrow s) = -\frac{\gamma}{r^2}\,\mathbf{r}^0$$

and

(γ) $$\gamma = \frac{4\pi^2 r^3}{T^2},$$

where γ represents a new physical concept operationally defined by means of (γ); (γ) defines a property of the sun which according to (g) is responsible for its power to attract the planets. Wenton called this property the *gravitational strength* of the sun.

Having obtained (g), Wenton changed the model conditions on the bare model sp as shown in Fig. 2. C1, C2, and C3 are retained.

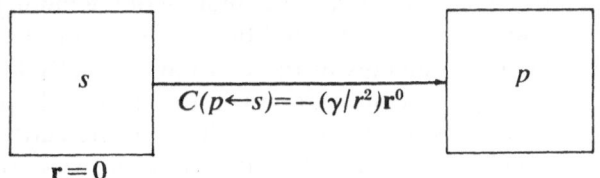

Fig. 2.

What are the possible motions of p? A simple calculus led Wenton to this answer, the set K_1,

(1) $\dfrac{1}{r} = \dfrac{\lambda}{h^2}\,(1{+}e\,\cos\,\vartheta).$

(2) $r^2\dot{\vartheta}{=}h.$

(3) $\gamma = \dfrac{4\pi^2 a^3}{T^2}.$

These model conditions were transformed into hypotheses by Pekler who tested them and confirmed them (= failed in his attempts to refute them). They are Kepler's laws.

The set of tested hypotheses K_1 at this stage differs from the original set of tested hypotheses K_0 in three respects:

1. *Accumulation:* K_1 contains more hypotheses than K_0. The hypothesis $K_1 3$ is a new hypothesis.

2. *Upgrading or refinement:* $K_1 1$ and $K_1 2$ may be regarded as descendants of $K_0 1$ and $K_0 2$, which also describe how planets move. However, $K_1 1$ and $K_1 2$ have a higher grade than $K_0 1$ and $K_0 2$ in this sense: Every test which fails to refute the Copernican hypotheses fails also to refute the Keplerian hypotheses, but there are tests which refute the Copernican but fail to refute the Keplerian hypotheses.

Remark: Upgrading means that predecessors of new hypotheses have to be removed from the body of scientific knowledge, which thus must possess both an entry and an exit.

3. *Organization:* The hypotheses in K_1 are organized in this sense: Every member of K_1 is represented by a model condition which is a theorem in a deductive system. The members of K_0, on the other hand, are organized in a very loose manner by means of S_0.

EXTENDING THE RANGE OF APPLICATION
OF THE GRAVITATIONAL MODEL CONDITION

There is a close analogy between the motion of the moon around the earth and the motion of a planet around the sun. Wenton was led by this analogy to assume that satellites of the earth are causally linked to the earth in the same way as planets are linked to the sun. He posited that falling bodies may also be regarded as satellites of the earth. So he assumed that the condition (g) may be imposed on a two-particle model which represents the earth and a satellite. He immediately obtained the following new model conditions:

$(\gamma' 1)$
$$\gamma' = gr^2$$

$(\gamma' 2)$
$$\gamma' = \frac{4\pi^2 R^3}{T^2}.$$

Condition $(\gamma' 1)$ refers to a falling body; g is the acceleration of a falling body in a vacuum near the surface of the earth; r is the radius of the earth. Condition $(\gamma' 2)$ refers to the moon. R is the distance to the moon and T is a lunar month.

Pekler confirmed that the right-hand side of $(\gamma' 2)$ has the same value as the right-hand side of $(\gamma' 1)$. The common value is the value of the gravitational strength γ' of the earth. Pekler's findings confirmed Wenton's assumption that (g) also applies to the satellite system of the earth in spite of the fact that the earth moves while the sun is at rest in the universe. (Wenton had not yet abandoned this belief.)

That the earth is coupled to its satellites in the same way as the sun is causally linked to its planets was a strong reason for Wenton to postulate that every object in the solar system has a gravitational strength and is coupled to every other object according to a formula such as (g). Wenton even went so far as to assume that all objects in the universe are coupled to each other pairwise in the same way. Condition (g) was thus extended to (g^*):

$$(g^*) \qquad\qquad \mathbf{a} = C(x \leftarrow y) = -\frac{\gamma}{r^2}\, \mathbf{r}^0,$$

where x and y range over mass-points representing any pair of objects in the universe.

REMOVAL OF AN INCONSISTENCY IN THE SET OF MODEL CONDITIONS

It is evident that the set of causal conditions C1, C2, C3 is incompatible with (g^*). The sun is, according to (g^*), affected by the planets, in conflict with what is asserted in C1, and the planets are causally coupled to each other. Bare models should therefore contain more than two mass-points. Figure 3 shows how Wenton removed this inconsistency.

We now have

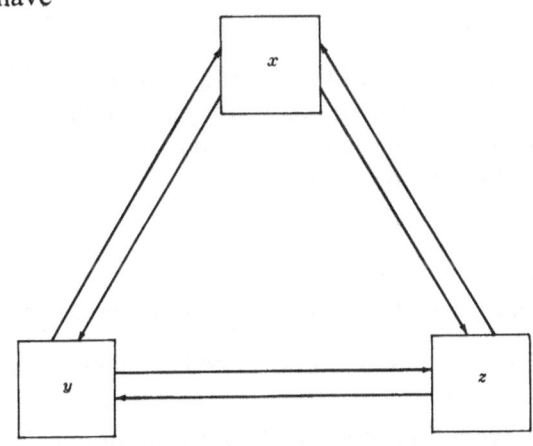

Diagram showing groups of one or more mass-points x, y, and z.
Fig. 3.

$$
\begin{aligned}
\text{C1a:} &\quad E(0|x) = 0 \\
\text{C1b:} &\quad E(x|x) = 0 \\
\text{C2:} &\quad E(x|y) = C(x \leftarrow y) \\
\text{C3a:} &\quad E(x|yz) = E(x|y) + E(x|z) \\
\text{C3b:} &\quad E(xy|z) = E(x|z) + E(y|z)
\end{aligned}
$$

C1a states that no effect takes place in an empty space. C1b states that no mass-point or group of mass-points is causally coupled to itself. C3a states that the effect which y and z together produce on x is the sum of the effects which each one of them alone produces on x. C3b is interpreted in an analogous way.

From these causal model conditions Wenton deduced the following theorems:

THEOREM 1a: $C(x \leftarrow yz) = C(x \leftarrow y) + C(x \leftarrow z)$.

THEOREM 1b: $C(xy \leftarrow z) = C(x \leftarrow z) + C(y \leftarrow z)$.
PROOF: Theorem 1 follows immediately from C2 and C3.

THEOREM 2a: $E(x|y) + E(y|x) = 0$.

THEOREM 2b: $C(x \leftarrow y) + C(y \leftarrow x) = 0$.
PROOF: Theorem 2b follows from Theorem 2a and C2. Theorem 2a is proved in this way:

$E(xy|xy) = 0$ (according to C1b) $= E(x|x) + E(x|y) + E(y|x) + E(y|y)$ (according to C3) $= E(x|y) = 0$ (according to C1b).

THEOREM 3: $E(x|0) = 0$.
PROOF: Theorem 3 follows immediately from Theorem 2a and C1a.

The earlier condition C3,$E(p|s)=$**a** is incompatible with Theorem 2a, because the sum of the acceleration of the sun and that of a planet is not zero. Wenton was therefore led to introduce a new concept, that of *inertia*, which he defined by the condition:

C4: $E(x|y)=$ma.

Making use of (g^*) in conjunction with the general causal model conditions, Wenton established the following important model condition:

THEOREM 4:   The inertia of a body is porportional to its gravitational strength.

PROOF: According to C4 and Theorem 2a $m_1\mathbf{a}_1 + m_2\mathbf{a}_2 = 0$. According to (g^*), $\mathbf{a}_1 = -\dfrac{\gamma_2\mathbf{r}^0}{r^2}$ and $\mathbf{a}_2 = +\dfrac{\gamma_1\mathbf{r}^0}{r^2}$, where $\mathbf{r}^0$ is a unit vector from Mass-point 2 to mass-point 1. It follows that $m_1\gamma_2 = m_2\gamma_1$. Hence $\gamma = fm$, where f is a constant of proportionality.

Remark: The new model conditions contain the so-called three fundamental laws of Newtonian mechanics.

APPLICATION OF THE NEW CAUSAL MODEL CONDITIONS

I will only report on one result among many which Wenton obtained.

From his causal principles in conjunction with (g^*) he obtained the following condition: $\mathbf{a} = -(\gamma + \gamma')\mathbf{r}^0/r^2$ for a two-particle model representing the sun and a planet. The acceleration of the planet is referred to the position of the sun. $\mathbf{r}^0$ is a unit vector from s to p. γ (γ') is the gravitational strength of the sun (planet).

From this condition Wenton deduced the following set of model conditions, the set K_2:

(1) $\quad \dfrac{1}{r} = \dfrac{(\gamma + \gamma')}{h^2}(1 + e \cos \theta).$

(2) $\quad r^2 \dot{\theta} = h.$

(3) $\quad \gamma + \gamma' = \dfrac{4\pi^2 a^3}{T^2}.$

The hypothesis corresponding to $K_2 3$ is an upgraded descendant of Kepler's third law. As the gravitational strength of the earth can be measured in a laboratory, $K_2 3$ can be used to measure the graviatational strength of the sun and all known planets. Upgrading or refinement of hypotheses can thus serve to make it possible to carry out measurements of significant physical magnitudes, which could not be measured before.

Remark: Wenton's causal model conditions can be grafted onto other models representing different kinds of physical systems. All such models exhibit Newtonian causality. We may call such models a family *of Newtonian models.*

GENERALIZATIONS FROM THE CASE STUDY

Development of knowledge about systems is characterized by the following general features:

1. There is a lineage of accepted hypotheses, such that descendants of an hypothesis have a higher grade: *upgrading or refinement.*

2. There is an *accumulation* of accepted hypotheses.

3. The accepted hypotheses are *represented* by model conditions which together with other model conditions form a deductive system.

Upgrading requires that the credibility control become more and more refined. This control may be described as a filter which becomes more and more fine-netted as the body of knowledge grows.

At any stage of the development of scientific knowledge there are *near* and *remote* model conditions. The former ones can be transformed into hypotheses by means of operational definitions. The remote model conditions are 'growing points' of scientific knowledge. They may have ancestors which are speculative hypotheses. Antispeculative positivists would manage to kill the growing points of science if they had their way.

What are the moves of the model builder?

1. He makes new model conditions to catch more features of the systems under investigation.

2. He responds to challenges. (a) A refuted hypothesis means that some model conditions must be changed. (b) New model conditions may conflict with previous model conditions.

The reactions to these types of challenge are subject to two rules: (1) Known positive analogy (similarity) between the model and the systems should be preserved or upgraded; and (2) negative analogy should be transformed into positive analogy. Net results: The positive analogy between the model and the systems under investigation increases.

DISCUSSION

YOURGRAU: I think this model is very interesting, but I do not believe that Newton or Kepler really went about in the way in which this model is developed here. In fact, I would say historically certainly not. Obviously, Newton did not talk model language. The only model was the point-mass, which was a fictitious entity; and he didn't even generally apply his own calculus.

Does this scheme help us to interpret Newton or Kepler better than we have been able to do before, either historically, mathematically, or by astronomy? Is this model a paradigm case for all discoveries in physics? I should think not. Now, my impression is that the active scientist or mathematician or logician does not employ this type of conceptual model. After all, there are so many entirely unpredictable, recalcitrant facts or data.

Men like Russell maintain that there is an element of creativity, an inspirational element which you can't account for; the model is only a very tedious, laborious way of 'showing' it. If I were Newton and I wanted to work on the methodology of my theory or theories, this might have been the model I would have chosen. But I wouldn't say that this model is an isomorphic presentation of what actually happens.

TÖRNEBOHM: In the first place, one reason I changed the names Newton and Kepler was to indicate that this model was not intended to be a true historical account of what actually happened.

Second, I don't presume that this extremely simple way of looking at the growth of scientific knowledge is such that it can function as a model of *all* processes of growth of scientific knowledge. I do say that this is a *way* of looking at knowledge which may be worthwhile pursuing. My model is one, so to speak, of peaceful evolution, not one of dramatic revolution. This makes it look almost deterministic. Wenton couldn't do anything else under the circumstances than what I tried to show he did.

The aims of this work are these. First, that a historian of science may wish to have something like a model showing how science may have grown. Second, I hope that a study of this kind may help to organize the various problems which are of interest to a methodologist.

TREDER: I think that in Törnebohm's model the scientists "Pekler" and "Wenton" are good and devoted Newtonian scholars. Wenton has learned his questions and research method from Newton. To the question, "What is the historical and psychological origin of the concepts used by Newton?", Törnebohm's model gives no answer.

TÖRNEBOHM: One thing that I have tried to do is to meet the needs of non-physicists who want to know how a physical theory may be generated and what makes it develop. A model of this kind may be of heuristic value to them.

HINTIKKA: Törnebohm indicated that his model might serve some of the purposes of a historian of science. Now I should very much like to see a closer comparison between Törnebohm's model and some of the material we actually have in the history of science, including the history of scientists' methodological views. I agree with Törnebohm (if I understand him correctly) that questions as to how science actually proceeds are often dismissed far too light-heartedly. Much more is involved in this progress than merely "creative jumps" to theories.

As a paradigm case, I would like to see a comparison between Törnebohm's model as developed in terms of some of Newton's discoveries and Newton's own methodological views. In addition to the methodological remarks in Newton's correspondence, there are some extremely interesting methodological remarks in some of Newton's books, for instance, at the end of *Opticks*. He certainly does not describe his own research as a series of creative jumps to theories. He compares his own method with what, at that time, was called geometrical analysis. He says that in the same way as we tackle difficult problems in geometry by analysis, we have to apply the same method of analysis to all difficult physical problems.

What is his method? This is partly a historical problem, but I submit that what happens is something like this: One takes a suitable particular case, for instance an experimental setup. One tries to make it as rich as is needed, in the same way as someone who practices geometrical "analysis" wants to include in his figure all the necessary "auxiliary constructions". If one succeeds in doing this, one can study the interrelations of these factors of the experimental situation in the same spirit as a geometer "analyzes" the interdependencies of the different parts of his figure. When they have been found, one simply generalizes from this single case to all similar cases without going through the detailed steps of Törnebohm's model or through the motions described in inductive logic. In other words, according to this model, what is essential is to find out that which is important is one single case, as in geometrical analysis, after which the generalization to other cases is more or less obvious.

I think my own model is intriguing, although I have not worked it out fully here. Perhaps it is even a more realistic sketch than Törnebohm's. It seems to me that a confrontation of Newton's own model with Törnebohm's might be very interesting both historically and topically.

Chapter V

EXISTENCE

Willard V. O. Quine
Harvard University

Ockham's razor applies to all science, including scientific philosophy, and it appeals to all scientists, including scientific philosophers. The minimizing of objects is a part of the drive for simplicity, which, in some sense, is central to science. Scientists are out to get the most for the least. They are out to drive a hard bargain with nature. It is significant that the minimizing of objects has its main appeal not as a minimizing of the mere number of objects, but as a minimizing of kinds in some sense. For it is the minimizing of kinds, not of number, that makes conspicuously for simplicity. Now the notion of a *kind*, what makes one big class count as a kind and another big class not, is in general unclear. Even so, this much is, I think, clear: if there is one difference of kind more drastic than another, it is the difference between particulars and universals. And so it is that the scientific philosopher, in driving his hard bargain, will allow no universals if he can help it.

Driving hard bargains makes for litigation. Our scientific philosopher wants to improve the ratio of utility to inventory, and one way is to quibble about what counts as inventory. Applied to universals, this maneuver consists in talking expressly of properties and propositions and other universals wherever convenient, and then appending a *caveat* to the effect that such talk is not to be taken to attribute existence to properties or propositions or other universals. Church cites examples from Ayer and Ryle. I shall limit myself to one, which is Ayer's: "... it makes sense to say, in a case where someone is believing or doubting, that there is something that he doubts or believes. But it does not follow... that something must exist to be doubted or believed."

If we can enjoy the benefits of unwelcome objects while disavowing the assuming of them, then disavowing and assuming are empty verbalism. No wonder, then, that some philosophers looked upon the question of the existence of universals as meaningless. Carnap did, and Lazerowitz in a way, and indeed Ayer himself. Still, that position is no easy resting place. We are bound to make sense of some questions of existence: existence of protons, pygmies, unicorns. When we get clearer on what it means to assume or repudiate such things, we are in a better position to see what sense is to be made of assuming or repudiating universals.

What we have to get clear on is when to consider that a sentence mentions certain things and thereby assumes their existence. Looking for names is no help, since anyone can disclaim assumption of the object by declaring that the alleged name is not a name. What counts, rather, are the variables. The objects assumed in a given discourse are the values of the variables. They are the things in the universe over which the variables are interpreted as ranging. If a man's assertions are such that protons, unicorns, or universals have to be among the values of the variables in order for the assertions to be true, then he has no business denying the assumption of protons or unicorns or universals.

We have the modern logic of quantification to thank for making evident the existential force of the variable. The existential quantifier "$(\exists x)$" is the distilled essence of existential talk. All imputations of existence can be put as existential quantifications. Moreover, all the other uses of variables — in universal quantification, in singular description, in class abstraction, in algebra — can be paraphrased in familiar ways so that the variables end up as variables of existential quantification.

This existential role belongs to variables only in the strict sense of the word; not to schematic letters. Schematic letters, such as the sentence letters of truth-function logic and the predicate letters of quantification theory, occur only in schemata and not in sentences, and they take only substitutions and not values. By variables I mean bound or bindable variables. These, if we construct our discourse along the lines of neo-classical logic, are the primary instrument of reference to objects. The objects are the values of the variables.

This referential primacy of variables can be made more evident with the help of Russell's theory of descriptions. For with its help we easily see that all names, indeed all singular terms except variables, are, in theory, dispensable. Applied to names the idea is just this: at the level of primitive notation you discard each name in favor of a primitive predicate which is true of just the named object. Then to serve the purpose of the discarded name, when desired, you can introduce a singular description by Russell's method of contextual definition on the basis of the primitive predicate. What this purely theoretical maneuver brings out is that the significant link between words and things is the link between variables and their values. Names are beside the point.

One use of a pronoun, in ordinary language, is as shorthand for a proper name or another definite singular term which stands as the pronoun's grammatical antecedent. This use is less important and less frequent than another use in which the grammatical antecedent is an indefinite singular term such as "a house" or "everybody". Now the pronoun in this latter use is what corresponds in ordinary language most nearly to the variable. In principle a clear thinker might have been able centuries ago to see the pronoun, and not the name, as the primary instrument of reference. But in fact we had to await quantification theory for the clarification it afforded of what had been, in its old pronominal form, too murky for easy appreication. We had to await the invention in 1879 of quantification theory by Frege (whom Yourgrau calls "old *Quantifex Maximus*"). And Russell's theory of descriptions was an additional clarificatory aid.

Still the clarification was not immediate. As early as 1905 Russell had pioneered in quantification theory and had created his theory of descriptions, but as late as 1940 we find him still failing to recognize that the objects assumed in a discourse are simply the values of the variables. In his *Inquiry into Meaning and Truth* we still find him arguing that if we try to eliminate universals by analyzing them in terms of resemblances of particulars, we are bound to be left with at least one universal, namely, similarity. I urge rather that if the universals can be analyzed out in terms of resemblances as Russell suggests, then we are not left with even the one universal, similarity, unless for some odd reason our analysis requires us to use similarity as a value of a variable. Otherwise we are left with a similarity predicate or symbol, but no similarity relation.

What I have said is not a plea for nominalism. I have no high hopes of getting an acceptable system of the world without admitting some universals, namely classes, among the values of the variables. But this is because I see no hope of analyzing out all the universals in terms of resemblances; it is not because of a surviving similarity relation.

Can we suppose that Russell felt some actual objection to equating existence with the value of variables? I think not. I must indeed recognize that in 1950 someone did object: Bergmann. But his objection to equating existence with the values of variables turned on his taking the term "existence", as philosophers sometimes do, in a narrower sense than "being" or "subsistence". Russell once favored that terminology too, but it is not concerned in his point about similarity, nor is it concerned in my doctrine. What I equate with the values of variables is the broadest sense of "being" that the user of the variables is to be seen as accepting. It is hard to picture anyone deliberately rejecting this account, since the quantifier "$(\exists x)$" is meant to mean precisely "there is something x such that." Such objections as I have noticed turn on misunderstandings. Ryle once objected on the ground that values of variables were expressions substitutable for variables, whereas there are things other than expressions. But this objection rested on a discrepant use of the phrase "value of a variable." G. P. Henderson objected that what values variables take is a question of language while what there *is* is a question of fact; but the answer to this is that in my view the purported values of the variables of a theory are merely what that theory takes them to be, and not what there really is, unless the theory happens to be true.

The doctrine must, I say, be obvious to anyone who reads it right and reflects how the existential quantifier is meant. It would seem too obvious to be worth propounding, were it not for the surprising and glaring cases of the need of it. One case is Russell's own position on similarity. Another is the case of Ayer noted earlier. I could cite others.

I referred to litigation. Can we hereafter depend upon each philosopher to own up to his ontic commitments, to everyone's satisfaction? Not so. If we have to translate a philosopher's discourse into quantifiers in order to assess his ontic commitments, it remains open to him to disown our translation. Or again, even if he seems to talk in quantifiers, he may plead some deviant, nonexistential intent for his apparent quantifiers. However, this is a predicament in which I can acquiesce. Quantification is an ontic idiom *par excellence*, and we may reasonably expect the ontic commitment of a theory to remain obscure to the extent that the way to translate the theory into quantificational terms is obscure. A man is free to be as obscure as he likes, and another man is free to despair of understanding him. Where an appreciation of the ontic role of variables is mainly beneficial is not in litigation, but in the home. If a man is interested in his own ontology and how to reduce it, then thinking in quantificational terms will prove helpful to him.

There are alternative logical styles, not using variables at all, that are just as good as the quantificational style for clarifying one's ontology. They are just as good because their intended translations into quantificational form are pat and evident. We have one such style in Schönfinkel's combinators. There we would no longer say that the objects assumed are the things over which the variables are interpreted as ranging; there are no variables. We would say rather that the objects assumed are the things that the functions are interpreted as admitting as arguments. They turn out, in the Schönfinkel scheme, to be a prodigal lot of objects, for the scheme encompasses a full set theory.

There are less prodigal styles that likewise avoid variables. They consist in subjecting one's predicates to a fixed battery of operators which generate further predicates. I have in mind here Tarski's cylindrical algebras. some related schemes of Halmos and Bernays, and one of my own. On these lines it is possible to frame a logic which, unlike Schönfinkel's exorbitant alternative, is precisely equivalent to the classical logic of quantification. And what are we to say of the ontic commitment of some particular theory that is formulated in this style of logic? No more than that the objects assumed are the things that are interpreted as fulfilling the predicates, simple and complex.

The translations which carry the ontic standard over from the quantificational style to these other styles are neither arbitrary nor mysterious, for they are the translations whereby the authors of these other styles explain the styles to us. The notation of quantifiers and variables is central to ontic commitment simply because it is the first clear regimentation of ordinary existence talk – of the locution "there is" and kindred ones. The ontic relevance of alternative logical styles is then a derivative and relative matter: it is relative to whatever translational correlations may have been stipulated or understood.

This is easy enough for our home-grown styles, but problems arise over remote cultures. What are we to say of the ontic commitment of a New Guinea aborigine – what are the purported values of his variables? How shall we translate his discourse into quantifiers and variables? I hold that there is in general no unique translation, not even unique in respect to ontic commitment, let alone logical style. I hold that our distinctively referential apparatus, comprising pronouns or variables and also identity, predication, and related devices, belongs in an essential respect to the theoretical part of our language: namely, it is underdetermined by all possible sensory stimulation. A result, or really another way of stating the point, is that our referential apparatus is subject to indeterminacy of translation. That is, translations not equivalent to each other could be reconciled with all behavior.

This does not mean we cannot translate such matters. We can and do, and on the basis of the translation we can impute ontologies. But I hold that the imputation is relative to the manual of translation adopted, and not purely a function of the aborigines' mental life or speech behavior. It may in this sense be said that ontological questions are parochial to our culture.

This is not to say that a thing may exist for one culture and be nonexistent for another. Existence is absolute, and those who talk of existence can say so. What is parochial is the talking of it. What to count as talk of existence in a remote language depends in part on accidental features of the construction of our translation manual.

This weakness of ontological communication between cultures must, I think, be acquiesced in. And besides, there is another limitation on ontological communication to be noted, a limitation that obtains even within the parochial setting of our home language. What I now have in mind is the predicament in which the proponent of the lesser of two ontologies finds himself when he tries to state the issue: he cannot say there are things which his opponent wrongly thinks there to be, without saying that there are things which are not. On this predicament I can only repeat my remark of 1948: that the disputant with the short ontology has to contrive to characterize the disagreement in linguistic rather than ontological terms, by citing his opponent's contested statements and dilating also upon the use his opponent makes elsewhere of the crucial terms of those sentences.

We have found that ontological discussion calls for talk about language in two connections. One connection we noted last: he who disbelieves in another man's objects has to talk about talk to report the disagreement. The other connection we noted earlier: which ontology to ascribe to a man depends on what he does or intends with his variables and quantifiers. This second appeal to language is no more to be wondered at than the first; for what is in question in both cases is not just what there really is, but what someone says or implies that there is. Nowhere in all this should there be any suggestion that what there is depends on language.

In Carnap's view, existence issues that are broad enough to count as philosophical are language-dependent, notably the problem of universals, while others, such as the existence of pygmies or unicorns or perhaps protons, are not. I disagree, and I think the illusion of a difference here between two sorts of issue is engendered by a certain strategy which I have called "semantic ascent": a strategy which is innocent and useful but apt to be misconstrued. This is a strategy of leaving the ontological plane and ascending to talk about language for neither of the two reasons noted in the preceding paragraph, but for a third reason: to be able to discuss very fundamental issues in comparatively neutral terms, and so to diminish the tendency to beg questions. Naturally the strategy proves especially useful for issues of a broadly philosophical sort, ontological or otherwise. But the philosophical truths, ontological and otherwise, are not for that reason more linguistic in content than are the more sharply focused truths of the special sciences. Between ontology and the more local existence statements I recognize no difference of kind.

I said that existence statements belong to the theoretical part of our language. Would this claim seem odd as applied to statements to the effect that there are sticks or stones or color patches? I think it is right even here, because of an inconspicuous but fundamental distinction that must be drawn between response and reference. The statement "Stone", or "That is a stone", is one that we may learn early to produce in response to appropriate sensory stimulation. Foreign observers of our response habits can determine what the appropriate range of stimulation is, moreover, and volunteer a sentence in their own language that corresponds to ours in this respect. "Stone" is in this clear behavioral sense an observation statement, rather than a theoretical statement. So far it is neutral with respect to reference, or objectification. The foreign observers might have an ontology not of stones but of feels and patches, or perhaps of fields of force, if we suppose for the moment that it makes sense to identify their ontology at all. Their translation of our observation statement "Stone" is predicated only on sameness of stimulating occasion, sameness in respect of what combinations of sensory receptors require activation; referentially it is neutral. When we go on to read a referential moment into our statement "Stone", on the other hand, or to affirm that there are stones, we are proceeding to integrate our stone-talk systematically into our peculiar referential apparatus of pronouns, identity, and predication in a certain way which amounts, in quantificational jargon, to taking stones as values of variables. This is a theoretical move, and one that can be simulated in the foreign language, only subject to the indeterminacy of translation.

What I have said of stones applies as well to color patches or anything else. Reification is a theoretical move, distinct from the observational component of an observation sentence.

I urged that between broad existence statements of philosophy and narrow ones of the special sciences there is no difference in kind. Granted, we have no uniform procedure for establishing the truth of existence statements broad and narrow. But neither is there such a procedure for the broad ones or the narrow ones alone. Existence statements differ in no way, epistemologically, from theoretical statements generally. They are parts of an inclusive theory whose overall claim to acceptance resides in the systematic simplicity, or something like that, with which the whole theory accommodates our observations. I am sorry that I have nothing new to say by way of illuminating this vague matter of acceptability of theories.

But it is easy to point to specific ways in which reifications, of universals, for instance, can contribute to systematic theory. I may begin with the old standby, 'ancestor'. Thus consider two two-place predicates,

'P' for parent and 'A' for ancestor. 'Pxy' means x is parent of y, and 'Azy' means that z is ancestor of y. Particulars, people, are to be assumed as values of our variables. Universals, parenthood, or ancestorhood for instance, are not here assumed as values of variables. But if we do assume certain universals as values of our variables, namely classes, then we can define 'A' in terms of 'P' and the two-place predicate 'ϵ' of class membership. As has been known since Frege and well known since Dedekind, we can explain 'Azy' as meaning that z is a member of every class that contains as members all parents of its own members and also y.

It would be a poor bargain, of course, to posit classes as new objects in our universe simply in order to reduce one pair of two-place predicates, 'P' and 'A', to another pair of two-place predicates, 'P' and 'ϵ'. The gain is seen rather in the fact that for every two-place predicate there is another two-place predicate, its *closed iterate* we might call it, that stands to it as ancestor stands to parent. Now the assumption of classes and 'ϵ' enables us to express, without further apparatus, the closed iterate of every available two-place predicate. This could not, presumably, have been accomplished by adding any one or several predicates or particulars to our vocabulary, but it is accomplished by adding the predicate 'ϵ' and assuming classes for it to apply to.

Another of the many systematic benefits of assuming classes can be seen in the avoidance of modality at certain points. Thus compare "All crows are black" and "All black crows are black".

$$(x) \ (Cx \supset Bx), \quad (x) \ (Bx \cdot Cx \cdot \supset Bx).$$

We might like to regard the first of these two statements as holding merely as a matter of fact and not by logical necessity. But the other patently holds by necessity, and we might like to record this trait by changing its material conditional to a strict one:

$$(x) \ (Bx \cdot Cx \cdot \ \rightarrow\!\!\!\!\!\dashv \ Bx).$$

On the other hand there are various reasons, having to do with something called "referential opacity" among other things, for preferring to avoid modal connectives. Now generality is often a satisfactory substitute for necessity. The need we felt to add a note of necessity to "All black crows are black" can perhaps be met well enough by just abstracting from the specificity of "black" and "crow" and saying that the rule holds generally; for any classes y and z, whatever belongs to both y and z belongs to y. Thus it is that the admission of classes as values of variables, and the adoption again of 'ϵ', can enable us to avoid modal logic. The need felt for modal logic cannot always be met thus by generality, but, as Russell has stressed, it often can.

There are also other ontological posits, beside that of classes, that can have the effect of resolving modality. Thus suppose, to return to "All black crows are black", that someone is not satisfied with the generalization over classes which was our substitute for necessity. An alternative line would be to keep the specific predicates "black" and "crow", and then get the force of necessity by quantifying over not just actual particulars but possible particulars: "All possible black crows are black". Here we see another example of resolving modality by enlarging the universe. This expedient does not really recommend itself, though, for unactualized particulars are in various ways an obscure and troublesome lot.

I turn finally to a more general question regarding the benefits that the positing of supplementary objects can confer. Might all supplementation of the range of values of variables be in fact dispensable early and late if not wanted in the end? Thus consider a statement whose variables range unrestrictedly over some broad universe, and suppose that the only consequences of this unrestricted statement that interest us in the end are restricted statements — statements whose variables are restricted to some narrower sub-universe. The question is whether, in general, a restricted statement can be constructed that has the same restricted consequences as the original unrestricted statement.

The answer is negative. This can perhaps be shown elegantly; I can only show it with the help of heavy artillery, as follows. The set theory of von Neumann and Bernays has a finite list of axioms, or, since we may conjoin them, a single axiom. Now we consider, in particular, from among the theorems of this set theory those that have all their variables restricted to a certain sub-universe. These are known to be precisely the theorems of another set theory: Zermelo's. So, if the answer to our question were affirmative, there would be a formula, with its variables restricted to the sub-universe, that would encompass precisely Zermelo's set theory. But it is known from the work of Wang and McNaughton that Zermelo's set theory is irreducible to a single axiom.

We have to conclude that multiplication of entities can make a substantive contribution to theory. It does not always contribute. Of itself, multiplication of entities should be seen as undesirable, conformably with Ockham's razor, and should be required to pay its way. Pad the universe with classes or what not, if that will get you a simpler, smoother over-all theory; otherwise don't. Simplicity is the thing, and ontological economy is one aspect of it, to be averaged in with others. We may fairly expect that some padding of the universe is in the interest of the overall net simplicity of our system of the world.

REFERENCES

1. Gustav Bergmann, "A Note on Ontology," *Philosophical Studies* 1 89-92, (1950).
2. Paul Bernays, Über eine natürliche Erweiterung des Relationenkalküls," in *Constructibility in Mathematics,* (A. Heyting, ed.), North-Holland, Publishing Co., 1959, pp. 1-14.
3. Rudolf Carnap, "Empiricism, Semantics, and Ontology," *Revue Internationale de Philosophie* 11, 20-40, (1950). Reprinted in *Meaning and Necessity,* 2d ed., University of Chicago, 1956.
4. Rudolf Carnap, "Replies and Systematic Exposition," in *The Philosophy of Rudolf Carnap* (P. A. Schilpp, ed.), Open Court, La Salle, 1963, pp. 859-1013, specifically p. 868.
5. Alonzo Church, "Ontological Commitment," *Journal of Philosophy* 55, 1008-1014, (1958).
6. George P. Henderson, "Intensional Entities and Ontology," *Proceedings of the Aristotelian Society* 58, 269-288 (1958).
7. Morris Lazerowitz, *The Structure of Metaphysics,* Routledge, London, 1955, Chapter 2.
8. Willard V. Quine, *Word and Object,* M.I.T. Press, 1960, Chapter 2.
9. Willard V. Quine, "On What There Is," *Review of Metaphysics* 2, 21-38, (1948). Reprinted in *From a Logical Point of View,* Harvard, Cambridge, 1953.
10. Willard V. Quine, "Variables Explained Away," *Proceedings of the American Philosophical Society* 104, 343-347, (1960). Reprinted in *The Ways of Paradox,* Random House, New York, 1966.
11. Moses Schönfinkel, "Über die Bausteine der mathematischen Logik", *Mathematische Annalen* 92, 305-316, (1924).

DISCUSSION

POPPER: Quine has shown clearly that different languages, different cultures, may have implicitly what he calls different ontological commitments. Now, a word about ontological commitments. I don't want to commit myself ontologically. If at all, I want to commit myself against commitments. So I think it should be stressed that the word 'commitment' as used by Quine really means something very relative to the language — or to the logical formalism — one adopts. With the current fashionable religious connotations of the word 'commitment' it could otherwise easily lead people to think: "Before I adopt a language I really have, first of all, to settle my account with God." But this is just what Quine does *not* want us to do. He definitely has opened a way to *not* settling our account with God before we adopt a language. So I think this point should be made clear.

QUINE: The word 'commitment' is unfortunate. It has a very solemn sound to it. I should stress that commitment can be momentary. In fact, you can have one commitment in one pocket and another in another, as one is likely to do more often than not in comparing theories and trying to decide which one to end up with.

KAPLAN: I don't understand Quine's point about the observational statements — "Stone", "That is a stone" — and their relation to the apparatus of reference which he says belongs to the theoretical part of language.

As I understand his criterion of ontological commitment, it amounts roughly to the following: that a sentence in some given language, a language which, indeed, must be interpreted, commits us to the existence of, say, unicorns, just in case the translation of that sentence into English logically implies (if I can talk about logical implication within English) the English sentence, "There exist unicorns." And I would take it that the example that he gives of the observational statement "Stone", which he paraphrases "That is a stone", using the demonstrative, would indeed imply, if we ever get the logic of demonstratives worked out, "There exists a stone". This, in a way similar to using a proper name such as "Pegasus" in the sentence, "Pegasus is a winged horse", does not logically imply "There exists a winged horse".

QUINE: On the matter of "stones" — yes, I suppose from the internal point of view within English it isn't too unreasonable to say "That is a stone" implies that there are stones. Actually, I am not sure even that the situation is quite as simple as that. Rather, I think it is in order (even from the point of view of our own language) for us to be clear on just what sort of thing we would want to take into our universe of discourse. Such a range of variables, for a given purpose, would require a certain amount of theory and not just a sentence. If I say "something is too bad", for instance, does it mean we are positing certain things which are bad and some among them that are too bad, and so on? The straight idiom doesn't indicate this, but even suppose it did, or suppose we had enough theory to do the job, my point about indeterminacy of translation is: that aspect of the theory (where those implications would lie, viz. the ontological aspect, the part that depends essentially on identity and pronominal cross-reference or quantified variables, and the like) is something that doesn't translate into another language except with the help of a partly arbitrary or accidental manual of translation. What we have that is not accidental at the start is translation by comparing simply the stimulatory situations which would give assent to this whole sentence "That's a stone", or "A stone is there", or "Stone", whatever we might think of it in terms of an ontology of our own that posits stones, stages of stones, or the abstract property of being a stone. All these sentences are alike, not in the sense of being synonymous in all respects or of having the same ontological consequences; they are alike in stimulus meaning, and it is only likeness in stimulus meaning that carries over when we are making a radical translation on a behavioral or empirical basis to another language.

Then the further sorts of meaning that we think of rather dimly and vaguely in ontological terms in the domestic situation are those which are theoretical. They don't have, in principle, any unique analogue in other languages. They are applicable to other languages only in relation to antecedently adopted manuals of translation between them.

YOURGRAU: I think Quine is concerned, like all of us, with the question of universals and particulars. I do not know any solution of that specific question and I agree that Ayer's rather flippant dismissal of that problem is not very satisfactory, but I haven't seen any other effective treatment, not even by Aristotle.

I am not an Aristotelian, so I do not even like the dichotomy between the universal and the particular. Yet the problem is still there; we simply have not been able to do anything about it.

I wish to point out in relation to Quine's use of the demonstrative pronoun, "this", in order to specify the particular, and then of the universal quantifier for the universal, that Frege, whom he quoted, made a very beautiful remark. He said, in effect, "I do not want to use the term 'universal'; I would like to use the term 'concept'." Now, how can one, by a trick, establish concept? I simply say *a* horse and not *the* horse. *The* horse is a particular, and *a* horse is a concept!

Russell's comment was: "You know, the trouble with Frege was that he didn't know any Latin, because *equus* in Latin is 'a horse' *and* 'the horse'. It is obvious that if Frege had known Latin, he could have solved the problem — if there is one."

This comment shows that the question of establishing universals in a precise manner is only possible if one is affected by quantification alone. Unfortunately, quantification seems to be the sole way of meaningfully approaching the topic of universals at present. Yet, I am not satisfied; because, as Kaplan pointed out, unicorns also have a beautiful resting place among universals, no less than stones and philosophers.

If I just quantify, then really nobody is capable of telling me whether this doesn't cover Pegasus, abstract entities, stones, physical entities, protons. There is no ontic commitment, because it covers *everything*: the abstract, the conceptual, the fictitious, the imaginary, as well as the real.

I wonder whether Quine would propose that one should introduce some sort of operator, say, "Here I talk in theoretical language and concepts, in scientific terms, as it were, and here I talk about stones and persons and objects in life". If that attitude is adopted we have to hypostatize most statements we utter, or we have to keep quiet.

Quine said, "Existence statements do not differ from theoretical statements in general." This is the only thing I am not very happy about. As a realist — and the *leitmotif* was realism — I must make such a sharp distinction.

QUINE: I take variables of quantification to range over *everything*. Not just things that exist in space-time, but also abstract objects, universals, if such there be.

Ordinarily you can't tell from a single quantified sentence what things are assumed to exist. You might ask the person who wrote it and thus get some further notion of what he would like to have among the values of his variables. If on the other hand he has given you not just a casual quantification but a substantial theory, then, by examining the existential quantifications which his theory implies to be true, you have better grounds on which to reckon his ontology.

Yourgrau proposes an operator which says, "Here I really mean these things exist". I would not even know how to interpret the words of this explanation, once they are dissociated thus from the ordinary old quantifier. But I suspect, here as in Russell's distinction between subsistence and existence, a bid to disavow unwelcome óbjects while still enjoying the convenience of quantifying over them. This kind of quibble is pointless, indeed unwelcome, insofar as one is interested in the question of what he is assuming as values of his variables; whether we call this "existence" or not is unimportant.

MERCIER: Now, Quine certainly is of the opinion that he spoke about an ontological subject matter. Suppose, then, one had to relate the approach of Gabriel Marcel to the ontological theme of Quine's presentation? What sort of connection would there be between the two?

QUINE: I tried it once, not with Marcel himself, but with a young colleague, Henry Bugbee, who is quite close to Marcel in his views. Well, I don't know how to build the bridge, if it can be done at all. I can't understand what the existentialists are up to.

HINTIKKA: It appears as if philosophers of logic, as distinguished from logicians and mathematicians who are doing technical logic, have overlooked an important task that can be defined fairly clearly in terms used by Quine. He spoke of the system of references that connect our symbols with whatever entities we are talking about. It seems to me that this system of references, or whatever you want to call it, is very often taken too much for granted. Some philosophers of logic have been primarily interested in mapping it, spelling out all the different ways in which symbols can stand for things out there in the world. However, these referential relations do not simply occur by nature. They are created by human activity. Undoubtedly there is a great deal to be done in saying exactly how this system of references comes about, what modes of behavior serve to establish and uphold this system.

Quine has in his writings implicitly assumed that we can always locate one such mode of behavior. Suppose we want to have a way of translating from what he has sometimes called "the jungle language" into our own mode of discourse. In explaining how propositional connectives are translated, Quine has made essential use of the notions of ascent and descent. I am not objecting to this. On the contrary, I am wondering whether we might take in something more. There might be other modes of behavior that a 'jungle linguist' could easily recognize and which he could make use of in his attempt to go beyond stimulus meaning. In this way we might perhaps eliminate some of the indeterminacy of translation. This is of course a very broad suggestion, but I am prepared to back it up with more detailed suggestions, relevant to the notions of existence and quantification.

QUINE: From my point of view this is a very welcome line of development, and it is conceivable that something could be done in the way of supplementing what evidence we can get from indeterminacy of translation. There would be quite a problem of making the criteria sufficiently sharp to be useful, but possibly there are some criteria in terms of various cultural manifestations of a non-linguistic kind. I would feel confident, even so, that the gap couldn't be closed, that there would still be indeterminacy of translation, and I would fully expect that in particular it would include the question of what, in our terms, the values of the variables are.

CASSERLEY: We are all aware that there is another way of using the word 'existence' to be found among a very large number of philosophers today. We might add to that a large number of other thinkers, for whom existence doesn't primarily refer to the existential import of universals, but very largely to the existence *per sua* of certain distinctive types of particulars. The result is to give us a doctrine of something like modalities of existence.

Well, I don't want to comment on that very much, just to say that there can be different games which seem to use the same kind of ball, for example, soccer and basketball. And, of course, if we are playing one game we play according to that rule, and if we are playing the other game, we play it according to the other rule. The really important thing, I claim, is for a large number of philosophers to get used to playing both games. They would play one game rather better than the other, but nevertheless if only more people would be accustomed to playing both games, bridges might be built. If I am going to ask Quine to play cricket with me, we ought to adhere to the rules of cricket, which, as everybody knows, is rather difficult. On the other hand, if Quine wants me to play tiddledywinks with him I will leave it to him to settle how many of his tiddledywinks equals one of my winks.

I try to play both games, though I must agree I play one game very much better than the other. But I attempt to keep up with both games, and it does seem that the real contribution which the thinkers (I won't use the difficult word 'existentialist' because nearly all the existentialists are claiming they are not existentialists, and nearly all the non-existentialists say that they are existentialists, and I am not quite sure what the word means by now) make to the whole problem of the relationship between the particular and the universal is that they have visualized and tried, perhaps in vain, to grasp the dimension in which we don't just think about the relationship between the particular and the universal, but actually experience it. The existentialist type of thought arose in the nineteenth century, as is known; we look at Kierkegaard, or Nietzsche, very largely as protesters against the threat of middle-class conformity. They saw a dilemma between being "this man" and "one of the men", and they experienced the pull on both sides.

If we could interpret "being one of the men" as the pull of the universal, and the insistence on being "this man" as the real meaning of the particular empirically, then we could say that here we have perhaps the one place which Plato and Aristotle and all the great traditional logicians are thinking about. I am not a crass empiricist, but it is always good to get a little bit of empiricism to help us deal with some of our more recondite problems.

Now, I am not sure we have this kind of bridge. I don't think the bridge is completed; I think there is a rather bad gap in the middle, but at least this is a contribution towards a bridge, for the existentialists rarely are bent on experiencing with enormous impact in depth precisely what the logicians have been talking about.

QUINE: Well, I really can't say anything more that would be at all constructive because I have failed to understand such things as I have tried to read in that vein. I must say I haven't made as much of an effort as I suppose I might have. There are all sorts of places that it would be well to have bridges to, and one has to be selective or follow hunches as to the possible fruitfulness of various other fields, things that might be related. I have felt up to now that mathematics, logic, physics, and behavioral psychology might be more relevant to considerations in in that part of philosophy having to do with universal discourse and values of variables than are large tracts of philosophy itself. I haven't found anything but discouragement in what efforts I have made to read not merely existentialists but related writers, such as phenomenologists and more distantly related ones, the whole post-Kantian idealistic tradition, to which I just don't respond. I find I am not getting illuminations from that tradition. The word 'existence' is the ball that happens to be used in the two games. That, of itself, wouldn't, I should think, be a sufficient reason for trying to find meaningful connections, especially in that direction rather than in another.

POPPER: I would think that logicians since Aristotle have wanted one thing: that logic should be the theory of inference and that we shouldn't be able to prove things in logic which are not true intuitively.

Well, this has not been easily achieved. In Aristotelian logic one can actually prove the existence of anything he wants to talk about, and this was really the first reason "existence" became a problem of logic. Logicians have been at great pains to build a logic in which one cannot prove the existence of everything. But they had great difficulty in succeeding in that. Obviously, a logic in which you can prove the existence of everything isn't a good instrument.

Now it seems to me that Quine has solved this problem. He has made certain suggestions for constructing a realistic logical system. By certain quite acceptable tricks—and his tricks, incidentally, are tricks to diminish the logical, linguistic distance between universals and particulars—he has succeeded in constructing a way of talking in which we do not necessarily anticipate the existence of everything we somehow are talking about. And the word 'about' is one of those words by which I actually commit myself to talking about everything in a certain way. But, anyway, that is not to be meant to be ordinary English and noncommittal.

This is why I don't want to commit myself ontologically, and that is exactly what I think Quine also did not want to do, and, incidentally, what Lejewski also did not want to do. His method is different and one which comes a little closer to Yourgrau's particular hopes. So, to put it in a simpler way: all these problems are to get a kind of austere logic, which is as noncommittal as possible, in order to speak about all those things which we think exist; in a way, to be able to discuss the existence of things without assuming in our discussion that they exist. That is the fundamental problem which logicians have been facing.

QUINE: I would agree. Of course, in order to work toward a system that is as weak in its ontology or ontological commitments as possible, we need also to have more clarity than we have had at some stages as to what it means to be committal. Indeed, that is a central part of what I have been trying to say.

Chapter VI

PROBLEMS OF UNIFYING COSMOLOGY WITH MICROPHYSICS

Dmitri D. Ivanenko
The Moscow State University

One must distinguish between the problems of a local unified theory of elementary particles, which are quite popular now because of the discovery of a multitude of particles and excited states — resonances — and the problems of a (super-) unified theory that tries to account for gravitation and cosmological features of the real universe. Up to now there have been only a few attempts in the second direction, chiefly because of the conviction (or superstition?) that the cosmological conditions of the universe as a whole are rather irrelevant for the processes of the microworld.

In contradistinction, I believe that a purely local theory of elementary particles is impossible. Let me point out that the great Einsteinian gravitodynamics, which constitutes the basis for making connections between space-time and (ordinary) matter, must anyhow be generalized in some respects. First there is a somewhat formal mathematical point arising from the fact that the interaction with gravitation of fermions (electrons, protons, neutrons, and so on) described by spinors goes *via* tetrads ($h_a{}^\nu$), that is, roughly speaking, through "square roots" of the components of the Riemann-Einstein metric tensor $g_{\mu\omega}$, and not by means of the Einstein potentials

$$g_{\mu\nu} = h_\mu{}^a h_\nu{}^a.$$

This fact, established some time ago (Fock—Ivanenko), requires a systematic revision of conventional general relativity with the eventual addition of new supplementary conditions (Möller, Rodičev, Schwinger-Deser, Treder).

The second essential point is connected with the quantization of the gravitational field, which among other things leads to the prediction of the transmutation of gravitons into electrons and positrons or photons, etc., and *vice versa* (Ivanenko, Wheeler, and Brill; also Piiz and Vladi-mirov); this must be taken into account for instance in considerations about the hot equilibrium and the origin of the universe (Zeldovič and Sakharov). The possibility of the transmutation of the curvature of space-time into ordinary matter links these three fundamental categories even more closely than before. Clearly, the conventional sharp distinction be-tween matter (ordinary, not gravitational) and geometry (space-time) arises from the relative smallness of such important transmutations under ordinary conditions of low energies or temperatures, or small values of curvature.

Moreover, the quantization of the metric leads to fluctuations of the latter (Wheeler), which can perhaps even produce topological changes or yield something like discrete geometrical structure, as has been dis-cussed independently by Ambarzumian, Ivanenko, Schild, Snyder, Koish, Finkelstein, and recently by Yukawa.

Thirdly, one must 'cosmologize' conventional solutions of the Ein-stein theory, which are valid essentially for distances of less than gal-actic order, say, by normalizing them, not, as is usual, on the flat Galilean metric at "infinity", but on the real Friedmann-like expanding metric. The centrally-symmetrical Schwarzschild solution has been generalized from this point of view by McVittie and by my collaborators Brejnev and Frolov. One may be inclined to consider the Dirac–Jordan hypothesis of the diminishing gravitation constant in connection with these cosmol-ogized solutions. Anyhow, one cannot ignore the fact (or the good joke) that, by applying Hubble's constant of cosmological expansion to earth-like dimensions, one obtains a value of the order of half a millimeter *per annum*, which coincides with our estimates (Frolov, Sagitov, and Ivan-enko; cf. also McDougall) and with the estimates of those geologists (Egy-ged and Wilson) who are inclined to support the idea of the earth's expan-sion.

Now let us try to take some very preliminary steps, however subtle they may be, in the direction of constructing bridges between cosmology and elementary particles, leaving aside our conventional, somewhat "geo-centric" or "galaxicentric", views that cosmological features cannot play any role in local phenomena (that is, inside our galaxy, inside our plane-tary system, or inside the microworld of elementary particles). On the contrary, let us make the hypothesis that the average cosmological con-ditions create a weak field, a specific ground state as an arena on which all local phenomena can be displayed. By the way, this may be considered as a modern version of Machian ideas!

Consider, to start with, the following fundamental properties of the real universe as explored thus far:

1. There is an overwhelming preponderance of baryons (protons, neutrons) over corresponding antiparticles.

2. This can also be interpreted as the gigantic cosmological value of hypercharge Y (which, for nucleons, coincides with the baryonic number B, although it is customary nowadays to treat Y and B in principle as independent quantities).

3. The difference between the number of protons and the number of neutrons seems to yield a large cosmological value of isotopic spin T or of its third component.

4. There exists a directed time arrow prescribed by the Friedmann expansion, a fact which does not depend on the details of various cosmological models (isotropic, anisotropic, rotating, and so on) or on controversies between exploding models (almost definitely established after the Penzias-Wilson discovery of relict photons) and the stationary world of Bondi-Gold (for the thermodynamic time arrow of Ne'eman).

5. It seems that the preponderance of electrons over positrons leads to a large cosmological value of the leptonic number, although the intervention of neutrinos makes this statement uncertain.

6. The eventual preponderance of (say, electronic) neutrinos can lead to a great cosmological value of a certain spirality.

7. The real universe is characterized by a specific value of the average density of all known kinds of matter which (not considering neutrinos, collapsed stars, and gravitons) is equal to $\approx 10^{-31}$ gm cm^{-3} and points to an open expanding Lobatchevskian world.

8. Space-time, both locally and in the average, is not flat, its curvature being connected with gravitation.

Surely all these phenomenological, fundamental, cosmological properties will in the future prove to be interrelated. We can now, for instance, conjecture about the connection between the expansion and the preponderance of baryons (or hypercharge), so that in a contracting universe one could expect an antibaryonic preponderance, which hypothesis should be taken into account in considerations about eventual relics.

All these specific features of our real universe, which distinguish it from any other possible worlds, for example a contracting one, a flat one, or a universe possessing preponderantly antibaryons, may eventually influence the behavior of elementary particles either directly or indirectly by constituting some average ground state or vacuum. By direct influence I mean, for instance, inclusion of the average cosmological, metrical (or tetradic) gravitational potentials in the equations of elementary particles or, preferably, in a fundamental nonlinear spinor

equation of some specific type (see below). Another possibility of direct
cosmological influence can arise if, with Kurdgelaidze, we make the hypothesis that the local wave functions of particles (antiparticles) are supplemented by factors $1 \mp \mathscr{F}$, where $\mathscr{F}$ is the Friedmann coefficient from
the theory of a nonstatic world. By the way, this latter conjecture gives
at once a qualitative explanation of combined CP-invariance or T-invari-
ance violation in K_2^0 decay, as these particles can be built from K and K:

$$\mathscr{F} \sim \frac{\rho}{4\rho_c}; \quad \rho_c = \frac{3H^2}{8\pi\kappa} \approx 10^{-29} \text{ gm cm}^{-3}; K_2^0 \rightarrow (K_2^0)_{\text{local}} + (K_1^0)_{\text{local}}.$$

At a more elaborate level there will be a rather severe test of such considerations, for the percentage of anomalous K_2^0 decays will be determined
by $\mathscr{F}$, that is, essentially by the average cosmological density of matter
or by Hubble's constant!

One could also try to introduce, in a somewhat analogous phenomenological manner, the influence of other effective, cosmological
forces corresponding to other fundamental cosmological properties.

However, there seems to be another method of accounting for
cosmological properties if one agrees that they are reflected in some
manner by the properties of the ground state (vacuum), viz. in peculiar
asymmetries or violations of some invariances and thus by violations
of some conservation laws, consequently noticed even in the local behavior of particles (even those due to isotropy?).

But the asymmetry of the ground state leads to the existence of a
Goldstone particle; for instance, the degeneracy and asymmetry of the
vacuum with respect to isospin leads to the existence of a "goldston"
(as we may call it for brevity instead of 'goldstoneon') equivalent to an
ordinary photon. (We shall neither discuss nor question the implication
of the Goldstone theorem treated especially by Salam and Weinberg,
and emphasized repeatedly by Heisenberg. He believes that it may
also be applied in the case of indefinite metrics in Hilbert space and
possibly also to discrete transformations, when it can yield goldstonic
fermions, not bosons as in the original version of the theorem; perhaps
even the zero-mass character of goldstons is not a general rule.) If so,
one can conjecture the existence of other goldstons having their origin
in other asymmetries of the vacuum, which — to repeat once more — are
somehow induced by cosmological asymmetries. In this way one must
expect above all the degeneracy of the vacuum with respect to hypercharge and the existence of a corresponding goldston, a hyper-photon,
say. These massless hyper-photons will be produced by sources possess-

ing some hypercharge value, e.g. by a galaxy. In this way one arrives again at the hypothesis of Cabibbo-Bell-Perring-Bernstein aimed at the explanation of K_2^0 decay, as K and K possess different hypercharges. But as I have previously remarked, one must sum over the whole expanding distribution of galaxies, otherwise one gets an infinite value (by summing over a static distribution), quite like the well-known paradox in Newtonian cosmology. One observes that in this way we also come quite close to our hypothesis of "direct" cosmological geometrical influence of the expansion on the K_2^0 decay.

The violation of Lorentz invariance due to space-time curvature can lead to gravitons as the corresponding goldstons. Indeed, the presence of gravitons (or photons) as real particles violates Lorentz (or isospin) invariance. Moreover, one is also attracted by the fact that both photons and gravitons are "compensons"; that is, the quanta of compensating fields, introduced by making the constant parameters of certain transformations (in iso-space or 4-space), depend locally on the coordinates. (For the theory of compensating fields, see Utiyama, also Sakurai, Kibble, Brodski-Ivanenko-Sokolik, Frolov, and Schwinger.) Are not goldstons generally compensons?

Using a different approach, one can try to bring into correspondence with asymmetries of the vacuum (and, on our hypothesis, also of cosmology) some other particles, for instance, according to Heisenberg, massless neutrinos (both electronic and muonic types), treating them as goldstons. One of the most important questions is the discovery of the goldston corresponding to a direct selected time-arrow — if this role is not already fulfilled by a hypercharge goldston. Or it may be that the expanding scale induces some specific asymmetry in the vacuum. This may cause the corresponding goldstons to reflect some complicated interplay of scale or a more general conformal group (the latter is advocated by Vigier and his collaborators as a fundamental group for particles). The very preliminary character of these considerations is clear enough and need not be stressed too much, but they seem to offer some new, reasonable, and stimulating possibilities in our persistent efforts to connect local quantum with global cosmological properties and may remind us of the impossibility of ever treating them separately.

As to a proper local unified theory of elementary particles, there have been, as is very well known, great successes in distributing (and predicting) the ordinary particles as well as their excited states, the resonances in octets, decuplets and so on, chiefly due to the discovery of the unitary SU_3 symmetry after previous rough classification on the basis of mass, spin, parity, isospin, and hypercharge (strangeness) value.

So far only hadrons (mesons and baryons) have been properly dealt with; the attempts to include leptons (Kiev and Nagoya symmetries of Sakata, massless quarks of Marshak) are not convincing. We can suggest here the application of the group $SU_3(J,L)$ to leptons in analogy to SU (J,B) for baryons (J is spin, L leptonic, and B baryonic, number). The latter, reflecting the empirical rule $2J + B = 0$ (mod 2), was — together with Gell-Mann-Ne'eman's famous $SU_3(T,Y)$ — rather successful (Sakita, Naumov, and De Cet) and must lead to other quarks! On this basis one can establish octets of massless (neutrinos and photons) and higher leptons (electrons, muons, hypothetical bi-photons) or work out a schema without bi-photons. Without going into the details of these schemata (Frolov, Ivanenko, Naumov, and Startsev), I should like to remark only that the relativistic generalization of $SU_3(J,L)$ to $SL(3,C)$ seems most plausible, like the generalization of SU_2 to $SL(2,C)$.

It is well known that the problem of relativistic generalization of SU_6 or SU_3 is a difficult one; we prefer to speak of the combination of internal and ordinary external 4-spaces, eventually invoking the form of a de Sitter model, in order to approach somewhat closer in this way the real cosmological conditions. (Cf. also Roman.)

Returning now to hadrons, let us admit a fundamental nonlinear equation as the basis for the description of 'pramatter'. In contradistinction to Heisenberg, who takes as a basic unit a Weyl (or Dirac) spinor whose components are iso-spinors, I prefer to choose unitary symmetry and, starting from quark pramatter, take for the fundamental unit a 12-component spinor (Dirac spinor, whose components are unitary spinors); then one can show that the non-linear term is of a vector type. Symbolically, $D\psi + \lambda(\psi^3) = 0$. [$D$ is the Dirac operator; $\lambda = hcr_0^2$, with r_0 the fundamental length giving the nonlinear interaction term; and ψ^3 expresses some nonlinearity of the type introduced first by Heisenberg and Ivanenko-Brodski]. Though the Lagrangian and fundamental equation may be SU_3 invariant, let us concede that the vacuum is degenerate and has only the SU_2 symmetry of strong interactions. Then, applying the rather powerful perturbation methods of Naumov (which are not as cumber some as the Tamm-Dancoff kind of calculations used for Heisenberg's group, but in known cases lead to practically equivalent results), one gets, for example, on account of the indefinite metric, mass splitting of the corresponding quark triplet. Consequently, the nucleonic quark proves to be 22% lighter than the Λ-quark, in good agreement with the phenomenological value in Zweig's theory.

I believe that the fundamental results of the recent nonlinear spinor unified theory, namely, the values of masses, magnetic moments, and coupling constants, are already close to quantitative agreement with experiment and are so impressive as to justify some degree of optimism in this respect. The theory can be considered as a most consistent development of the fusion theory of de Broglie, of the phenomenological schemas of Fermi-Yang and Sakata, and of various treatments of classical nonlinear extensions of relativistic equations (Kurdgelaidze, Ivanenko, Rodičev, Petiau, Finkelstein) obtained mostly by Heisenberg and his collaborators Dürr, Mitter, Jamazaki, and others, and supplemented by Nambu-Jonas-Lasinio, Marshak-Okubo, Naumov, *et al*. Perhaps some skilful combination of nonlinear spinor theory and cosmological considerations will bring us in the not too remote future nearer to the elaboration of a unified theory of physical reality.

REFERENCES

1. D. Ivanenko, in *Proceedings of the Galilean Jubilee Conference*, Florence, 1964, in *Nuovo Cimento, Suppl.*, 1965.
2. D. Ivanenko, in *Proceedings of the 2nd Soviet Conf. on Gravitation*, Tiflis, April, 1965.
3. D. Ivanenko, in *Proceedings of the Einstein Symposium*, Berlin, November, 1965: *Entstehung, Entwicklung und Perspektiven der Einsteinschen Gravitationstheorie*, Akademie-Verlag, Berlin, 1966.
4. D. Ivanenko, in *Progr. Theoret. Phys., Suppl., Commemoration Issue for the Thirtieth Anniversary of the Meson Theory by Dr. H. Yukawa*, 1965, pp. 161-166.
5. J.A. Wheeler, *Geometrodynamics*, Academic Press, New York, 1962.
6. J.A. Wheeler, in *Proceedings of the Einstein Symposium*, Berlin, November, 1965: *Entstehung, Entwicklung und Perspektiven der Einsteinschen Gravitationstheorie*, Akademie-Verlag, Berlin, 1966.
7. J.A. Wheeler, in *Gravitation, Groups, Topology*. Gordon Breach, New York, 1964.
8. C. Möller, in *Proceedings of the Galilean Conference*, Florence, 1964, in *Nuovo Cimento, Suppl.* 1965.
9. C. Möller, in *Proceedings of the Einstein Symposium*, Berlin, November, 1965: *Entstehung, Entwicklung und Perspektiven der Einsteinschen Gravitationstheorie*, Akademie-Verlag, Berlin, 1966.
10. W. Heisenberg, in *Proceedings of the Symposium on the Non-Linear Theory*, Munich, 1965.
11. W. Heisenberg, *Introduction to the Unified Field Theory of Elementary Particles*, Interscience, New York, 1966.

DISCUSSION

MERCIER: Ivanenko's paper attacks the problem of unification from a standpoint which is different from that taken by Einstein and which has become traditional. His views have been made known, as he says, on several occasions. I have myself independently discussed unification at various times, and I have taken a critical position against the traditional Einsteinian approach, which has the disadvantage of ignoring quantization altogether. Ivanenko seems rather optimistic about unification. After having attempted years ago with E. Schaffhauser and J. Schaer to elaborate a unified field theory in the Einsteinian sense based on considerations of Finsler's geometry (which, like many attempts by other authors, did not succeed), I became rather pessimistic about the possible final success of such attempts. Still, in the book which J. Schaer and I have published, *The Idea of a Unified Theory*, the epistemological and methodological advantages of unification are explained at length.

It seems at this time as if each irreducible interaction might need an appropriate or proper dynamics. Einsteinian gravitodynamics (or its Newtonian approximation) is, strictly speaking, suited for gravitation only and is not quite appropriate for dealing with other interactions. This has been discussed in detail in my paper at the Einstein Jubilee in Berlin.

It is clear, however, that at least some links exist between the various dynamics developed so far, and they produce what might be considered a quasi-unification. Physics appears to be in a steady state of quasi-unification which, however, never becomes total. If a total unification were possible, the following alternatives are thinkable: either it would still assume the form of a dynamics, or it would avoid dynamics altogether. C. F. von Weizsäcker has suggested that dynamics might collapse; in that case, the complete description of the world would only need a few universal constants like α — and an Eddingtonian picture of nature would be attained. Current physics does not seem to advocate this picture, for the theory of particles suggests more and more the possibility of an infinite regress in looking for "smaller" and "smaller" structures, leading from molecules to atoms, from atoms to nuclei and electrons, from nuclei to baryons, accompanied by all the particles known as leptons, mesons, and heavy particles, then from such particles to hypothetical quarks, and so on. And if at the present time we think of only three quarks, it may well be that in the future there will be lots of them, just as the proton-electron pair of the twenties has been extended to dozens of particles in our day. Of course, the classification of particles by means of groups of symmetries is in a way very successful and can be interpreted as suggesting the final disappearance of dynamics at the "elementary particle" level. But these symmetries must be broken or violated if the individual particles are to be accounted for. As far as I can see, this violation will finally require one or several dynamical schemata.

If, however, the unification of physics remains a question of dynamics, it will presumably have to be done with the help of a variational principle, or the like, containing a Lagrangian L. This Lagrangian will have to be uniquely determined, either by the fundamental principles establishing the unification (the theory of the "only possible world"), or by the choice of one final L among possible Lagrangians (the Leibnizian idea of the "best of all possible worlds").

TREDER: I think Ivanenko is right in implying a strong asymmetry of the universe with reference to the new quantum numbers and the (selected) directed time arrow.

I am also fascinated with the idea of perhaps interpreting the gravitons as Goldstone particles. (At this moment I am working on a development of general relativity in which the Goldstone philosophy is — implicitly — a main point of argument.) But I wonder what must compensate for the destruction of symmetries in orthodox general relativity. In reference to the tetrads (or the metrical spin vectors), the curved Riemannian space permits the same general local Lorentz group as the flat Minkowski space. The curvature of the space-time does not break the Lorentz symmetry, but a tele-parallelism of space-time breaks this symmetry and permits rigid Lorentz rotations alone. That is, the "compensating gravitational fields" described by Utiyama, Kibble, Ivanenko, etc., are compensating nothing in general relativity, because the postulated defect of Lorentz symmetry for a curved space doesn't exist. I wish to give a simple example:

In the Minkowski space of special relativity, the Dirac equation (written in the form that uses the Dirac-Pauli spin matrices) is covariant with rigid Lorentz rotations alone. But, if we are writing the Dirac equation with the tetrads used by Ivanenko and Fock, we can give to this equation a general Lorentz-covariant form. In the common presentation Dirac's equation is:

$$\gamma^\nu \delta^i_\nu \psi_{,i} + K\psi = 0, \tag{1}$$

with the special Cartesian tetrads $h^i_\nu = \delta^i_\nu$. In the explicit Lorentz covariant form of Ivanenko and Fock, the above equation becomes

$$\gamma^\nu (h^i_\nu \psi_{,i} - h^i_\nu \Gamma_i \psi) + K\psi = 0, \tag{2}$$

with the orthonormalization:

$$h_i{}^\nu h_{k\nu} = \eta_{ik}. \tag{3}$$

Each fixation of the tetrads $h^i_\nu(x^l)$ which is compatible with (3) gives a specialization of the general form (2), and this special form of (2) is covariant with rigid Lorentz rotations again. The single difference between the Minkowski space and a curved space is the orthonormalization rule for the tetrads:

$$h^\nu_i h_{k\nu} = g_{ik}(x^l). \tag{4}$$

If we choose one tetrad-system compatible with (4), then the equations which we have in reference to this fixed tetrad-system are covariant with rigid rotations only. (This choice might be arbitrary or the result of a gravitation theory with tele-parallelism.) All this is independent of the curvature of space.

I think that the proposition "the gravitons are Goldstone particles" is right for a gravitation theory with tele-parallelism only, because in such a theory the gravitons can compensate for a real defect of Lorentz symmetry. However, according to this theory, the Goldstone gravitons are particles with spin 1, because these gravitons are the quanta of the 4-tetrad vector fields.

YOURGRAU: Since Ivanenko published his beautiful book on classical field theory, the pundits have been once more deeply impressed by his superb knowledge of certain actual domains in theoretical physics. And it was in field theory where he and Heisenberg established a nonlinear wave equation which made it possible to treat the mass spectrum of strongly interacting particles, if only in a qualitative manner. In brief, Ivanenko's published research findings, especially in nuclear physics—his well-known nuclear model, the Ivanenko-Tamm theory of specific nuclear forces, his prediction (together with Pomeranchuk and

Sokolov) of electromagnetic radiation emitted by synchroton accelerators, his linear matrix geometry, and so on — testify to the brilliant achievements of our friend Dmitri; they are of the highest level, often original, always displaying his competence as a theoretical physicist, and certainly never dull.

These preliminary remarks are made in order to protect myself against any accusation of bias and unfair comments. Now, the gist of Ivanenko's paper can be summarized in two words: wishful thinking. For example, gravitons — are they supposed to be bosons? — are dealt with as if they *were* elementary particles, such as electrons, positrons, neutrons, mesons, etc., etc. To quote verbatim: "...the existence of a corresponding goldston, a hyper-photon, say. These massless hyper-photons will be produced...e.g. by a galaxy." I find conjectures of that kind at best 'accurate exaggerations'. Some years ago, Ivanenko suggested that gravitons can transform into neutrino-antineutrino, electron-positron, and other similar pairs. So far we are still without any experimental evidence for the physical existence of gravitons.

It is only fair to mention that J. Weber (1961) and K. W. Ford (1963) treated gravitons, at least theoretically, as legitimate physical entities. Both authors were convinced of the desirability and necessity of such particles, but they were not very optimistic as to the chances for experimental (observational) evidence of those massless, "relativistic" particles. Their scepticism, as far as direct observation is concerned, was partially based upon the fact that the gravitational force operating between an electron and a proton is by a factor of c. 10^{40} weaker than the electrical force. Hence, my scepticism is shared, so far, by at least two strong advocates of the 'physical reality' of gravitons.

In this connection I also recall that Pirani treated his so-called gravitinos like the other massless particles, viz. photons and neutrinos. He compared them to neutrinos of negative energy. Of course, we consider photons and neutrinos as fundamental particles and theoretically, at least, the existence of gravitons is definitely feasible. However, the consequences of such a discovery, i.e. of gravitons, would be so tremendous that we would have to rewrite whole chapters of physics, re-examine the now-accepted belief in an expanding universe, and look at quantum mechanics in a novel manner. Still, perhaps the experimentalist will have the last word in this matter and the theorist may be forced to pray for the arrival of another Schrödinger or Einstein. By the way, I recommend strongly to our colleagues here (who are interested in this field) to read again the theory of Synge who postulated the possible existence of repulsive and attractive impulses many years ago. Perhaps the graviton may also possess either negative or positive energy. But all this is highly speculative and *in statu nascendi*.

Many of Ivanenko's ideas in this paper are most stimulating and definitely support those attempts that entice us with their beauty and have the elegance of some highly speculative 'quasi-theories' — until we succumb to Circe's alluring efforts and accept a or *the* unified theory. He knows that Einstein's endeavors regarding unification of quantum theory and general relativity were a failure. Heisenberg's recent work on this very subject is certainly more plausible than Einstein's; still, it is in no way conclusive. Ivanenko too is haunted by the spectre that a unified theory is a logical *and* physical necessity.

Well, he is undoubtedly in good company. But my fear is that whenever our 'extrapolative leaps' are too great, even the most persuasive conjectures will ultimately paraphrase out into interesting, though merely speculative, pseudo-answers to genuine physical questions.

Chapter VII

THE EVOLUTION OF EVOLUTION

Julian V. Langmead Casserley
Seabury-Western Theological Seminary

One of the most characteristic evolutionary or cosmogenetic processes is a gradual transition from one mode of development to another. The time has now come, speaking from a de Chardinesque point of view rather than merely reflecting and relying upon Teilhard's views, to compare and contrast the dominant modes of growth and development in the different spheres of emergence which supervene upon and enfold within themselves their predecessors in the evolutionary hierarchy.

For example, we may compare and contrast the dominant types of causality to be found in the geosphere, the biosphere, and the noosphere. Even in the geosphere causality has a certain ambiguity. The discovery that on a miscroscopic level causality in the geosphere is ambiguous rather than rigidly deterministic is connected with the work of the great physicist Heisenberg in the 1920's and with the principle of indeterminacy which he propounded. However, by the use of statistical techniques it was possible to show that this microscopic ambiguity had almost no tangible and ponderable consequences on even the most modest macroscopic level. In fact, the principle of indeterminacy, though it might bother old-fashioned and deterministic physicists, made very little difference to any actual state of affairs. Certainly some philosophical interpreters of the principle erred by equating indeterminacy with freedom.

In the biosphere, however, we note a considerable intensification of the ambiguity of causality, and the consequences of the ambiguity become more significant and effective. For example, the enzymes serve as catalytic agents which apparently perform no other function than that of facilitating certain types of biological process. However, it is always possible that an enzyme which more usually or even normally facilitates one type of process may on some rare occasion facilitate another. Some researchers, for example, think that this may be very relevant to the problem of the causes and possible cures of cancer.

Obviously, a deterministic, "one-one" conception of causality means that if we have defined and described the cause with sufficient accuracy then we shall recognize that only one effect is possible. Ambiguous causality means that each cause must be understood in relation to a whole spectrum of possible effects, of which some, of course, are no doubt very much more probable than others. What we are saying is that in the geosphere this element of ambiguity is indeed present and traceable, but that it has no effective impact and makes no significant difference in at least the overwhelming majority of overt occasions. In the biosphere, however, it becomes increasingly impossible to ignore the ambiguity of causality, for now it is a factor the consequences of which can be traced in the record of what overtly occurs.

In the noosphere this ambiguity of causality becomes increasingly more noticeable. For the conscious beings who inhabit the noosphere find in their immediate experience of existence so much evidence of the indeterminism of their situation that they are sometimes tempted towards an almost dogmatically indeterministic interpretation of themselves, which is gravely mistaken. There are indeed deterministic elements in the conscious experience of life in the noosphere, but they are so increasingly ambiguous that they are often hardly recognized as deterministic at all.

In secularized philosophy the problem of freedom has for many centuries been diagnosed and discussed as a question of the relationship between an alleged entity called the "free will" and an alleged type of process which we may entitle absolute determinism. In Christian theology, however, as we can see in Augustine, Boethius, and many other classical writers, the controversy about freedom takes a very different form. In this context the question is not whether man is determined or not, but rather whether he is determined by impersonal forms of process that operate very effectively at levels beneath him in the hierarchy of being—evolution is only our version of the antique hierarchy of being hypothesis—or whether, on the contrary, he is capable of allowing himself to be determined by personal forms of process which prevail absolutely at levels of being superior to his own. In other words, whereas the problem of freedom is visualized by the purely secular philosophy as the free will vs. determinism controversy, the theologian interprets it as a divine providence vs. fixed fate controversy. It certainly appears to be the case that, although there is not and cannot be an absolutely indeterministic situation, the developing creation of the world renders causality increasingly ambiguous and the condition of creatures an increasingly flexible one, permitting in the noosphere a real measure of limited self-direction and autonomy. The consequence is that in this ambiguous situation men tend either apathetically to permit themselves to be determined by the forces inferior to them deriving from lower levels of the cosmogenetic process, or actively to adapt and devote themselves to processes superior to them that derive sometimes from further extensions of cosmogenetic development which are hardly as yet clearly discerned, and sometimes — at least conceivably — from the absolute reality that transcends the cosmogenetic process altogether. To be determined entirely by the latter is what the theologian calls "freedom". To be determined absolutely by the former he thinks of as slavery. The question of freedom is thus not one of what we want to do or of whether we are able to do what we want to do, but rather a question of the cosmogenetic status of the forces to which our conduct and decision submit and conform. Are they, in other words, the forces that preside and predominate in our past evolutionary history, or the forces that will be completely victorious only in some future consummation of our destiny? Thus we may reformulate the question of freedom, in a language which secularizes our theological interpretation of it as a controversy between providence and fate, by transforming it into a debate between history and destiny, between the past and the future. Any conception of evolution or cosmogenesis of the type that predominates in the writings of people like Samuel Alexander, Teilhard de Chardin, or perhaps even Bergson, inevitably tends toward the view that the future represents a conditioning agent, and in some sense even a causal force, that is more powerful and decisive than the past. The inherent cosmic conservatism of conventional science is very largely due to the way in which, being so much more familiar with the scientific scrutiny of events in the geosphere than with those in the biosphere and the noosphere, it tends always to lean too heavily on the causality of the past and to ignore the increasingly evident causality of the future.

Second only in importance to the increasing ambiguity of causality, as one emergent form succeeds another in cosmogenesis, is the heightening of the significance of individuality. Of course, the distinction between essence and existence, between being merely "that kind of thing" and being "this particular one" can be traced at every cosmogenetic level. Thus, each stone in a heap of stones is merely one of the stones, like all the others, and yet at the same time "this one" which is precisely what no other stone can possibly be. Nevertheless, at the level of stones, the distinction between essence and existence is hardly significant. There is apparently no tension or conflict between essence and existence in the case of stones, and who would ever speak of the alienation of one particular stony existence from its stony essence! In the biosphere, although we might conceivably be able to trace something analogous to the existential conflict in, for example, the more developed anthropoids, it can hardly be claimed, except perhaps where there is some particular love of an animal pet, or exceptional grace and beauty in the form and motion of some unique individual, that the individual is more significant or more important than the type. Certainly, however, in the noosphere we encounter a situation in which the individual human existence, is, at all events for other individuals closely associated with him, far more significant than the general human type—more significant not merely in the order of value, but even to an increasing extent more significant for the explanation of action.

The process of evolution tends toward inverting the primordial relationship between the type and the individual. The latter, barely significant although nevertheless present and traceable in the earlier stages of the process, has become by the end of it far more important than the former, so that by the end of the process it is the type that has become barely significant although no doubt still traceable. All this is not necessarily to endorse the kind of social philosophy that is called individualism and supposed wrongly, I believe, to be essential to democracy. On the contrary, the inversion of the relationship between the individual and the type is the essential step toward setting up and emphasizing a novel possibility which is uniquely characteristic of what I will call the Christosphere, the phase of evolutionary development which lies beyond the noosphere. For now we aim at a unity that will intensify the freedom, autonomy, and dignity of the individuals entering into it and will in no sense annul them. Such a unity is the most important of all mystical concepts. It is classically set forth theologically in the doctrine of the Trinity. It is characteristic of the Christosphere that it is an area in which more and more mystical processes of this type begin to operate and even to prevail.

In the debates characteristic of recent centuries it must be allowed that the totalitarians have collected at least some of the points. They would insist, for example, that it is only in the area of social collaboration originally based on the division of labor that the most characteristic values of culture are produced and flourish. The individualist, however, is right to reply that these values, though they could not emerge except out of some distinctive species of society, would nevertheless be impossible without the occurrence of what might be called creative and charismatic individuals. This is of course true, but it is even more important to insist that, although cultural values cannot arise except in society, yet it is only by individuals that they can be consciously valued and enjoyed. The individual is thus essential to the recognition and enjoyment of even the most purely social values. Nevertheless, to defend and justify the traditional emphasis upon the individual in democratic western society in such terms as these is at the same time to recognize the necessity of society. There is a polarity between individual and society and an energizing tension between them, but either would be inconceivable without the other. It is only in society that real individuals actually appear, and only in a very good society that they are able to wax rounded and rich.

Another error closely associated with the conventional individualism that so many western thinkers have employed to bolster up the case for a democratic society is the notion that freedom is a purely individual value and that society with its inevitable authority and power is the chief enemy of freedom. In fact, it is every bit as important to speak in terms of the freedom of society to develop in the direction of its evolutionary destiny, as to speak of the freedom of the individual from the more undesirable manifestations of social power. Government is neither necessarily the enemy nor the inevitable assassin of freedom. On the contrary, in a truly free society both society itself and social organs like government are free to function, develop, and carry out on an increasing scale and with intensified efficiency their duties and responsibilities toward the freedom and welfare of all the people. A free society is inevitably one in which government is big enough to do its job properly. The fears and prejudices which would cut it down to too small a scale inevitably restrict its capacity to grow to a size that makes possible the orderly performance of its functions. Thus, we must speak not only of the necessary freedom of the individual to be utterly himself, but also of the freedom of the social order to which he belongs to become effectively itself.

This is perhaps in the last analysis a mystical conception, for mysticism is not a purely religious process. Granted, the word is usually used in a highly specialized sense in which it refers to the relationship between God and man, or at all events to man's spiritual behavior in a narrow and exclusive sense of the word 'spiritual', but in truth the whole life of man, insofar as it involves love, friendship, responsibility, cooperation, marriage, parenthood, and other forms of intimate exchange between people, involves mysticism. For where persons acknowledge rather than merely use each other, creating a unity which is posterior although not necessarily prior to their individual being, there indeed we meet with phenomena of the kind that is called mystical. Religion may, indeed, raise the mystical to a kind of zenith at which it proclaims itself in all its fullness, but more generally, and less purely experienced, mysticism is part of life, and profound interpersonal relationships are impossible without it. The goal of social development, it may thus be claimed, from the present point of view, is an ultimate free choice, on the part of many, of some one life in mystical union which will defend and intensify the freedom and autonomy of those who enter into and freely bind themselves by the bonds of union.

As man in the noosphere, therefore, the individual persistently tends to become more and more evidently and effectively of greater significance than the category of beings to which he belongs. In the biosphere any one crocodile is less important than the crocodilian race, but a man is now much more significant than the human race. Those who see a man's posterity as his only true immortality are retrogressive biophysical types who are unable to perceive — we will not say the difference between a man and a crocodile — at least the difference between a man's relationship to the human race and a crocodile's relationship to the crocodilian race. Thus Hitler supposed that a German was a mere part of Germany and apparently a subordinate and expendable part at that. When we deny indignantly that a German is a mere part of Germany, do not let us do so because we are mere liberal humanitarians who think that a German is, like anyone else, merely a part of the human race. Let us denounce the Nazi heresy in the name of an authentic humanism that knows that a person is not *merely* a part of anything at all, but a complete and bottomless world of meaning and value in himself. For such a humanism the transition from a doctrine of development, which takes the form of a merely or purely racial evolution into a doctrine of personal development culminating not in the persistence of the type but in a rich measure of eternal life for the person, is indeed the mark of a more profoundly informed concept of evolution and a more secure grasp of the scope and direction of cosmogenesis.

Both the increasing ambiguity of causality and the inversion of the primordial relation between type and individual illustrate the way in which the modes of evolution themselves evolve. The process may be described as one of inversion. Dominant processes become recessive, and processes so shy at first as to be barely perceptible become dramatically operative and effective on an almost universal scale. Thus dawning consciousness and intelligence in the higher anthropoids and the first men make almost no difference to what actually happens. Some processes have at this time become conscious of themselves, but they are very much the same processes that were formerly unconscious of themselves. Consciousness has made very little difference as yet to what there is for it to be conscious of. Contrast this with the present situation of the world only a few hundred thousand years later. Certainly consciousness has not yet drastically modified everything, but it must be said that it has enormously transformed most of the things that are objectively there for it to be conscious of. We are most of all conscious of the vivid effects of consciousness (e.g. through technology). Perhaps the older philosophical idea confined itself too much to the way in which the individual consciousness interpenetrated the individual experience of which it was conscious. Upon a vaster and more cosmic scale, the case for a substantially idealistic doctrine according to which consciousness, as it expands and intensifies, modifies and transforms the being of which it is conscious, would seem to be undeniable. We might call this a historical as opposed to a merely epistemological idealism.

At first a new emergent, for example, the vitality of the first living creatures, seems like a mere peripheral excrescence or growth upon the great body of what preceded it. Who would dare to prophesy in the age of the amoeba and similar unicellular organisms that one day the organic would be vastly more important than the inorganic and indeed revolutionize and transform the latter's structure? Similarly, the first symptoms of intelligence and consciousness in the age of the biosphere would hardly prompt the observer to suppose that the day would come when the noosphere would become vastly more impressive in its achievements and growing control than the forces let loose in the biosphere. Living as we do in the first stumbling days of the Christosphere, it may still seem to many almost impertinent to suppose that one day the mystical forces let loose in the Christosphere will become dominant in world history and be revealed as even vital to man's very survival; yet that is what is beginning to be made plain to us even now by this age in which God is patently increasing the numbers of the human race and at the same time intensifying their proximity. By 2000 A.D., the population of the earth may have been doubled and dumped in everybody's back yard. The necessity for techniques of mystical love and unity making possible a common life based upon felt fraternity, unimpeachable justice, and active tolerance and charity will then become so great as to be undeniable and even unquestionable. The Christian may well say that God who gave men the gospel of love seems bent upon creating a world in which it is more manifestly relevant than ever before.

Thus evolution itself evolves by the inversion of its modes. Many of the hackneyed science vs. religion problems are based upon the failure to recognize the extent to which such inversions have taken place and are appearing even now. For example, the stress upon the struggle for existence which led many conventional Darwinians to the conclusion that conflict was the most creative thing in the universe was due to a failure to realize than an evolutionary mode dominant in the early biosphere is no longer dominant, although still important, in the noosphere, or even in the later biosphere. The use of the Darwinian thesis by Sumner to justify cut-throat capitalism and free-for-all economic competition, or by Gumplowicz to defend and whitewash aggressive warfare, was clearly the result of supposing that, despite all the evolution that has taken place since, we are still in the early stages of the biosphere. Similarly, the view that social psychology grasps and comprehends the profoundest social and historic processes, or that the psychology of religion understands what it essentially means to be religious, are clear examples of the mistake of supposing that now, in the age of the Christosphere, we are still in the noosphere. Actually, of course, the inversion between the psychological processes characteristic of the noosphere and the mystical processes characteristic of the Christosphere has already begun and is indeed beginning to make some, at least, of its consequences visible and significant.

Thus, we have observed in the passage from the geosphere to the biosphere through the noosphere to the Christosphere three characteristic modes of development: (a) the increasing ambiguity of causality, (b) the significance and effectiveness of the individual, and (c) increasing relevance of 'tough' mystical love. Each is an example of this inversion, typical of the relationship between once dominant and once recessive types of process, so that what appeared at first to be merely excrescent and peripheral more and more tends to assume the central role, gradually banishing processes of a type formerly dominant to a new periphery grouped around a new center.

DISCUSSION

TENNESSEN: The suggestion by Casserley with which I am chiefly concerned is the (implied) suggestion that we should aim at something that would embrace both analytical and existential philosophy, and this will come to something like classic metaphysics. I'm not sure whether this is an argument for or against the embracement, but in any case I shall maintain that the suggestion is not only movingly naive, touchingly optimistic, but also pernicious for almost all possible interpretations of it. The existentialist and the analytical philosopher accuse each other of not knowing what they are talking about, and the remarkable thing is that they are both right, in at least one true sense of knowing what one is talking about.

There is one sense of not knowing what one is talking about, when e.g. one sticks to a very, very low level of preciseness: as when I say that this book is heavier than that, and I go ahead without having made it clear whether I mean to say (1) that one of the books weighs more than the other, or (2) that it is more difficult to understand, it is heavier reading. One would have to decide which interpretation one attempts to be dealing with now; this is particularly obvious in the case of existentialism, as is exemplified in the very famous criticism by Carnap of Heidegger.

Carnap convincingly takes apart what I have called "the general continental grandiloquence and the particularly pompous teutonic turgidity in Heidegger's high-flown, glutinated, conglomerate of bombastic neologisms. Its stylistic rhythm has the peculiarly mesmerizing effect of kettledrums finding a short-cut, as it were, from the receiver's tympanum directly to his volitional layers." So much for existentialists not knowing what they are talking about.

Now, what is more interesting is to consider the existentialist criticism of the analytical philosopher. I think something like the following example might help. The Finnish philosopher and psychologist Eino Kaila uses it somewhere in one of his books to the effect that at the time of the Finnish-Russian war a spy was caught behind the Finnish line. It is an obvious case, a clear-cut one; the spy is going to be executed at dawn, and he knows this as well as anyone can know anything. If one gave him a questionnaire: "Are you going to die?", he would check the questionnaire in the box for "yes", and would not consider "no", "don't know", etc. Therefore he remains absolutely stoic in court. He knows the outcome. However, he can't help himself, he gets involved. The peculiar thing happens that as the trial goes on and the death sentence is pronounced, the spy collapses. What on earth happened? He couldn't know the outcome better than he did. What indeed could have caused the collapse?

The explanation, according to Heidegger, is that his 'knowledge perfection' has increased along a new dimension. He now knows, as it were, in bones and marrow, what he earlier knew in a very superficial way. And the existentialist aspect of all this? Well, we all go about quite complacently, feeling quite comfortable and believing that everything is orderly and fine and that there is nothing to worry about. And if we were presented with the same questionnaire, we would all know which box to check. But do we know in the existentialist manner about imminent death in the way in which the spy knew it *before* the death sentence was pronounced, or do we really know it in the way *he* knew it *after* the death sentence was pronounced?

One of the obstacles we encounter when we intend to convey that kind of information is precisely *preciseness*, the ordinary language of prose and discourse. It seems there is an inherent difficulty here: if we try to be terribly precise we lose out on integration, on knowing in this bones and marrow business.

So, as Artaud has said, we would have to "break through language to touch life". We have to turn ourselves as communicators into victims burning at the stake, signalling to each other through the flames.

This is, then, what is achieved by the use of poetry and drama. In a simpler way one could try to impress upon oneself the fact that every life is like falling out of a tall building. Suddenly you realize the yawning abyss below you, and you are saying to yourself, "What can I do about it?" There isn't anything you can do, but you can choose any kind of attitude you want. You can say, "Well, this is very comfortable as long as it lasts," and try to make the best of it. You take out your transistor radio—two transistor radios, so you have stereophonic music—you know, one has to live a full life. Or one can decide to be terribly serious about life, one's beingness (*Daseinsweise*), and start counting the windows. Then one would have the satisfaction, when one makes the final "hit on Broadway", that life wasn't totally wasted.

Or one can do fantastic things, like having an enormous ostrich feather with which to tickle oneself under the arm as one goes down, accompanied by Dionysian laughter.

CASSERLEY: Tennessen accuses me of being vague. Of course, this charge may be made against a statement either on account of the literary form in which it is presented, or on account of the ideas expressed. I certainly want to stress the necessity of a prolonged period of discussion in which a given set of ideas gradually articulate themselves with greater clarity. In general we may say that vagueness is often the by-product of the comprehensiveness and generality of an idea. Whenever a hypothesis appears to cover too much, it will always tend to be accused of a lack of clarity and distinctive application. Thus, for example, evolutionary ideas, as we now call them, in the centuries before Darwin were so general and universal in their sphere of application that they could plausibly be described as vague. Perhaps I should say also that evolutionary ideas prior to Darwin were to be found among a very few minor classical writers, but occur much more frequently from St. Paul onwards, particularly in the Greek fathers and Augustine. They also crop up more rarely in the Latin scholastics, especially St. Bonaventure and in one celebrated phrase of Aquinas.

Of course, in none of these writers is there any idea that an evolutionary process has actually taken place. St. Thomas Aquinas speculated about the possibility, but for him, as for all these writers, the difficulty was the lack of time. We are very familiar with the way in which early Christian writers supposed that the world was coming to an end quite shortly. What we observe less often is that for these same writers it began only quite recently. An evolutionary process requires an enormous amount of time. Darwin was able to think of an evolutionary process as having really occurred because the geologists had greatly increased their estimate of the age of the earth. Why, then, do I credit the Greek fathers, Aquinas, and some other Latin scholastics with evolutionary ideas? I do so because they gave an account of the earth in hierarchical terms like those of the post-Darwinian emergent evolutionists, and they attempt, for example in the development of the theory of the plurality of forms, to show how a species having one characteristic form might acquire another without altogether forfeiting the characteristic form which it possessed before. St. Bonaventure indeed says that the lower form stretches out its arms to embrace the higher form. The idea of an actual evolutionary process they tend to put aside regretfully as an attractive but unevidenced speculation. St. Thomas employs the phrase *"Magis placet"*, i.e. it would be much more pleasing to the theological mind to accept a theory of creation by evolution, but, of course, an honest theologian cannot accept any theory merely because it is much more pleasing to the theological mind.

Darwin gave greater precision to this rather vague mass of speculative ideas by narrowing the range of their application and by insisting on one particular method of evolutionary progression. He confined the hypothesis of evolution to the development of species, and he interpreted all evolutionary processes whatsoever in terms of his own central conception of survival and the struggle for existence. Since Darwin's time, however, the general tendency of evolutionary thought has been toward a recovery of some of that generality which we find in the earlier writers. We may think of philosophers of evolution like Bergson and Alexander, or even A. N. Whitehead in some moods, of Julian Huxley, Teilhard de Chardin, and many others. In different ways, and whether they know it or not, they all represent a return to the outlook characteristic of the Greek fathers. They provide us with the outlines of a theory of creation by evolution, and they describe the world in hierarchical terms governed by the stage in the process at which different levels of reality emerged and began to flourish. In particular, they would insist on giving an evolutionary account of the development of the inorganic world and at the other end of the story, so to speak, an evolutionary account of the development of human personality toward its destiny or goal. Unfortunately, as Tennessen has correctly observed, in restoring the generality of the hypothesis they had to sacrifice much of Darwin's clarity and precision. What we are now engaged upon is a process of intellectual experimentation out of which — so we hope — there will emerge a broader evolutionary hypothesis than Darwin's, which, although more complex than his because more general in its field of application, will yet be scientifically verifiable. Teilhard de Chardin, upon whose writings my remarks were chiefly based, made an important contribution tending in this direction, but we cannot claim that he in any way completed the process. If the vague applications of classical writers and theologians, chiefly interested in the doctrine of creation, are to be transformed into a viable scientific hypothesis, very much remains to be done. And, no doubt, until then some people like Tennessen will complain of an oppressive atmosphere of vagueness, while others, like myself, will experience a bracing climate of hope.

YOURGRAU: I must say I do not know what Casserley believes evolution tells us. If he means that man is biologically developing, I have no objection, though I prefer his present form. If he means Nietzsche's superman, it is certain that his interpretation of that term will hardly coincide with mine. I hope he means man's developing toward further intellectual proficiency and excellence and superiority. So if he means mental growth, I accept it. But if he tries to "sell" Christology, or let us say, the tenets of traditional Christian interpretation, then he would definitely meet with tremendous and often impolite counter-arguments.

I noticed he committed a blunder which non-physicists or non-scientists often succumb to. It is preposterous to apply Heisenberg's uncertainty principle either to metaphysics or ethical theory. Granted, Casserley is in good company, because Eddington and Jeans made the same mistake. But this is no excuse. The uncertainty principle is still subject to many inane interpretations, so many indeed, that we can ascribe practically any metaphysical nonsense whatsoever to it.

I appreciate that questions have been raised concerning the contrast between analytic and existentialist theories; and I think if we must accept the dichotomy at all, I agree that analytic philosophy — if treated in isolation — is empty, is arid. However, I do not propose at all that because analytic philosophy or linguistic philosophy can be painfully wrong, existential philosophy could ever be correct.

CASSERLEY: To my mind, the great excellence of analytic philosophy is that it is a way of doing philosophy which those who have some kind of existentialist attitude toward life could make good use of. It would introduce into their thought a great many clarifications, like a ray of light in a fog, and that is the kind of thing I am looking for.

BONDI: I found both the paper and its subject matter peculiarly repulsive. I find it so on two grounds. One, on the general ground that in our present state of knowledge it is quite unduly ambitious to attempt a synthesis of this kind. What we all want to do is to chip away and learn a little more, so that one day as a result such a synthesis can be attempted.

I want to make it quite clear that in opposing any such present attempt at synthesis I am in no way opposed to inter-disciplinary colloquia, at which we can certainly learn to understand each other's language; but this doesn't mean that any one of us is fit to do the job of all of us.

Secondly, there is to me the peculiarly repulsive habit of some (but fortunately by no means all) Christian theologians of thinking that theirs is the only valid method of thought, without recognizing that the majority of mankind and the majority of intelligent, sincere people do not hold Christian views. There are, of course, I recognize, those who do hold Christian views with intelligence and sincerity, and do not automatically assume that all others are in error.

KEYES: Casserley has left open the possibility of a creative synthesis involving the positive contributions of both the existential and analytical. The nature of the questioning of this proposition has been up to now primarily on the anthropoligical level. Is it not possible to open that question on the level of ontology? If so, what notion of being will express the creative synthesis when we have, on one hand — to borrow an expression from Gilson — "existential neutrality", and, on the other hand, the notion of being as "to be" or *Sein*, to borrow an expression from an unpopular source, namely Heidegger?

CASSERLEY: Bondi's criticism is one that not only I but even Teilhard would entirely agree with. Teilhard in many places says that he was trying to issue some kind of trial balance, which might be helpful, but he was as aware as Bondi is, as I am sure we all must be, that the moment has not come for dogmatism and finality in such matters.

However, I cannot recollect saying anything which deserved the charge of thinking that Christian theologians are the only serious people around. On the contrary; for example, I carefully balanced Nietzsche and Kierkegaard with precisely that in mind, because the astonishing thing about them is that despite their diametrically opposed religious views, they did manage to say very much the same thing, and I would certainly agree with Bondi altogether that any attempt to use religious philosophy for theologians only and ignore the others would indeed be, as he said, repulsive. I cannot believe that I allowed myself to use any expression which really can, from any objective point of view, have that meaning.

BONDI: While I was not accusing Casserley of displaying Christian superiority, the very name of 'Christosphere' shows that Teilhard is wide open to this criticism. I would like to make that clear. Secondly, I castigated no dogmatism, but the very making of the attempt at a synthesis.

BRECK: Teilhard de Chardin, and Casserley after him, have tried to build the very sort of bridges we need. Indeed, they have added another dimension, that of theology, to the domain of the humanist and the scientist, or rather, they have seen the world of nature and the world of people, as it were, *sub specie universitatis*. One might anticipate the fact that their conclusions, as well as their methods, would not be entirely acceptable to scholars whose point of departure is the material universe. For the thoughtfulness and thought-provoking nature of papers such as Casserley's, we have much to be grateful.

CASSERLEY: What I have expounded in this paper is at the moment what I might call a "naive conjecture", a description of a certain way of experiencing life and enjoying a vision of its meaning. It could only be turned into a fully developed and sophisticated theory by a further process of analysis and criticism. Some theorists have proposed that every sophisticated theory begins with a naive conjecture which is transformed into a tenable and verifiable hypothesis by a critical process, in which we more expertly exclude errors and misinterpretations, and in which we enrich the original conjecture by incorporating into it other elements that might at first blush have appeared antithetical to it.

We are concerned with a cosmological vision of the entire evolutionary process that came to a man who was at the same time a gifted and richly creative palaeontologist and a Christian mystic of a deeply inspired and inspiring type, Père Teilhard de Chardin. Even this brief and preliminary introduction calls for the more careful definition of at least two of the terms that we have employed. Why is the original conjecture described as "naive"? Or rather, in what does its naiveté consist? The naiveté of the original conjecture is due to the way in which it merely describes an initial hunch or vision and yet employs the vision as though it were in fact a prescriptive theory. This passage from naive conjecture to sophisticated theory may well be a characteristic of all scientific and intellectual processes whatsoever. It is certainly characteristic of most of them. The naiveté of the original conjecture is to be found simply in treating the vision even in the original conjectural form as a sophisticated theory when it is as yet nothing of the kind.

The second word I want to define is 'mysticism'. In most philosophical and scientific processes of thought we seek to interpret and make sense of what we experience. In mysticism, however, men seek not to think their experience but rather to experience what they have thought. Thus mysticism in metaphysics and theology is rather like verification in science, and not altogether unlike the concept of verification in old-fashioned logical positivism. Its effort is to experience what life has compelled us conceptually and somewhat abstractly to think, and if mysticism usually acknowledges a considerable measure of failure, it also claims an at least not inconsiderable measure of success.

Chapter VIII

A PLEA FOR PLURALISM IN PHILOSOPHY AND PHYSICS

Arne Naess
University of Oslo

The impact of science upon society and upon the individual is today of such an order, that *any view or vision of the world and mankind* labeled unscientific by authoritative scientists of philosophers of science has little chance of being enjoyed, expressed, or being made an object of serious logical and empirical research. Among such world views, too easily ignored by admirers and philosophers of science, I have particularly in mind those conventionally classed as not empirical or not rational, including a list of philosophies inspired by the Dane, Sören Kierkegaard, and by the Germans, Hegel and Karl Marx. But I also have artistic visions in mind, such as those of the best science fiction.

To people with different visions, one and the same scientific result *means* something different. They agree completely on an "object level", but may disagree completely on the "meta-levels". The differences might be part of the domain for serious metascientific research, but little has been done so far. It is a great task to help verbalize and conceptualize such visions so that others can see man, the universe, and world history with the coloring specific to a particular vision. Without deep engagement and serious research, we tend toward eclecticism, traditionalism, incongruence, "greyness", and incompleteness (partiality) in our views.

My plea for multiplicity of precisely formulated views in the philosophy of science is not an invitation to physicists to engage in vague speculation or to take seriously what this or that philosopher, e.g. Marx, Hegel, or Kierkegaard, has said about science or what they have intended to be answers to scientific problems. It is first a plea for extreme vigilance in distinguishing intersubjective, intercultural results of physical research proper from interpretations of these results within a framework larger than professional, technical physical science. Secondly, it is a plea that

all different mutually inconsistent interpretations such as are suggested —
but practically never more than suggested — by competent physicists
and philosophers of physics be elaborated with painful clearness and in
detail. An opposite plea would be for immediate decision about what is
true and what is false, using e.g. traditional vague formulations of
so-called operationalism, pragmatism, rationalism, dialectical materialism, idealism, and so on.

The world of personalities, of consistent personal perspective, is
today still a colorful world, immensely satisfying to contemplate in its
unbelievable variety. It would be disastrous to use the prestige of science to lay down limits and exclude some world views not refuted by
science, or to call them irrational or intellectually dishonest because
they are inconsistent with a definite 'correct' philosophy of science
proclaimed to be the only valid one. The same holds good for metascientific views considered testable, but which are not yet tested by genuine
interpersonal and intercultural methods.

After all, the carefully formulated *results* of genuine scientific research are largely neutral toward differences in world views as long as
these concern fundamentals: there is always room for differences in ultimate rules, valuations, premisses, postulates. As long as we do not have
any established criterion for testing different metascientific theories
concerning the completeness of a theory (e.g. quantum mechanics),
why not work out different views of completeness in all preciseness and
detail instead of trying to promote just one view that is not even stated
clearly?

Summing up what I wish to communicate so far: comparability of
world views requires precise formulation. Precisely formulated, basic
positions are mostly seen to involve different assumptions, postulates.
Attempts to point out the "true", "correct", or "valid" are futile in such
cases. Pseudo-refutations promote conformity, leaving us with colorless
world pictures devoid of inspiration from visions.

Now some words about the use of the term 'fundamental' or
'basic'.

The degree of fundamentality of a position is relative to the status
of the discussion and research at any given moment. Roughly, those propositions, rules, or norms will be ultimate, which make up the last links in
argumentations. If mathematics, as suggested by Lakatos, has no foundations outside itself, certain purely mathematical propositions and rules
will be fundamental to any comprehensive position within mathematics.
But philosophy of mathematics will then take over, asking, "*Why* does not
mathematics have any foundation outside itself?" If the answer is con-

sidered a satisfying one, the questioner comes to a rest. But there are questions leading further, namely, what is a satisfactory explanation? Which are the criteria of a true or good explanation, and are there ways of testing its truth or goodness? Now we certainly are outside physics, mathematics, or history, and our argumentations lead us to take up positions which I would call ultimate, basic, or fundamental—without in any way implying that there is any eternal or absolute resting point in questioning.

The tolerant and liberal attitude in philosophy of science serves the interest of keeping alive and working out in detail views which as yet cannot be intersubjectively and decisively tested.

Let T_1 and T_2 be different theories of explanation, with both descriptive and prescriptive components. What would constitute tolerance and liberalism in relation to them? If P is a researcher in the field of metatheory and he has a conceptual frame of reference in which he can state, compare, and test the truth or validity of T_1 and T_2, considerations of tolerance and liberalism do not enter in any other way than if he would have to do with two theories, e.g. theories about the relation between molecular weight and the boiling point of alcohols or any other particular scientific domain.

But if T_1 and T_2 are not very superficial, they are likely to reach conceptual layers where we today do not have proper intersubjective and intercultural methods for testing. This holds especially for the prescriptive component. The metascientific researcher is apt to succumb to illusions of comparability and closes the door to research, insisting that he has performed a test with decisive results in favor of T_1 (or T_2).

Among the scientifically incomparable theories of explanation, it is enough here to mention those tending towards seeing a *reduction* of the thing explained (the *explanandum*) to the *explanans*, and those tending towards seeing a mere *shorthand description* of observational results.

Let then T_1 and T_2 be two different theories of testing, confirming, disconfirming, corroborating, and related activities. As they appear today, they have both descriptive and prescriptive features. There are at present no intersubjective and intercultural acknowledged ways of testin theories of testing! If P_1 believes in T_1, he will normally use it to test T_2, in spite of T_2 containing rules or descriptions in conflict with T_1, and T_2 will do the same with T_1. This circularity of procedure is, in some cases, inevitable, and perhaps only extreme rationalists find the situation repugnant. But it is at least clear that metatheories of testing are not today scientifically testable.

In fields in which convictions are strong and research is weak, tolerance means adhering to intellectually decent ways of debate: not to pretend to disprove a theory without having bothered to state it clearly; not to pretend to have refuted a theory or point of view, when one has allowed oneself to take for granted certain premises, presuppositions, or postulates which are negated in that theory or point of view.

What is most needed in these fields is an intense fight for explicitness. It is only if certain levels of explicitness are reached that serious search for methods of comparison and testing can be instigated. Only then can truth or falsity be discerned — if ever.

Research attitude here involves cooperation within different schools of thought to help make each other's point of view clearer. It is often easier for researcher P_1 adhering to T_1 to explicate a hidden assumption in the rival theory T_2 than it is for P_2, who himself for years has worked with T_2 and believes in T_2.

Hertz, Poincaré, Duhem, Planck, Einstein, and Bohr, to mention only a few outstanding names in the realm of philosophy of science, had different over-all pictures of science, of its basic conceptual framework, and its methodology. The differences are parts of differences in total views, made more or less explicit in their non-scientific writings and sayings.

Any effort to restrict the variety of such total views by pretending some are "unscientific" is not only unempirical, but goes against the kind of intellectual honesty and open-mindedness which is the greatest gift of research to mankind.

If two positions cannot be compared as to truth, if they cannot be refuted in the sense of Popper, the one does not, in relation to available conceptual frameworks, have a greater validity than the other. Neither can we say they both are lacking in validity, because that would leave us without anything to start out with. I propose the following way of putting it:

All non-contradictory fundamental positions (points of view) *have the same non-zero status of validity.*

Open-minded constructive research, having such positions as an object, consists of clarifying and increasing the scope of them rather than trying to reduce their number. Reduction in number does not occur only by pointing out inconsistencies. This is indeed sometimes difficult, because concepts of consistency show variation with variation of position. It is also difficult because pointing out inconsistency requires a semantics, and today we have not one single semantic system, but various incomplete, competing ones.

The term *"validity"* needs (even in this preliminary discussion) some elucidation:

Truth—as agreement with reality—is here taken to be *a kind of* validity. The kind of validity of a fundamental position is not truth; it belongs to the class of validities such that p and non-p may both be valid, though p may not be both valid and not valid.

The logic of this concept of validity is closely related to that of tenability: a view is tenable if not refuted. If neither p nor non-p is refuted, they are *both* tenable, but, of course, not within one unit of argumentation. The term "refuted" as used here does, of course, need some elucidation, but I am afraid I must stop here.

The use of the term "reality" in philosophy of science (e.g. as accepted by Mercier) immediately suggests questions about the criteria of being real and the various positions concerning reality and appearance — going all the way back to the dialogues of Plato. Most of our dicussions in philosophy of science certainly need not, even *should* not, go that far. My plea for pluralism is — in respect to the debate on concepts and criteria of reality and appearance—the modest one of proposing that no door leading to the various ultimate positions should be shut by restrictive clauses.

The trend today is unfortunately against pluralism because it is more and more plain that Eddington, Heisenberg, Rosenfeld, and others have misled many physicists, and made them adopt strange, philosophically interesting but unphysical views. But a cry "back to normality" or "back to realism" will not, I feel confident, close the doors to new approaches, even if we agree with Landé that to maintain that the electron has *no* position between two measurements is rhetoric rather than a revolution in the theory of knowledge. The antipluralist trend is sound as a reaction against pluralism "by invasion"—by invading "physics itself" and constructing unnecessary and strange philosophical interpretations within this very discipline. (But what is "physics itself"?)

It is part of the business of the historian to try to make us experience from the inside (by *Einfühlung*) extremely different views of man and the universe. But how is this possible? It seems to rest upon the assumption that the reader of a historical or philosophical work portraying the differences has a conceptual framework wide enough to cover the most extreme differences. If this were the case, we are in a sense back to monism, since a universal conceptual framework would be the only ultimate or basic one. I shall argue that we need not assume we have such an ultimate *conceptual* framework in order to make pluralism understandable and defensible at a non-technical level. I need for this argumentation to introduce some semantic concepts.

If somebody utters a sentence T_0 with truth – or validity – claim, the definiteness (not necessarily the "depth") of cognitive meaning is limited by the set of discriminations he makes. The network of discriminations in the form of distinctions in meaning is not a stable one. Thus, if I say that this ship is of 10,000 *tons* or that π and h (Planck) are *constants*, I may have a very crude idea of what I intend to say, but it may be definite enough *for the purpose at hand.* An expert at Lloyd's Register of Shipping, for example, will, in his professional capacity, have a high or sharp definiteness of intention. It can be conveyed to outsiders only by means of perhaps 500 words. As regards the term "constant", one's network or grading of discriminations may be at least temporarily refined by reading articles such as those of Yourgrau about different usages of the term "constant", or by trying to compete with Quine, Church, or Mates introducing the term "constant" in mathematical logic, or by trying to prove some fairly general theorems about constants. Degrees of definiteness of intention may be, and have been, experimentally measured and compared, but this is a complicated affair.

Economy of thought requires that we work with a definiteness of intention commensurate with the requirements of the task, of the problems confronting us at any time. Problems in quantum physics confronting physicists who do not aim at making radical advances, do not require a high definiteness of intention regarding the significance of the symbols in, let us say, the Heisenberg equations. It is, therefore, misleading to say, as many do, that most physicists subscribe to the Copenhagen interpretation. In so far as the so-called Copenhagen interpretation is formulated as a *specific* interpretation, contrasted with others, only a small fraction of researchers and teachers in physics in western countries seem to discriminate. A high percentage at least *say* they do not discriminate. Their definiteness of intention is too low to reach relevant distinctions.

So much about the concept of definiteness of intention. The other concept, "preciseness", can be introduced as follows: A sentence T_1 is *more precise than* a sentence T_0, if there is at least one interpretation which T_0 admits, but T_1 does not, and there is no interpretation admitted by T_1 which is not also admitted by T_0.

If T_0 and T_1 are sentences in the philosophy of science, an interesting criterion of admittance is the actual ways of interpreting T_0 and T_1 within a competency group defined, e.g. by having training both in physical and in philosophical research.

Thus, when Leon Rosenfeld says to an audience of philosophers of science, "The type of causality of classical physics is determinism", the level of preciseness in communication will be measured roughly by mapping out the diversity of interpretations among the audience at hand. This level will be dependent, but of course not entirely dependent, upon the diversity of interpretations of the terms "causality", "classical physics", and "determinism".

That T_1 is more precise that T_0 may also be defined by saying that the range of differences in interpretation of T_1 falls within the range of interpretation of T_0, or that the range of interpretation of T_1 is a subclass of interpretations of T_0.

Applying these concepts to the pluralist postulate, I suggest that we place any *talking about* wide systems, the metasystematic utterances, at a T_0-level. It is a relatively neutral level, not because of wideness of scope, vastness or abstractness of conceptions, but because of its low level of discriminations (in relation to systematic conceptualizations). Any sufficiently vigorous effort to exact a delimitation of the pluralism postulate inevitably plunges it into the arms of a definite system or family of systems.

This is easily seen, considering the fact that the above formulation of pluralism includes the words "non-contradictory", "fundamental", "position", and "validity". Any fairly precise account of what might be intended by these vague and ambiguous words must reveal some world-view idiosyncracies of the author, or some of his basic methodological positions. This ruins the communicability of the pluralism, making it only understandable within a definite camp. Pluralism is, therefore, in some sense only an *ad hoc* and rough position, "exposed to wind and weather" and awaiting its destruction. But what is not *ad hoc*?

Pluralism does not *rule out* that ultimately there must be one truth. But except in matters of little concern, or in practical affairs, many of us are never able to satisfy ourselves for any reasonable length of time with any definite solution to even one major theoretical question. And why not let this color one's stand toward ultimate positions?

Let me make a digression on the plurality of "embryonic" world-views among non-philosophers.

It is an almost universal belief among philosophers that non-philosophers (or more specifically, men of common sense and, of course, youngsters who have not yet heard of philosophy) are naive realists in ontology, that they think truth consists in agreement with reality, etc. Taking "naive realism", "ontology", "truth", and "agreement with reality" as vague and ambiguous words on the T_0-level, they can be used as starting points of preciseness which lead us ultimately to interesting differences in world-views in general, and metascientific views in particular.

If empirical evidence is considered of any importance in this field of easy speculation, it supports an opposite conclusion. When they are directly or indirectly stimulated toward formulating philosophical opinions, I have found that youths from 14 to 18 years old express in a crude way, with low definiteness of intention, very different ontologies, epistemologies, and other positions of fundamental import.

In environments where certain trends of philosophy dominate, gifted young students tend to adopt the current opinions and attitudes, although an impressive teacher may induce some of the students to accept his views, even if they are looked down upon within the dominating circle. But this is the exception. In any case, the *narrowing down of variation* is not due to any intellectual inferiority of certain basic views, and certainly not to clear-cut falsification. Intellectually, there seems to be a decline in variation due to absence of systematic development of various intensively incompatible views on the professional level (the "monolithic" tendency). The "amateurs", kept isolated from authoritarian adults, show a far greater tolerance of ambiguity (as this term is used in psychology), and also the courage to leave debates on fundamentals open.

These are, of course, empirical hypotheses, and they have only in part been subjected to research. But results obtained point in that direction.

Finally, what is the relation of philosophical pluralism to the contemporary discussion of physical reality?

Listening to what some physicists authoritatively tell us, pluralists get into trouble: to accept as pure physics what they tell us *must* today be accepted – and not as conceptions derived from some basic conceptions – entails accepting certain fundamental positions as the only possible ones. Thus, Leon Rosenfeld *insists* that the development of physics entails certain views in the logic of concepts. If this logic (which is more akin to ontology in the usual sense) is expanded, it fits Hegelian basic positions, not others. Those of us who are not physicists are used to, and inclined to, accept at face value what we are told is pure physics, and we are tempted to look at certain philosophies as falsified by physics. This means giving up pluralism. Listening to other physicists, however, we begin to suspect that physicists have succumbed to a gigantic *non sequitur,* and are offering us positions on false grounds. Hence we shall look with interest for evidence that different groups of contemporary physicists, all presumably very competent, have incorporated different positions in their so-called physics. This is happily the case. We should, therefore, be in a position to discriminate "pure" physics from "philosophical physics", looking for pure physics in what is agreed upon by all physicists today. Philosophical physics would be physics explicitly developed within the frame of reference of a fundamental position.

Pure, unphilosophical physics is of course, strictly speaking, nonexistent; it is a fiction. However, a position akin to Pierre Duhem's may well be developed—*akin* to Duhem, because his doctrine that the succession of good physical theories makes them approximate a natural classification of real objects cannot, if accepted at face value, avoid coloring the physicist's criterion of a good theory. And this makes him take a kind of realist philosophical position, thus *leaving* his "pure" physics.

The pluralist in me is interested in the further elaboration and clarification of the subjectivist interpretation worked out by Heisenberg. Eddington can be radicalized in the direction of Berkeley's idealism. Of value to pluralism, too, is the idea of Leon Rosenfeld and others, that there is something dialectical, in the Hegelian or Marxian sense, in the doctrine of complementarity. In his famous Tokyo lecture (1960) this eminent, vehemently antipluralist physicist, Rosenfeld, made quantum physics part of a far from trivial metaphysics. He there said, among other things: "Complementarity denotes the logical relation, of quite a new type, between concepts which *are mutually exclusive,* and which therefore cannot be considered at the same time because that would lead to logical mistakes, but which, nevertheless, must both be used in order to give a complete description of the situation." Logicians have not, as far as I know, been inspired to work on this quite new type of logical relation. The main reason, I think, is that the environment of logicians (in the West) is unhegelian or even antihegelian: The notions of "concept" and of "logic" implicit in Rosenfeld's views do not belong within the mainstream of formal logic. They do belong to the Hegelian framework of Rosenfeld's philosophical physics.

Neither the Heisenberg nor the Rosenfeld philosophical theory has as yet the preciseness required for univocal location within the network of fundamental positions, but I hope that some philosophically trained Copenhagen people will take up the problem of how to find careful, precise formulations. This will not, however, be of any consequence if Landé is right: the point of departure *inside physics* of the Heisenberg and the Rosenfeld theory does not at all warrant any interpretations different from older particle physics. A quantum physics without particle-wave dualism cuts out any special quantum philosophy of the Heisenberg and Rosenfeld varieties.

But what, now, has the development of physics in the last decades to do with philosophical pluralism?

The developments have convinced me, first, that fundamental advances in physics are made by physicists for whom physics is not a formidable set of tricks of the trade, but whose thinking proceeds within the framework of ultimate positions, of philosophical interpretations of the terms and formulae used in physics. With time, the philosophical theory of these physicists is "rubbed off", because physical practice does not require preciseness in fundamentals.

Second, I am convinced that the positions among creative physicists are and will continue to be mutually inconsistent, that efforts to stifle the sources of diverse philosophical inspiration constitute not only a methodological but a general cultural evil. The fight between supporters of so-called idealist and realist conceptions is barren except for increasing the explicitness, comprehensiveness, and consistency of each kind of conception.

The pluralist Bernard d'Espagnat says in his carefully written *Conceptions de la Physique Contemporaine* (1965) that there are as many original interpretations of physics as there are "possible conceptions of the relation between man and the world" (p. 11). This is a happy formulation, I think, if we are permitted to add "basic" to his term "relation between man and the world", and "at least" before "as many". In the matter of particulars, or non-basic problems, most views (even among the consistent ones) are eliminated by research, or will soon be eliminated as improbable, badly corroborated, etc. Pluralism based on keeping alive refuted hypotheses is, of course, uninteresting to us as researchers.

Let me append to this pro-pluralist sermon a remark that might (mistakenly) be taken as anti-pluralist:

Vigier embraced the idea of a theory of hidden variables before he could describe a single possibility of experimental confirmation. He has been unjustifiably criticized for this among some physicists, but, on behalf of all pluralist philosophers of the world, I would thank him; we have, as philosophers, little or no chance at all of creating alternatives in physics, and are, thus, rather helpless when physicists point to certain interpretations as inevitable and definitive. Courageous physicists who suggest new paths even before it is seen where (or what) they might lead to experimentally are, therefore, especially welcome. Vigier was inspired, however, by a form of dialectical materialism, in a way which would scarcely be possible if he were a pluralist in philosophy. This underlines the curious fact that it is difficult simultaneously to promote unconditionally both pluralism and swift, radical advancement in science.

REFERENCES

The least unsatisfactory *survey* of different world views is that of Karl Jaspers in his *Psychologie der Weltanschauungen*, first edition Berlin, 1919. The least unsatisfactory recent arguments for a monolithic rather than pluralist view is offered by P. F. Strawson in his *Individuals*, London, 1959. For semantics and theory of argumentation used to establish the (relative) neutrality of ordinary language and of physical research towards a variety or world views, see my short *Communication and Argument*, Oslo and Allen and Unwin, 1966, or my long *Interpretation and Preciseness*, Oslo University Press, 1953. That *even Hegel's* world color can be reexperienced today, and *even in England*, is exemplified by F. N. Findlay in his *Language, Mind and Value*, London, 1963 (see particularly p. 230).

DISCUSSION

YOURGRAU: Naess seems to plead for some sort of tremendous tolerance and liberalism, as far as philosophical schools are concerned. He is saying, it seems to me, that all philosophical schools are more or less equivalent. According to my opinion, they can't all be valid. Having the sheep with the lions, the existentialists and the analytical philosophers, the rationalists and the realists, the pragmatists and the metaphysicians, all in one happy paradise – I just don't believe in the probability of such a paradise.

In fact, it can be proven that some views held by a given philosophical school are in no way consistent with other views. So why not draw the most plausible conclusions? If they are wrong, we make corrections by trial and error. Philosophy, physics, mathematics – they all have learned from errors. You can't afford the luxury of having no viewpoint at all and simply say, "Well, I listen to you and you listen to me". Philosophically I find such an attitude slightly . . . repulsive.

In mathematics we have three famous schools. I grant that this *is* pluralism in a way, but even though one may contend that the intuitionists or the logicists or the formalists have each a very good case, one still has to say, "I am either a formalist or an intuitionist or a logicist". I don't think one can be creative in three *generically* different scores like a gifted musician.

When it comes to logic, although I have heard about many "logics", I think it has been very beautifully shown by Quine and later by other thinkers that most findings of modal logic can be reduced to nonmodal logic. These are interesting adventures, but on the whole I think we have a definitely unique viewpoint even among logicians, although one may pay tribute, or at least lip service, to other logicians with whom one doesn't agree.

I would like to stress that (although I agree the limits of definiteness are not very sharp) our aim must be, as scholars, to sharpen our concepts *even at the risk of antagonizing our colleagues.* There are situations where one just cannot build bridges even within his own field. Bridges to another discipline are one thing; illusory bridges among your own fraternity are an entirely different thing.

Naess' viewpoint on contemporary physics is interesting, but unfortunately I can't agree with it. For me, any sincere physicist with academic integrity tries to tell us something about the world, sometimes in very involved conceptual language, sometimes using a mathematical apparatus which is intricate and abstract. But still, he is for me only a physicist and not a mathematician if he really wants to know something about reality. And there I think pluralism is an approach I would not recommend to my students as *the* ideal recipe.

Finally, Vigier is responsible together with Bohm, *et al.,* for a very interesting new model of the elementary particle and an aesthetically fascinating theory. Regrettably, as long as I haven't been shown for his precious model and theory even an iota[13] of experimental evidence, I have to consider the adherents to those "original" ideas as quixotic. But let me confess that the very moment when I see an ever so slight experential support for Vigier's contentions (and those of his collaborators), I shall become converted and eschew all my previous commitments in that particular domain.

NAESS: Let me restate the definition of the term "pluralism" as I used it: All explicit non-contradictory, fundamental positions have the same non-zero argumentational status of validity. Explicitness and fundamentality require preciseness. This rules out the standard statements of philosophical schools, as far as I can see. But behind the vague slogans of schools there are differences of basic views. These should be explicated.

At the lowest level of preciseness, the T_0-level, contradictions do not appear, nor are interesting differences in premises, postulates, rules of inference statable. To be able correctly to say to somebody, "You stated a contradiction", is a compliment because it means he operates above a certain minimum level of preciseness, and this is what I, of course, will require of an explicit fundamental position. If we cannot locate where you stand in the landscape, how can we argue with you?

I am not excessively liberal and tolerant toward so-called schools of philosophy as we find them "defined" in dictionaries and bad histories of philosophy. What is asked for is not "tremendous tolerance and liberalism", but rather abstinence from totalitarian, sectarian, and conformist attitudes, and willingness to take part in teamwork of serious research across ideological boundaries.

It would of course be queer to proclaim that "all philosophical schools are more or less equivalent" and their premises, assumptions, presuppositions, and postulates all "valid". Most school formulations are not precise enough even to make exact comparisons or to search for inconsistencies. Mists do not collide. And note: physicists are responsible for a lot of this mist!

Researchers do not like to be labeled by names of so-called "schools", and this holds also good of philosophers. A formulation said to express an assertion or rule characteristic of pragmatism *as a school* is mostly a formulation nobody would accept or take seriously, because of vagueness and ambiguity. The same holds true for formulations pretending somehow to catch the essence — "existentialism", "Hegelianism", "ordinary language philosophy", "rationalism", etc. When Yourgrau states that "all philosophical schools" cannot be "valid", he may have ism-formulations in mind, and I agree. "Validity" or "invalidity" are terms we both would reserve for formulations of statements or rules which show a minimum level of interpersonal preciseness. I have only such formulations in mind when advocating pluralism.

If, on the other hand, we take carefully selected formulations from, e.g. the so-called pragmatist C. I. Lewis or Quine or from the anti-pragmatist Gottlob Frege, or from a Hegelian like Grenness, some may have the character of fundamentality which make them "equivalid", "acceptable one at the time". That is, there is no sufficiently wide conceptual framework, including semantics and methodology, within which they can be scientifically compared and tested. Maybe future generations will see such frameworks; today it seems best to travel on different roads. To invoke trial and error as a test-method here would be naïve.

In the current philosophy of physics there are a great number of metascientific sentences on "completeness of a theory", "explanation", "observation", "levels of physical reality", "interaction of subject and object", "path of a particle". What is needed more than anything else is more precise and elaborate formulation of the positions suggested but not clarified by those sentences. Cooperation between people trained in philosophy and physics is needed in order to make use of the suggestions by Heisenberg, Rosenfeld, Bohr, Fock, Bohm, Vigier, Landé, Weiszäcker, de Broglie, and others. The school terms "idealism", "pragmatism", "dialectical materialism", etc. are apt to thicken rather than dispel the haze. And the stress on one position being true and the other false is premature; it leads away from constructive metaresearch and the difficult work of finding hidden premises and of formulating, as exactly as possible, differences in postulates, fundamental assumptions, and styles of research. Contrasting positions must be sharpened, kept alive until if possible, refuted in an intellectually honest way. The emphasis on multiplicity of mutually inconsistent views is *not* eclecticism. The eclectic is rather a monist than a pluralist, rather an irrationalist in the "Popperian" sense[1] than a rationalist.

VIGIER: I wholeheartedly agree with Yourgrau. I think the real problems now, which are philosophical problems, are being fought on quite a different battlefield: the battlefield of science. The great periods of philosophy were precisely the periods where the problems were tied to an explosion of scientific knowledge such as in Greece, the Renaissance, etc. Now in a sense we live in a similar but even more exciting period. Progress in the last forty years has been more important than in the last two thousand years; nine out of ten scientists who have ever lived — are alive at present.

If one reflects on the nature of the big problems to which our knowledge has been brought to bear in the last ten years, we see we are on the way to answering problems which have been discussed on a purely rhetorical level for the last two thousand years between, for instance, the materialists and the idealists.

[1]*Cf.* his stress on "rational discussion", which I take as stress on *pro et contra,* and on contra-arguments against his own tentative conclusions. This introduces necessarily pluralism, i.e. at least *two* alternatives, two mutually inconsistent ways of reasoning on any subject. *(The Logic of Scientific Discovery,* Preface to 1958 edition.)

Consider, for example, the problem of whether life can be explained in terms of a possible behavior of matter. The answer to that problem is being given now, not by philosophers but by the scientists themselves. It rests on the discovery of DNA and the introduction of the feedback ideas of Wiener and many other scientists into the theory of life. This is the beginning of a real definition of the problem of life.

There is the same thing in the problems of soul and consciousness. Thinking machines are beginning to clarify certain elements of what "thought" is. This means that the corresponding problems are moved from the level of words to the level of the laboratory. I think the whole issue, of course, is that of synthesis within scientific knowledge.

I am against eclecticism. All ideas should be allowed to be expressed, but in science they are fought and settled on a very special level, that of scientific practice, by the results of experiments. I don't think the discussion with the Copenhagen school is a merely rhetorical one. I am certain it is an issue that is going to be settled on the scientific level. A lot of so-called philosophy appears to me to be just sophistry. People like Teilhard de Chardin pretend to answer questions which they have no right to answer because all this talk about spheres is just words. The people who are doing the actual work are the biologists, for they are really approaching, on a precise, controllable level, the fundamental problems raised by our knowledge. Hence, questions concerning physics or the laws of physics should be settled by the physicists themselves on the level of experiment. I want to chase the ideologists away from knowledge because the truly fundamental revolution of our time is precisely the construction of a synthesis out of scientific knowledge.

Now, this does not mean that I think there is a unique philosophy which is going to issue forth from the discoveries of scientific knowledge. I hold that we will observe in the following years new scientific revolutions—this is not a Teilhard de Chardin-like prediction—and a new qualitative synthesis will arise because science works by successive revolutions. There are long periods of accumulation of knowledge and then periods of breakthrough which change the whole traditional point of view.

If one reflects on the nature of modern science, one sees that all the barriers between the different branches of knowledge are falling down. Consider, for instance, the impact of the work of a man like Norbert Wiener. I think that Wiener's contribution will appear, in time, as important perhaps as that of Descartes. The ideas of cybernetics, the notion of feedback, have transformed completely our ideas about the behavior of nature and about causality. They have opened the way to a new scientific explanation of what is life and what is soul.

What is really happening is the explosion of science. Now, instead of having long periods where different branches of knowledge were independently deadlocked, any revolution in one branch changes the outlook in all other branches.

This is a fantastic change and explosion, and of course it is imperative to bring scientists and philosophers together. But, as I say, I have a deep conviction that the most advanced thought, the solution of the most advanced problems, originate in the laboratories themselves.

TREDER: I think one of the best remarks Einstein has made about philosophical problems is that all physicists are philosophical opportunists. They do not have a philosophical system that will work. A physicist might be in one respect a realist and in another a positivist. The conceptions of the philosophers are only working instruments for the physicists. Of course, some physicists may or may not lean toward some special philosophical systems. A relevant criterion of the physical meaning of the works of these physicists is the invariance of their results according to different philosophical points of view.

NAESS: As I was trained in psychology in general and social psychology in particular, the interview technique is natural to me; so I asked physicists doing research in quantum physics whether they apply such a method to the Copenhagen or any other interpretation. The answers mostly amounted to a denial that they have had to take a stand. Their work is neutral toward crucial differences of interpretation, and they often succeed, as they admit, by "mere" tricks of the trade. And if their work happens to get them into close contact with burning questions of metascience and of fundamental views in general, they often are, or act as, "opportunists", as pointed out by Treder. In other, perhaps more rare, cases physicists do not act in this way, as exemplified by Niels Bohr and Einstein, for whom there was no clear distinction between a scientifically "pure" physics and a physics interpreted within a broader framework of metascientific and fundamental issues. In the early 1920's so-called "pragmatist" and "positivist" attitudes were perhaps the *most* opportune ideas for "getting quantum physics moving". Einstein could not leave his fundamental positions, however, and Heisenberg did one of the crucial jobs. ("What is a physical theory?", "What is physical reality?" were burning questions affecting their conceptions *in* as well as *about* physics.) In still other cases, metascientific issues *dominate* the physicist in his work, as was the case with Pierre Duhem. The result was sometimes a happy one, sometimes a deplorable one, from the point of view of heuristics.

All this I try to elucidate by using my concept of "depth of intention". The varieties of fundamental positions arise from a great depth of intention in any extraordinary situation. For most researchers in most of their work, great depth and the corresponding fine-grained set of discriminations (to use a phrase from psycho-physics) is a luxury.

Recent significant advances in molecular biology have brought to the front just those problems which cannot be solved in the laboratory alone, but which require going back to premises, postulates, and basic human policies. The "social" sciences in their turn require us to pose as precisely as possible the values we shall defend. They do not operate automatically, but require "input". When ordinary people ask scientists to state the implications of the discovery of DNA, to show what questions of eugenics, ethics, politics are implied, an answer must be stated in ordinary language, not in that of the biological laboratory. The scientist must be conscious of his own position in the vast metascientific field of rational debate.

BONDI: On the historical side, I think Naess' remark about great advances in physics coming from physicists with a philosophical bent of mind is true in some cases and untrue in others. The greatest experimental physicist of our century, Rutherford, was perhaps the most non-philosophical thinker who ever lived. Dirac made many great contributions, but I think what most would regard as his

greatest, viz. the equations of relativistic quantum mechanics, were essentially
a trick of the trade, a usable one, but no more than that. So I cannot entirely agree
there.

On the other hand, when Naess pleads for pluralism, then it seems to me very
clear that pluralism is of the very essence of science. To me, it is the greatest
glory of science that people of different religions and different ideologies can work
together effectively. It shows that in science there are certain methods on which
we can agree, in which experiment forces agreement, even if there are others
on which we do not agree.

Next, there is the point where I am perhaps a little more 'Vigierist' than
Vigier himself. There is science, yes, and in this we have theories that are testable in the "Popperian" sense. But we do not know (until we have explored alternatives) which part of the framework is essential and which is not. For eighty
years after Maxwell proposed his theory, the field concept was thought to be
a vital part of it, and the tests of electromagnetic theories were thought to show
that the field concept was right. Then Wheeler and Feynman demonstrated that
we could arrive at Maxwell's equations from an "action at a distance" concept,
leading to the same experimental tests, and so we now know that the field concept
is only *one* alternative basis. It has made the field part of the Maxwell theory
move out of science just because it is no longer testable. But removing it from
science strengthens the demarcation and is thus a true scientific contribution. If
the Bohm-Vigier theory does not lead to any possible experimental discrimination, it will yet show that different approaches are possible. Thereby, it will have
removed from science part of the usual foundations of quantum theory as not
being testable. Therefore, it will have made a valuable contribution.

My last point will merely be to support the plea for pedestrian work. As
may be known, I have occasionally done some speculative work. I recall giving
a lecture to students on some particularly speculative matters, and they became
most interested and fascinated. And so I closed with the advice, "I am delighted
if I have interested you in this, but do not do any work in such speculative fields
until you have done some good pedestrian work so people know you are a real
scientist and not just a crank."

NAESS: As Bondi indicated, sometimes philosophy of science is more of a burden
than an asset for the creative scientist. I note not with reluctance but with de-
light that in science sometimes very unphilosophical minds do excellent things.
In the case of physics, I venture to suggest that in certain years, from 1920 on,
those who just said, "Let us have more fun, doing some really good mathematical tricks", could do it more easily because they had *not* the world-view of
Einstein. But they had a background of positivism and pragmatism I would not
be right in saying that they were mainly philosophically inspired, but 'unphiloso-
phicalness' may be of a philosophical kind, an *implicit* pragmatism, or practica-
lism.

POPKIN: I think the history of ideas thrives on the sort of pluralism Naess is
advocating. If there hadn't been this sort of pluralism, things would be very dull
because we would be just studying the same idea over and over again. Fortunately, there has been this pluralism, but I find in studying it there is always a tension between the sort of tolerant pluralism which Naess is advocating and the
dogmatic pluralism that has gone on all the time. The contribution of the many

views that have existed in the past seems to depend not on the fact that they tolerate each other but on the fact that they don't. But the dogmatism, the fact that people believe these things, really take them so seriously, claim they have the unique way to truth, has led them to produce the novel ideas that they have. If they were as tolerant as Naess would like them to be, I wonder whether they would really make this sort of contribution.

Pascal observed that Pyrrhonian scepticism is true as long as there are dogmatists, but as soon as there are no longer any dogmatists, this scepticism will become false.

TENNESSEN: Vigier remarks that the solutions of the most advanced problems are given by the scientists themselves, achieved in the laboratories, and he mentions one problem to which, he claims, the solution is now coming from laboratories, viz. the problem: "What is life?"

First, I want to say that this model is not adequate. It gives the impression that one is either here or there, whereas the point is that in most cases we are focused somewhere in the middle, and that we shift, oscillating from one situation to another. The model overstates our definiteness of intention. This is misleading, as the whole thing is more like one enormous porridge of meanings and intentions.

Let's take the problem of immortality, for instance. If you say man is immortal, you can work out a formula. You can write, "This is the maximum of tenability and this is the consolation-value of your theory". But in my opinion, it goes something like this. The more tenability, the less consolation-value. But what do people do with "immortality"? One may, on the one hand, give it a definition which it is very likely to be *true*. Then, on the other hand, one could at the same time hold an idea of immortality which would be much different, but which has great *consolation value*. Thus it becomes a porridge of intentions and meanings; and one oscillates imperceptibly between both extremes, which gives one the very comfortable feeling that both ideas are true: we are immortal, and we have the consolation. We are utilizing the vagueness and ambiguity of "immortality" to comfort ourselves.

Well, I would maintain that it is the same thing Vigier does. He is assuming unlimited definiteness of intention. Possibly his intention is, at one point, A, but he is tacitly assuming that it is, at another point, say, B. Therefore, when he is talking about laboratory experiments as being relevant, they are only relevant to someone who has the level of intention A; but this is totally irrelevant for anyone who has the level or depth of intention B. And of course, somewhere between A and B there would be mixed feelings, and these are the mixed feelings I wanted to express.

VIGIER: The question of life is a precise question which can be split into a series of problems and has indeed been put by biologists in quite a precise set of single problems.

Now we understand how the cell works, because we know that the DNA is a code and the code gives instructions for all cellular chemical reactions. The living cell can be defined as a memory which reproduces itself.

This is the way in which a problem passes from the level of vague words into a precise formulation, and that is the way knowledge moves forward. Look at the way, for example, in which knowledge moved during the Renaissance, the vague way in which questions in astronomy and in physics were expressed before Galileo and other thinkers, and the precise way in which they left these problems.

NAESS: If all fundamental views or theories about physical reality were to be considered in a critical atmosphere, it could only be done in an environment of tolerant pluralism. As regards the chances of professional maximal (or optimal) achievement in an atmosphere of tolerant pluralism, I share Popkin's pessimism. Dogmatism, narrow-mindedness, cocksureness, and fanaticism have sometimes been conducive to great achievements because they have made a researcher work out the consequences of a bright idea with greater energy. Intolerance does not only protect weaklings. Generosity and wide perspective is sometimes bad heuristics but always good cultural philosophy.

Is our great problem today lack of perfect achievements? Is it not rather the drying up of sources of a colorful multiplicity of views and attitudes due to the world-wide integration of technical, "scientific" culture?

Vigier says that the problem of life is a series of precise problems. But I would say there is no precise series. A hundred and fifty years ago one of the central questions concerning life was the question whether organic matter could be explored and explained successfully by chemists and physicists. Tremendous scientific victories in organic chemistry, since Wöhler in 1829 succeeded in obtaining urea from ammonium cyanate, have decided that issue. It is not a living issue any more. There has been a shift of problems — many times.

Only specialists on Aristotle can make us aware of his frame of reference in *his* "physics". As long as we did not go deep enough, we interpreted it in terms of Galileo and the decadent "Aristotelian" physics of his time. Today — to our astonishment — we see a "physics" of Aristotle that is capable of a rejuvenescence, not as a physics in the contemporary sense, but as a view of physical reality and of the logic of "things". The discovery is analogous to that of the independence of stoic logic from the conceptual frame of — the same incredible genius — Aristotle. Both discoveries are feats of tolerance of diversity and careful inspection of fundamental assumptions, premises, postulates.

NEF: I would only note that to distinguish is not necessarily to tear apart. To distinguish can be a means of bringing about agreement on a higher level and concerning matters that are more vital to man than those about which we disagree.

Chapter IX

ON SEMANTIC INFORMATION

Jaakko Hintikka
University of Helsinki and Stanford University

In the last couple of decades, a logician or a philosopher has run a risk whenever he has put the term "information" into the title of one of his papers. In these days, the term "information" often creates an expectation that the paper has something to do with that impressive body of results in communication theory which was first known as *theory of transmission of information* but which now is elliptically called *information theory* (in the United States at least).[1] For the purposes of this paper, I shall speak of it as *statistical information theory*. I want to begin by making it clear that I have nothing to contribute to this statistical information theory as it is usually developed.

Some comments on it are nevertheless in order. One interesting question that arises here is why a logician or a philosopher of language should want to go beyond statistical information theory or to approach it from a novel point of view. The reason usually given is that this theory seems to have little to say of information in the most important sense of the word, viz. in the sense in which it is used of whatever it is that meaningful sentences and other comparable combinations of symbols convey to one who understands them. It has been pointed out repeatedly that many of the applications of statistical information theory have nothing to do with information in this basic sense. It has also been argued that the use of the term "information" in such contexts is apt to create misunderstanding and unrealistic hopes as to what statistical information theory can offer to a logician or a philosopher of language.

This, at any rate, is why a few philosophers, notably Carnap and Bar-Hillel, outlined in the early fifties a theory which sought to catch some of the distinctive features of information in the proper sense of the term (see [3-5]). This kind of information was called by them *semantic information*, and the theory they started might therefore be called a theory of

---

[1]For this theory, see, e.g. [1]. A popular survey is given in [2].

semantic information. It is this sense of information that I shall be dealing with in the present paper.

The relation of this theory of semantic information to statistical information theory is not very clear.[2] On the formal level, the two theories have a certain amount of ground in common. In both theories, information is defined, or can be defined, in terms of a suitable concept of probability. The basic connection between probability and information (in at least one of its senses) can also be taken to be one and the same in the two cases[3]:

$$\inf(h) = -\log p(h), \tag{1}$$

where p is the probability-measure in question. From (1) we obtain at once the familiar entropy expression,

$$-\sum_i p_i \log p_i, \tag{2}$$

for the expectation of information in a situation where we have a number of pairwise exclusive alternatives with the probabilities p_i ($i = 1,2,...$). This expression can thus occur in both kinds of information theory. In general, the two theories can be said to have in common a certain calculus based on (1) and on the usual probability calculus.

It is more difficult to say what the relation of the two theories is on the interpretative level after all the obvious misunderstandings and category-mistakes have been cleared away. It is sometimes said that there is a difference between the concepts of probability that are involved in them. In statistical information theory, a frequency interpretation of probability is presupposed, while in a theory of semantic information we are presupposing a purely logical interpretation of probability. The difference between the two theories will on this view be in effect the difference between Carnap's probability$_2$ and probability$_1$.

It seems to me that this way of making the distinction rests on serious over-simplifications. I shall make a few comments on them at the end of this paper.

It is nevertheless clear that this explanation catches some of the difference in emphasis between the two theories. In statistical information theory, one is typically interested in what happens in the long run in certain types of uncertainty situations that can be repeated again and again.[4]

[2]On this subject, cf. [6], Chapters 15-18; also [2], pp. 229-255.

[3]E.g.[2], p. 242.

[4]It is therefore natural that such expected values or estimates of information as (2) should play an important role in this theory. Its statistical character is not the only reason, however, for the prevalence of expressions similar to (2); they have a place in the semantic theory, too, as I have argued in another paper.

In a theory of semantic information, we are primarily interested in the different alternatives which one can distinguish from each other by means of the resources of expression we have at our disposal. The more of these alternatives a sentence admits of, the more probable it is in some "purely logical" sense of the word. Conversely, the fewer of these alternatives a sentence admits of, i.e. the more narrowly it restricts the possibilities that it leaves open, the more informative it clearly is. It is also obvious that this sense of information has nothing to do with the question which of these possibilities we believe or know to be actually true. It is completely obvious that the sentence (h & g) can be said to be more informative than (h v g) even if we know that both of them are in fact true.

Semantic information theory accordingly arises when an attempt is made to interpret the probability measure p that underlies (1) as being this kind of "purely logical" probability. In this paper, I shall examine, in the light of recent work, what happens when such an attempt is seriously made.[5]

The basic problem is to find some suitable symmetry principles which enable us to distinguish from each other the different possibilities that certain given resources of expression offer us. It seems to me that insofar as we have an unambiguous concept of (semantic) information of which serious theoretical use can be made, it must be based on such distinctions between different cases, perhaps together with a system of "weights" (probabilities) given to these different possibilities.

This will become clearer when we consider certain simple examples. The simplest case is undoubtedly that given to us by the resources of expression used in propositional logic. These resources of expression are the following:

1. A finite number of unanalyzed atomic statements $A_1, A_2, \ldots, A_K$;
2. Propositional connectives $\sim$ (not), & (and), v (or), plus whatever other connectives can be defined in terms of them (e.g. material implication or $\supset$).

Here the different possibilities which we can distinguish from each other, as far as the world of which the atomic statements are about is concerned, are perfectly obvious. They are given by the statements which we shall call *constituents*.[6] These could be said to be of the form

$$(\pm) A_1 \& (\pm) A_2 \& \ldots \& (\pm) A_K, \tag{3}$$

where each of the symbols $(\pm)$ is to be replaced by $\sim$ or by nothing at all in all the different combinations. The number of the different constituents will be 2^K.

[5] I shall draw on the works I have published elsewhere. See [7-12].
[6] The term (and the idea) goes back all the way to Boole.

The sense in which the constituents give us all the possible alternatives which can be described by the given atomic statements and connectives should be obvious. For instance, if $K = 2$, $A_1 =$ it is raining and $A_2 =$ it is blowing, the constituents will be the following:

$$
\begin{array}{lll}
A_1 \,\&\, A_2 & = & \text{it is raining and blowing} \\
A_1 \,\&\, {\sim}A_2 & = & \text{it is raining but not blowing} \\
{\sim}A_1 \,\&\, A_2 & = & \text{it is blowing but not raining} \\
{\sim}A_1 \,\&\, {\sim}A_2 & = & \text{it is neither raining nor blowing.}
\end{array}
$$

Each statement considered in propositional logic admits some of the alternatives described by the constituents and excludes the rest of them. It is true if one of the admitted alternatives is materialized, false otherwise. It can thus be represented in the form of a disjunction of some (perhaps all) of the constituents, provided that it admits at least one of them (i.e. is consistent):

$$h = C_1 \lor C_2 \lor \ldots \lor C_{w(h)}. \tag{4}$$

Here h is the statement we are considering and $w(h)$ is called its width. For an inconsistent h, we may put $w(h) = 0$.

In this simple case it is fairly straightforward to define that "logical" concept of probability which goes together with suitable measures of information. Obviously, constituents are what represent the different symmetric "atomic events" on which we have to base our measures of information. From this point of view, a statement h is the more probable the more alternatives represented by the constituents it admits of, i.e. the greater its width $w(h)$ is. Obvious symmetry considerations suggest the following definition:

$$p(h) = \frac{w(h)}{2^k}. \tag{5}$$

It is readily seen that this in fact creates a finite probability field. The above definition of information (1) then yields the following measure for the information of h:

$$\inf(h) = -\log p(h) = -\log \frac{w(h)}{2^K} = K - \log w(h), \tag{6}$$

where the base of our logarithms is assumed to be 2. It is of some interest to see that (6) is not the only possible definition of information in the propositional case. As was mentioned above, the basic feature of the situation we are considering is clearly that the more alternatives a statement excludes, the more informative it is. (This idea was spelled out for the first time in a general form in Popper[13]. If the reader does not appreciate it at once, he should try thinking of the different alternatives as so many contingencies he has to be prepared for. The more narrowly he can restrict their range, the more one clearly can say he knows about them.) This basic idea does not imply that the exclusiveness has to be measured as in (6), although (6) clearly gives one possible way of measuring it. An even more direct way of carrying out this basic idea might seem to be to use as the measure of information the relative number of alternatives it excludes. The notion so defined will be called the content of a statement:

$$\text{cont}(h) = \frac{2^K - w(h)}{2^K} = 1 - p(h). \tag{7}$$

This is in fact a perfectly reasonable measure of the information that h conveys. It is of course related in many ways to the earlier measure inf. One such connection is given by the equation

$$\inf(h) = \log\{1/[1 - \text{cont}(h)]\}. \tag{8}$$

The relation of the two measures of information is a rather interesting one. It has been pointed out that our naive views of (semantic) information have sometimes to be explicated in terms of one of them and at other times in terms of the other. It has been suggested that "cont" might be viewed as a measure of the substantial information a statement carries, while "inf" might be considered as a measure of its surprise value, i.e. of the unexpectedness of its truth.[7] Some feeling for this difference between the functions "cont" and "inf" is perhaps given by the following simple results:

$$\text{cont}(h \ \& \ g) = \text{cont}(h) + \text{cont}(g) \tag{9}$$

if and only if $(h \ \text{v} \ g)$ is logically true,

$$\inf(h \ \& \ g) = \inf(h) + \inf(g) \tag{10}$$

if and only if h and g are independent with respect to the probability measure p defined by (5);

$$\inf(h) = \text{cont}(h) = 0 \tag{11}$$

if and only if h is logically true.

[7]Cf. [5], p. 307.

Of these, (11) shows the rationale of the choice of the term "tautology" for the logical truths of propositional logic. (10) shows the basic reason why "inf" is for many purposes more suitable a measure of information than "cont". It is natural to require that information be additive only in the case of statements that are probabilistically independent of each other; and this is precisely what (10) says. In fact, from (10) we can easily derive definition (1) if we make a few simple additional assumptions concerning the differentiability and other formal properties of "inf".

Further results are obtained in terms of relative measures of information. Their definitions are obvious:

$$\text{cont}(h/e) = \text{cont}(h \ \& \ e) - \text{cont}(e)$$

$$\text{inf}(h/e) = \text{inf}(h \ \& \ e) - \text{inf}(e).$$

The following results are then forthcoming:

$$\text{cont}(h/e) = \text{cont}(e \supset h) \tag{12}$$

$$\text{inf}(h/e) = -\log p(h/e), \tag{13}$$

where $p(h/e)$ is the usual conditional probability. Both (12) and (13) are entirely natural. [The naturalness of (13) is brought out by a comparison with (1).] It is interesting to see, in view of this naturalness, that they contain different measures of information and that they cannot be satisfied by one and the same measure.

Simple though these results are, they already give rise to (and material for) several more general considerations. Independently of whether we are using the measure "inf" or the measure "cont" for our concept of information, the information of an arbitrary statement h will be the smaller the more probable it is. This inverse relation of information to probability has been one of the cornerstones of the views Popper has put forward concerning the nature of scientific method and scientific knowledge. He has criticized attempts to think of scientific methods primarily as methods of verifying or of "probabilifying" or confirming hypotheses and theories. He has emphasized that the true aim, or at least one of the true aims, of the scientific enterprise is high information. Since this was seen to go together with low probability, at least in one of its senses, it seems to be more accurate to say that science aims at highly *improbable* hypotheses and theories than to say that it aims at highly probable ones. Attempted falsification is for Popper a more important idea than verification or confirmation.[8]

[8] [13], *passim.*

In the extremely simple situation we studied, these facts are reflected by the observation that one could normally make a hypothesis h more probable by adding more disjuncts to its expansion (4). It is obvious, however, that the adjunction of such new alternatives would only result in the h's becoming more permissive, hence flabbier and hence worse as a scientific hypothesis.

Popper is undoubtedly right in stressing the importance of information as an aim of the scientific procedure. It is not quite obvious, however, how this search for high information-content should be conceived. It is also not clear whether it has to be conceived in a way that excludes high probability as another *desideratum* of science. Already at the simple-minded level on which we have so far been moving, we can say this much: The information of a hypothesis or a theory has obviously to be defined by reference to its *prior* probability (as was already hinted at above). It is prior probability that is inversely related to information in the way explained above. Now the high probability that a scientist desires for his hypotheses and theories is clearly *posterior* probability, probability on evidence. There is no good general reason to expect that information and posterior probability are inversely related in the same way as information and prior probability are. In some simple cases, it can be shown that they are not so related.

One appealing idea here is to balance one's desire for high (prior) information and one's desire for high (posterior) probability in the same way in which we typically balance our values (or "utilities") and the hard realities of probability, viz. by considering the *expected* values of utility and by trying to maximize them. Here information would be the sole relevant utility. The appeal of this "Bayesian" approach to scientific method is witnessed by the number of scholars that have suggested it.[9] Unfortunately, attempts to carry out this program have not got very far as yet.[10] I believe that this is basically due to unfortunate choices of the underlying measure of information.[11]

In order to find more suitable measures for cases that are at least in principle more interesting than propositional logic, let us consider some more complicated cases of this kind.

The next more complicated case considered by logicians is called monadic first-order logic. The languages considered under this heading employ the following resources of expression:

[9]See, e.g., [14, 15].
[10]See, e.g., [16, 17].
[11]This is in effect surmised by Hempel in [15], p. 77.

1. A number of one-place predicates (properties) $P_1(x), P_2(x), \ldots,$ $P_k(x)$.
2. Names or other free singular terms $a_1, a_2, \ldots, b_1, b_2, \ldots.$
3. Connectives $\&, \mathbf{v}, \sim, \supset,$ etc.
4. Quantifiers (Ex) ("there is at least one individual, call it x, such that") and (x) ("each individual, call it x, is such that") together with the variables bound to them (in this instance x).

Out of these we can form atomic statements concerning individuals of the form $P_i(a_j)$, propositional combinations of these, results of generalizing upon these, propositional combinations of these generalizations and atomic statements, etc.

What are the different basic symmetric cases out of which we can construct suitable measures of information?

This question is an easy one to answer—far too easy, in fact. We can give several entirely different answers to it which yield several entirely different measures of information. We shall begin by considering some such measures of information and probability known from literature.[12]

1. **Carnap's p⁺.** This is obtained by extending the above considerations to monadic first-order logic without any modifications. It is assumed that every individual in our domain ("universe of discourse") has a name a_j. Let us form all the atomic statements $P_i(a_j)$ and deal with them precisely in the same way as with the atomic statements of propositional logic. The constituents of propositional logic will then become as many *state-descriptions* (as they have been called by Carnap). They are of the form

$$(\pm)\, P_1(a_1) \;\&\; (\pm)\, P_2(a_1) \;\&\ldots\&\; (\pm)\, P_k(a_1) \;\&$$
$$(\pm)\, P_1(a_2) \;\&\; (\pm)\, P_2(a_2) \;\&\ldots\&\; (\pm)\, P_k(a_2) \;\&$$
$$\cdot \quad \cdot \quad \cdot \quad \cdot \quad \cdot \tag{14}$$

The probability-measure p^+ is obtained by giving all these state-descriptions an equal probability (*a priori*).

In this way we can of course treat finite universes of discourses only. Infinite domains have to be dealt with as limits of sequences of finite ones.

Relative to a given domain, every statement in the monadic first-order logic is a disjunction of a number of state-descriptions. It is true if one of these is true, and false otherwise. It may be said to admit all these state-descriptions and to exclude the rest. The problem of assign-

---

[12]The pioneering work in this area has been done by Rudolf Carnap. See [18, 19]. Cf. also [20], especially Carnap's own contribution, pp. 966-998.

ing *a priori* probabilities to statements of the monadic first-order logic therefore reduces in any case to the problem of specifying the probabilities of the state-descriptions.

But are state-descriptions really those comparable and symmetric cases between which probabilities should be distributed evenly? This does not seem to be the case. Indications of this are the strange consequences of basing a definition of information on p^+. For instance, suppose we have observed a thousand individuals and found that all of them without exception have the property P_i. Then it is easily seen that the probability $p(P_i(a_{1001})/e)$ that the next individual a_{1001} has the same property (e being the evidence given by our observations) is still equal to 1/2, which was the probability *a priori* that the first observed individual a_1 be of this kind. Thus we have $\inf(P_i(a_{1001})/e) = 1 = \inf(P_i(a_1))$. In other words, we should be as surprised to find that $P_i(a_{1001})$ as we were to find that $P_i(a_1)$. There is obviously something amiss in such a measure of information (or surprise).

2. **Carnap's p***. Suggestions for improvements are not hard to find. Perhaps the most concise diagnosis of what is wrong with p^+ and with the associated measures of information is to say that what we are interested in here are not the different alternatives concerning the world we are talking about, but rather the different *kinds of* alternatives. It is the latter, and not the former, that should be weighted equally, i.e. given an equal *a priori* probability.

This idea of a different kind of world was in effect interpreted by Carnap as meaning a *structurally* different kind of world.[13] On this view, what are to be given equal probabilities are not state-descriptions but disjunctions of all structurally similar or isomorphic state-descriptions. Two state-descriptions are isomorphic in the intended sense if and only if they can be obtained from one another by permuting names of individuals. Disjunctions of all such isomorphic state-descriptions were called by Carnap, not surprisingly, *structure-descriptions*. Each structure-description can now be given an equal *a priori* probability which is then divided evenly among all state-descriptions compatible with it (i.e. occurring in it as disjuncts). Two state-descriptions thus receive the same probability if they are members of the same structure-descriptions, but rarely otherwise. The resulting probability-measure is called p^*.[14]

---

[13]The line of thought presented here is not Carnap's.

[14]This measure was for a while preferred by Carnap to other probability-measures, and it appears to be an especially simple one in certain respects.

On the basis of this probability-measure we obtain measures of information. For statements concerning individuals these measures of information are not unnatural, but rather seem to catch fairly well some of the features of our native idea of information. However, for general statements (statements containing quantifiers) the resulting measures of probability and information have highly undesirable features. It turns out, for instance, that in an infinitely large domain every statement beginning with a universal quantifier has a zero probability independently of evidence, provided that it is not logically true. Conversely, in an infinite domain every existential statement has zero information independently of evidence, provided it is not logically false. These results are approximated by what we find in large finite universes. They seem to be especially perverse in view of the fact that in a large universe it would seem very difficult to find individuals to exemplify existential statements, which suggests that in a large universe statements which nevertheless assert the existence of such individuals should have a high degree of information rather than a very low one.

   3. **Equal Distribution among Constituents.** For statements containing quantifiers we therefore need better measures of logical probability and of information. They can easily be come by if we analyze the structure of the monadic first-order logic a little further. It turns out that we can find another explication for the idea of a kind of alternative, different from the one we saw Carnap offering for it.

   Basically, the situation here is as follows. By means of the given basic predicates we can form a complete system of classification, i.e. a partition of all possible individuals into pairwise exclusive and collectively exhaustive classes. These classes are defined by what Carnap has called *Q-predicates*. An arbitrary Q-predicate can be said to be of the form

$$(\pm)P_1(x) \ \& \ (\pm)P_2(x) \ \& \ \ldots \ \& \ (\pm)P_k(x), \tag{15}$$

where each symbol $(\pm)$ may again be replaced either by $\sim$ or by nothing at all in all the different combinations. The number of Q-predicates is therefore $2^k = K$. An arbitrary Q-predicate will be called $Ct_i(x)$ where $i = 1, 2, \ldots, K$.

   Q-predicates may be said to specify all the different kinds of individuals that can be defined by means of the resources of expression that were given to us. By means of them, each state-description can be expressed in the following form:

$$Ct_{i1}(a_1) \ \& \ Ct_{i2}(a_2) \ \& \ \ldots \ \& \ Ct_{ij}(a_j) \ \& \ \ldots. \tag{16}$$

In other words, all that a state-description does is to place each of the individuals in our domain into one of the classes defined by the Q-predicates. A structure-description says less than this: it tells us *how many* individuals belong to each of the different Q-predicates.

By means of the Q-predicates we can form descriptions of all the different kinds of worlds that we can specify by means of our resources of expression without speaking of any particular individuals. In order to give such a description, all we have to do is to run through the list of all the different Q-predicates and to indicate for each of them whether it is exemplified in our domain or not. In other words, these descriptions are given by statements of the following form:

$$(\pm)(Ex)Ct_1(x) \ \& \ (\pm)(Ex)Ct_2(x) \ \& \ldots \& \ (\pm)(Ex)Ct_K(x) \tag{17}$$

The number of these is obviously $2^K = 2^{2^k}$. It is also obvious how (17) could be rewritten in a somewhat more perspicuous form. Instead of listing both the Q-predicates that are instantiated and those that are not, it suffices to indicate which Q-predicates are exemplified and then to add that those are all the Q-predicates that are exemplified, i.e. that each individual has to have one of the Q-predicates of the first kind. In other words, each statement of form (17) can be rewritten so as to be of form

$$\begin{aligned} &(Ex) \, Ct_{i_1}(x) \ \& \ (Ex) \, Ct_{i_2}(x) \ \& \ldots \& \ (Ex) \, Ct_{i_w} \\ &\& \ (x) \, [Ct_{i_1}(x) \lor Ct_{i_2}(x) \lor \ldots \lor Ct_{i_w}(x)] \end{aligned} \tag{18}$$

for some suitable subset $\{Ct_{i_1}, \ldots, Ct_{i_w}\}$ of the set of all Q-predicates.

This form is handier than (17) for many purposes. For one thing, (18) is often shorter than (17).

The statements (18) will be called constituents of monadic first-order logic. Among those statements of this part of logic which contain no names (or other free singular terms) they play exactly the same role as the constituents of propositional logic played there. These statements are said to be closed. Each consistent closed statement can now be represented as a disjunction of some of the constituents. It admits all these constituents, excludes the rest. Constituents (18) are thus the strongest closed statements of monadic first-order logic that are not logically false. For this reason, they might be called *strong generalizations*. The difference between constituents and structure-descriptions should be obvious by this time: A structure-description says of each Q-predicate *how many* individuals have it, whereas a constituent only tells you whether it is *empty* or not.

The analogy between the constituents of propositional logic and the constituents of monadic first-order logic suggests using the latter in the same way the former were used. We may give each of them an equal *a priori* probability and study the resulting measures of information. As far as the probabilities of state-descriptions are concerned, we have more than one course open for us. We may distribute the *a priori* probability that a constituent receives evenly among all the state-descriptions that make it true.[15] Alternatively, we may first distribute the probability-mass of each constituent evenly among all the structure-descriptions compatible with it, and only secondly distribute the probability of each structure-description evenly among all the state-descriptions whose disjunction it is. Since the latter method combines some of the features of the first one with those of Carnap's p^*, we shall call it the "combined method".[16]

Both methods yield measures of probability and information which are in certain respects more natural than those proposed by Carnap and Bar-Hillel, especially when they are applied to closed statements. It is of a special interest here to see how probabilities are affected by increasing evidence. Let us assume that we have observed a number of individuals $a_1, a_2, \ldots, a_n$ which have been found to exemplify a number of Q-predicates. Let these Q-predicates be

$$Ct_{i_1}(x), Ct_{i_2}(x), \ldots, Ct_{i_c}(x). \tag{19}$$

This is clearly the general form of a body of evidence consisting of completely observed individuals.

Some constituents are ruled out (falsified) by this evidence while others are compatible with it. Clearly, (18) is compatible with it if and only if $w \geq c$. The important question is what happens when we obtain more and more evidence compatible with these constituents, i.e. what happens if c remains constant while n grows? Without adding any details I shall say simply that a closer analysis shows that in this case the posterior probability of precisely one constituent grows more and more and approaches one as its limit, while the posterior probability of all the other constituents converges to zero. This unique constituent which gets more and more highly confirmed is the unique simplest constituent compatible with our evidence. It is the unique constituent which says that only those Q-predicates are instantiated in the whole universe which are already instantiated in our sample. If our sample is rather large (large absolutely

[15]This was the course followed in [7].
[16]It was proposed in [8].

speaking, though it might nevertheless be small as compared with the whole universe), this constituent clearly represents on intuitive grounds too the most rational guess one can make concerning the whole universe. In other words, in monadic first-order logic the posterior probabilities of our constituents lead us to prefer that unique constituent among them in the long run which is the only rationally preferable constituent in the case of a large sample on obvious intuitive grounds. Asymptotically at least, our results in this case therefore accord with our inductive good sense.

This kind of argument is in a sense completely general, for when we run through all the individuals of our universe, the number c of observed Q-predicates will have to stop at some finite value $1 \leq c \leq K$.

As far as comparisons between constituents (strong generalizations) are concerned, we thus obtain an asymptotically realistic model of the scientific procedure according to which we are guided in our examination of hypotheses by high posterior probability, and by high posterior probability only.

This cannot be the whole story concerning our choice of hypotheses and theories, however, not even in the artificially simple test case of monadic first-order logic. For one thing, it was already noted that if we consider generalizations other than strong ones, we can always raise their probability by adjoining to their expansions (into disjunctions of constituents) "irrelevant" new disjuncts. If our choice is not restricted to constituents, high posterior probability cannot therefore be the guide of scientific life, conceived of as a choice between different generalizations (closed statements).

A better guide is given to us by the Bayesian decision principle, according to which one should strive to maximize expected utility. What results does this principle lead to here?[17]

Let us ask whether we should accept or reject a hypothesis h, which is assumed to be of the nature of a generalization (closed statement). The expected utility of accepting it will be

$$p(h/e) \cdot u(h) + p(\sim h/e) \cdot u(\sim h), \tag{20}$$

where e is the evidence on which we are basing our decision and $u(h)$ and $u(\sim h)$ are, respectively, the utilities resulting from our decision when h is true and when it is false. Since we want to consider information as the sole utility here, we must obviously put $u(h) = \inf(h)$ or $u(h) = \text{cont}(h)$.

---

[17]This question was asked by Hintikka and Pietarinen [11], whose line of thought we shall here follow.

In other words, what we have gained by accepting h is the information it gives us. Conversely, if h is false, what we have lost is the information we would have had if we had accepted $\sim h$ and not h. In other words, we have to put $u(\sim h) = -\inf(\sim h)$ or $u(\sim h) = -\text{cont}(\sim h)$.[18]

It turns out that the choice between our two information-measures "inf" and "cont" does not make any essential difference here. For simplicity, we may consider what happens if "cont" is used. Its intuitive meaning as "substantial information" seems to make it preferable here in any case.

In this case we have $u(h) = 1 - p(h)$ and $u(\sim h) = p(h)$. Substituting these into (20) and simplifying, we obtain for the expected utility of our decision the very simple expression

$$p(h/e) - p(h). \tag{21}$$

What this represents is simply the net gain in the probability of h that is brought about by our evidence e.

If h is a generalization (closed statement), it can be represented as a disjunction of constituents as in (4). For (21), we shall then obtain

$$\sum_{i=1}^{w(h)} p(C_i/e) - \sum_{i=1}^{w(h)} p(C_i) = \sum_{i=1}^{w(h)} [p(C_i/e) - p(C_i)]. \tag{22}$$

In the right-hand sum the different terms are the net gains that our evidence induces in the probabilities of the different constituents compatible with h. We already know how they behave asymptotically. Since the probability of one and only one constituent approaches one while the probabilities of all the others approach zero, asymptotically there is one and only one constituent whose posterior probability is greater than its probability *a priori*. This result is independent of our decision to give equal *a priori* probabilities to constituents; it holds as soon as all of them have any nonzero probabilities *a priori*.

In other words, in the last sum of (22) at most one term is positive when the number of observed individuals is large enough. Let the constituent which gives rise to this member C_c. Then what we have to do in order to maximize (22) is to choose h to be identical with this one constituent. According to the Bayesian approach that I have sketched, this

---

[18]The choice of the utility-function u here is a tricky matter which deserves further discussion. The present choice can be defended, but this necessitates spelling out more fully our *desiderata*. I hope to return to these problems in another paper.

will be our preferred hypothesis asymptotically. We have already seen that this hypothesis seems to be the most reasonable thing we can in these circumstances surmise about the universe in any case.

Simple though this result is, it is very suggestive because it is, to the best of my knowledge, the first clear-cut positive result which has been obtained by considering induction (choice of scientific hypotheses) as a process of maximizing expected utility. It is, in other words, the first clear-cut application of decision theory to the philosophical theory of induction.

Other considerations are also possible here. Instead of looking for a generalization which maximizes expected information we can ask: Which kind of individual would contribute most to our knowledge if we could observe it? Asking this is tantamount to asking: What Q-predicate yields the highest information value $\inf[Ct(b)/e]$ to the statement $Ct_i(b)$ when b is an individual which we have not yet observed? Again, easy calculations give us a simple answer. The information of $Ct_i(b)$ in relation to the kind of evidence e which we have been considering is maximal when $Ct_i(x)$ is a Q-predicate which has not yet been exemplified in the evidence e, provided that this evidence is large enough. An individual b having such a Q-predicate would falsify the previously preferred strong generalization, for this was found to assert that only those Q-predicates are instantiated in the whole universe that are already exemplified in a sufficiently large e. In other words, the most informative new observations will be those which falsify the previously preferred generalization. This simple observation may go some way toward explaining and justifying Popper's emphasis on the role of falsification in scientific procedure. On the view we have reached here, however, the emphasis on falsification is not important solely because it guards us against false hypotheses, as Popper sometimes seems to suggest. It is also a consequence of our search for the most informative observations.

All these results seem rather important philosophically. Their theoretical foundation is somewhat shaky, however, as long as we do not better understand the reasons for the choice of the underlying probability measures on which our measures of information are based. Furthermore, it must be pointed out that although the results we have obtained are often very natural asymptotically, in almost all the cases I have so far discussed they are not very natural quantitatively. For small values of n (= the number of observed individuals) they give too high probabilities to generalizations and thus correspond to a wildly over-optimistic inductive behavior with respect to inductive generalization from small samples. Furthermore, in some cases (especially when the *a priori* probabilities are divided evenly among state-descriptions) we do not obtain reasonable results for singular inductive inferences.[19]

4. Probability-Measures Dependent on Parameters.

All these difficulties can be eliminated by taking a somewhat more general view of the situation. It seems to me that we have so far overlooked an important factor that affects our ideas of information. For it seems to me obvious that the information that statements concerning individuals (they may be thought of as observation-statements) are supposed to have depends heavily on the amount of order or regularity that we know (or, for a subjectivist, believe) prevail in the world. For instance, if we know or believe that the universe is perfectly regular in the sense that all its members have the same Q-predicate (though we do not know which), one single observation-statement of the form $Ct_i(a)$ gives us all the information we might want to have (insofar as it can be expressed by means of the resources we have at our disposal in monadic first-order logic). Taking a less extreme case, if we know that almost all the individuals of our domain have one and the same Q-predicate, the observation of two individuals having the same Q-predicate tells us more about the probable distribution of individuals into the different Q-predicates than what it would do if we knew that individuals are likely to be distributed more or less evenly among several Q-predicates.[20]

Our probability-judgements are apt to exhibit the same dependence on our knowledge of the amount of order or regularity in the universe. If we know (or believe) that the universe is very regular, we are apt to follow the indications of observations much more boldly than we would do if we knew that it were fairly irregular. In the latter case, our estimates of the distribution of unobserved individuals between the Q-predicates will be much more conservative than in the former and also much more dependent on *a priori* probabilities. Observed regularities would not be followed up as quickly as in the former case because they would be expected to be due to chance rather than the actual regularities that obtain in the universe at large. Thus the amount of regularity we think there is in the universe is reflected by the rate at which observations affect our probabilities.

---

[19]See the last few pages of [7].

[20]It is not very clear, however, how these entirely plausible uses of the concept of information are to be looked upon from the point of view of our formal definitions of information. It would take us too far to explain this matter here; I shall return to it in another connection. Suffice it to say that we have to distinguish between the information certain observation-statements convey concerning their own subject matter and the information they convey concerning the universe at large.

This effect has to be taken into account in defining measures of information. It seems to me that it has to be taken into account on two different levels. We have to consider the amount of regularity or irregularity that there is in the world (or is thought of as being there) as far as the distribution of individuals between those Q-predicates is concerned that are instantiated in the universe. Secondly, we have to consider how regularly the individuals are apt to be concentrated into few Q-predicates rather than to be spread out among all of them.[21]

Let us consider these problems in order. The former effect is shown by the probability that the next individual will exemplify a certain given Q-predicate $Ct_i(x)$, when we know that a certain definite constituent C_w which allows w Q-predicates to be instantiated (among them $Ct_i(x)$) is true. Let us assume that we have observed n individuals of which q exemplify $Ct_i(x)$. It is natural to assume that the desired probability should depend on the numbers w, n, and q only. From this assumption it follows, together with certain other rather plausible assumptions, that the desired probability is expressed by

$$\frac{q+\lambda/w}{n+\lambda}, \tag{23}$$

where λ is a free parameter, $0 < \lambda < \infty$.

This expression (23) may be taken to be a kind of weighted average of the *a priori* factor $1/w$ and the purely empirical (*a posteriori*) factor q/n which is nothing but the observed relative frequency. In this weighted average, λ is the weight of the *a priori* factor. For instance, if $\lambda = 0$, we have the most purely empirical rule of induction ("straight rule") which tells us to follow observed relative frequencies directly.

According to what was said earlier, λ is thus also a measure of the *dis*order or *ir*regularity that we know or believe to obtain in our universe of discourse, as far as the distribution of individuals between the different Q-predicates that C_w to be allows instantiated are concerned.

In any finite universe, the choice of λ determines together with (23) the probabilities *a priori* of all our state-descriptions. This is perhaps seen most easily by pointing out that because each state-description can be written in the form (16), its probability can be expressed in the following form:

$$p[Ct_{i_1}(a_1)] \cdot p[Ct_{i_2}(a_2)/Ct_{i_1}(a_1)] \cdot p[Ct_{i_3}(a_3)/Ct_{i_1}(a_1) \& Ct_{i_2}(a_2)] \ldots . \tag{24}$$

[21]For the following, cf. [9, 19].

In order to find a completely defined distribution of *a priori* probabilities to state-descriptions (and therefore to everything else too) we thus have but to specify the *a priori* probabilities of the different constituents, for everything we have said in the last couple of paragraphs has been relative to the truth of one of them.

These *a priori* probabilities of constituents are precisely what is at stake when we consider the other kind of regularity that there may or may not be in the universe, for this kind of regularity was what determines how likely it is that some of the Q-predicates, or even several of them, are empty. A convenient way of telling how likely it is for a constituent C_w to be true, i.e. for certain general laws to hold in our universe, is to compare the *a priori* probability $p(C_w)$ of C_w with the probability of obtaining a sample compatible with it in a universe whose individuals have been distributed "randomly" between the different Q-predicates in accordance with (23) except that w is replaced by K. Such a universe might be called an *atomistic* one. My suggestion for specifying the probabilities *a priori* that the different general laws hold in our universe is to say that they are proportionally as great (or small) as the probabilities of obtaining samples of a fixed size (say containing α individuals) compatible with these laws in an atomistic universe. The larger α is, the smaller these probabilities obviously are. In fact, for a constituent C_w admitting w Q-predicates to be instantiated, the probability of obtaining an atomistic sample of size α compatible with it is by (23)

$$\frac{\lambda w/K}{\lambda} \frac{1 + \lambda w/K}{1 + \lambda} \frac{2 + \lambda w/K}{2 + \lambda} \cdots \frac{\alpha - 1 + \lambda w/K}{\alpha - 1 + \lambda} \tag{25}$$

which generalizes as

$$\frac{\Gamma[\alpha + (\lambda w/K)] \cdot \Gamma(\lambda)}{\Gamma(\alpha + \lambda) \cdot \Gamma(\lambda w/K)}, \tag{26}$$

where Γ is the familiar gamma function of analysis and α an arbitrary non-negative real number. My suggestion is to make the *a priori* probability of C_w proportional to (25) or (26).

Together with (23), these *a priori* probabilities suffice to specify our probability distribution. Strictly speaking, this works without further explanations only in the case of an infinite universe, while in a finite universe certain further explanations are needed. They are not pertinent to our present purposes, however.

The parameter α which occurs in (25) and (26) is a kind of measure of the amount of disorder (irregularity) we expect, or are entitled to expect, to obtain in the universe as far as general laws are concerned. According to what was said earlier, it therefore plays the role of an *index of caution*: it is a kind of weight of the *a priori* factor in inductive generalization, somewhat in the same way λ was a kind of weight attached to the *a priori* factor in singular inductive inference. However, in the case of generalizations, *a priori* considerations tend to discourage neat inferences. If we are to go by symmetry considerations alone, we must obviously say that since all Q-predicates are pretty much on a par even after large universe all of them are likely to be exemplified, which means that no nontrivial general laws hold.

All the different ways of defining measures of probability and information which were discussed above (and a number of others) may be taken to be special cases of our two-dimensional continuum of probability-measures depending on α and λ. Carnap's p^+ results as a limiting case by letting $\alpha \to \infty$, $\lambda \to \infty$. His p^* results when $\alpha \to \infty$, $\lambda = K$. More generally, the λ-continuum of inductive methods considered by Carnap in some of his works results when $\alpha \to \infty$. The two-stage procedure of first distributing *a priori* probabilities evenly among constituents and then as evenly among state-descriptions as is compatible with the first distribution amounts to letting $\alpha = 0$, $\lambda \to \infty$. The "combined method" is tantamount to putting $\alpha = 0$, $\lambda = w$.

All the undesirable features of these particular probability-measures and of the associated measures of information can now be seen to be consequences of the particular choices of α and λ that underlie them. If $\alpha = \infty$, inductive generalization are judged on *a priori* grounds only. Unsurprisingly, it then turns out that all the different Q-predicates are, in a large universe, highly likely to be instantiated, and that a statement to the effect that some of them are instantiated is therefore highly uninformative. It is not any more surprising that a system with $\lambda = \infty$ should fail to yield a reasonable treatment of singular inductive inference, for this choice of λ amounts to treating singular inductive inferences completely on *a priori* grounds. Likewise, it is now seen to be inevitable that the systems in which probabilities are distributed evenly among constituents should give wildly over-optimistic results as far as inductive generalizations are concerned, for they are based on the choice $\alpha = 0$, which means that we are expecting a great deal of uniformity in the world as far as generalizations are concerned. All these "mistakes" can be corrected by choosing α (or λ) differently.

Independently of these choices, the asymptotic results mentioned above hold as long as α and λ are both non-negative finite real numbers and $\lambda \neq 0$. Furthermore, they remain valid when $\lambda \rightarrow \infty$. Hence, in all these cases, the connection between reasonable inductive generalization and the maximization of expected information will remain the same. Here we cannot study in detail the consequences of the different choices of λ and α in other respects.

A matter of principle should be taken up here, however. It seems to me natural and inevitable that the factors which our parameters α and λ bring out should influence one's measures of semantic information. At the same time, it seems to me impossible to justify any particular choice of these parameters by means of purely logical considerations. For what they express is, from an objectivistic point of view, the amount of disorder that there probably is in the universe, or, from the subjectivistic point of view, one's expectations concerning the amount of this disorder. An objectivist or a subjectivist can thus understand what goes into these two parameters, but it does not seem to be possible even to associate a clear sense to them on the basis of a strictly logical interpretation of probability. Insofar as the considerations that are presented in this paper concerning the relation of information and probability and concerning the role of the parameters α and λ are acceptable, they therefore suggest rather strongly that a strictly logical interpretation of probability cannot be fully carried out, at least not in the simple languages we have been considering. Rather, the force of circumstance (more accurately speaking, the role of the idea of order or regularity in our judgments of information and probability) pushes one inevitably toward a much more "Bayesian" position in the sense that we are forced to consider the dependence of our *a priori* probabilities on non-logical factors and also to make them much more flexible than a hard-core defender of a logical interpretation of probability is likely to accept. It also seems to me that a closer investigation of the factors affecting our judgments of semantic information and probability will strengthen this view and perhaps necessitate an even greater flexibility in our distribution of *a priori* probabilities.

Among other things, this implies that the relation of a theory of semantic information to statistical information theory is more problematic than earlier discussions of this relation are likely to suggest. This relation cannot be identified with any clear-cut difference between different senses of probability.

The distinctive feature of a theory of semantic information and of the associated theory of probability will perhaps turn out not to be any special sense of probability but rather the need of developing methods for discussing the probabilities and information-contents of *generalizations (closed statements)*. For I do not see how one could avoid speaking of the information that different generalizations convey to us in any serious theory of semantic information. And if this is to be understood along the lines of any concept of information that is related to a suitable notion of probability in a natural way, we need a concept of probability which can assign finite nonzero probabilities to generalizations in all domains. Above, some such senses of probability were examined in the context of certain especially simple languages. The task of studying these senses of probability and similar ones seems especially important from a subjectivistic point of view, for it seems to me obvious that people do in fact associate degrees of belief and of information with general statements. They even seem to be able to compare these degrees of belief with the degrees of their belief in various singular statements.

REFERENCES

1. C. E. Shannon and W. Weaver, *The Mathematical Theory of Communication*, The University of Illinois Press, Urbana, Illinois, 1949; A. I. Khinchin, *Mathematical Foundations of Information Theory*, translated by R. A. Silverman and M. D. Friedman, Dover, New York, 1957.
2. C. Cherry, *On Human Communication*, M.I.T. Press and John Wiley, 1957.
3. Y. Bar-Hillel and R. Carnap, "Semantic Information," *British Journal for the Philosophy of Science* 4, 147-157 (1953).
4. R. Carnap and Y. Bar-Hillel, "An Outline of a Theory of Semantic Information," *Technical Report No. 247*. M.I.T. Research Laboratory in Electronics, 1952; reprinted in Y. Bar-Hillel, *Language and Information*, Addison Wesley and The Jerusalem Academic Press, 1964, Chapter 15.
5. J. G. Kemeny, "A Logical Measure Function," *Journal of Symbolic Logic* 18, 289-308 (1953).
6. Y. Bar-Hillel, *Language and Information*, Addison Wesley and The Jerusalem Academic Press, 1964.
7. J. Hintikka, "Towards a Theory of Inductive Generalization," in *Proceedings of the 1964 International Congress for Logic, Methodology, and Philosophy of Science* (Y. Bar-Hillel, ed.), North-Holland Publishing Co., Amsterdam, 1965, pp. 274-288.
8. J. Hintikka, "On a Combined System of Inductive Logic," in *Studia Logico-Mathematica et Philosophica in honorem Rolf Nevanlinna, Acta Philosophica Fennica* 18, 21-30 (1965).
9. J. Hintikka, "A Two-Dimensional Continuum of Inductive Methods," in *Aspects of Inductive Logic* (J. Hintikka and P. Suppes, eds.), North-Holland Publishing Co., Amsterdam, 1966, pp. 98-117.
10. J. Hintikka and R. Hilpinen, "Knowledge, Acceptance, and Inductive Logic," in *Aspects of Inductive Logic* (J. Hintikka and P. Suppes, eds.), North-Holland Publishing Co., Amsterdam, 1966, pp. 1-20.
11. J. Hintikka and J. Pietarinen, "Semantic Information and Inductive Logic," in *Aspects of Inductive Logic* (J. Hintikka and P. Suppes, eds.) North-Holland Publishing Co., Amsterdam, 1966, pp. 81-97.
12. J. Hintikka, "Induction by Enumeration and Induction by Elimination," in *The Problem of Inductive Logic, Proceedings of the International Collquium in the*

Philosophy of Science, London, 1965, (I. Lakatos, ed.), Volume 2, North-Holland Publishing Co., Amsterdam, 1967.

13. K. Popper, *The Logic of Scientific Discovery*, Hutchinson, London, 1959. The bulk of this work consists of a translation of *Logik der Forschung*, Springer, Vienna, 1934.

14. C. G. Hempel, "Deductive-nomological vs. Statistical Explanation," in *Minnesota Studies in the Philosophy of Science*, Volume 3 (H. Feigl and G. Maxwell, eds.), University of Minnesota Press, Minneapolis, 1962, pp. 98-169, especially pp. 153-156.

15. C. G. Hempel, "Inductive Inconsistencies," *Synthèse* 12, 439-469 (1960); reprinted in C. G. Hempel, *Aspects of Scientific Exploration and Other Essays in the Philosophy of Science*, The Free Press, New York, 1965, pp. 53-79, especially pp. 73-78.

16. R. D. Luce and H. Raiffa, *Games and Decisions*, John Wiley, New York, 1957, pp. 318-324 and especially p. 324.

17. I. Levi, "Decision Theory and Confirmation," *Journal of Philosophy* 58, 614-624 (1961).

18. R. Carnap, *Logical Foundations of Probability*, University of Chicago Press, Chicago, 1950; 2nd ed., 1963.

19. R. Carnap, *Continuum of Inductive Methods*, University of Chicago Press, Chicago, 1952.

20. P. A. Schilpp (ed.), *The Philosophy of Rudolf Carnap*, The Library of Living Philosophers, La Salle, Illinois, 1963.

DISCUSSION

HINTIKKA: What can a philosopher do in order to help us understand the scientific enterprise? The aims of scientific inquiry are somehow to be characterized. There are, of course, many different ways of doing so. One way is to sketch suitable simple models of the scientific enterprise, and then to try to define as precisely as we can how the success of the enterprise is to be measured. Our grasp of how science actually proceeds gives us a pretty good idea as to what we are doing when we are sketching these models. On the other hand, such models help us to clarify our own ideas of the scientific enterprise.

In this philosophical program, the concept of information seems to me to play a crucial role. It is a relatively familiar and clear idea. Hence we have some ways of telling what it is and how it is to be defined. On the other hand, it is a very handy concept for the purpose of straightening out our ideas of science and the progress of science. I think Popper's use of this concept is a good case in point. I do not see anything wrong in attempting to do this either from a Carnapian or from a Popperian point of view. It seems to me a rather promising attempt, as philosophical programs go.

To forestall the use of certain types of Popperian argument against my paper, I would like to suggest that it is misleading to say that probability is always (on all its interpretations) a measure of safety. Let us not be misled by the well-known connection between probability theory and betting theory. What it shows is — very roughly speaking — that violations of the axioms of probability calculus are unsafe. It does not say that everything that satisfies these axioms measures safety in any sense. There seems to be a very precise sense in which high probability goes together in my model with the absence of safety. The more heavily our probability-mass is concentrated on one constituent, the more violently can a new observation upset our (*a posteriori*) probability distribution. Even the quest of high *a posteriori* probability thus cannot automatically be identified with the kind of quest for safety Popper has justly criticized.

In particular, safety seems to have little to do with the fact that rational acceptability and high probability sometimes go together. In my model they go together in the case of the posterior probability of constituents, but not in general, and I do not see any deep significance in the fact that they sometimes point in the same direction.

More than anything else, I want to get the record straight on the following point: What is maximized in the simple model I sketched is not probability at all, but information. Speaking more precisely, I suggested that the success of a generalization be measured by the expected information of the generalization in question; and this turned out to give us in fact a nice model. One reason why this seems to me important is that it is a special case of the basic ideas of decision theory, where we have quite a few interesting and important ideas available for further work. More work remains to be done to see how useful they are for general philosophical purposes.

POPPER: I would just like to summarize a few of the things on which Hintikka's paper is based. There is an idea, which is perfectly legitimate, the idea that a good hypothesis is a probable hypothesis. But then there are *other* ideas of probability which have been formulated into the probability calculus, and according to the probability calculus, if I bet on two horses it is more improbable that I shall win than if I bet only on one horse. Of course, I *may* win very much more money if I bet on two horses than if I bet on one horse, if I win. But I am very much more likely to lose my money. Or if I twice toss a penny, it is more improbable that both times will come up heads than that if I once toss a penny, it will come up heads.

Now, the second kind of idea has been developed into a mathematical calculus, the so-called probability calculus. The first idea, that it is good to have a probable hypothesis, has been by almost everybody held to be very closely related to that probability calculus. That is to say, most people have thought that if they speak of a probable hypothesis as a good hypothesis, then they mean that this hypothesis is probable in the same sense in which we speak of the probability of other games of chance.

Actually, this is implicit in the work of quite a number of logicians, although they very rarely say so explicitly.

Now, I pointed out long ago something terribly simple, namely this: In the sense of the probability calculus it is clear that *the more a hypothesis says, the less probable it is*. A hypothesis which says, "These two horses will win consecutive races" says more, and tells us more, than a hypothesis which says only of one horse (among those which will start) that it will win. A hypothesis which says, "I will toss heads twice with a penny", does tell us more than a hypothesis which says, "Oh, perhaps I will toss heads with a penny, at least among two or three or many sets of tosses."

Thus, the more I am saying, the less probable is what I am saying. I therefore suggested that we could measure the content, the informative content, as I called it (when I published this idea in 1934). I suggested that we can measure the informative content of what we say by the improbability of what we are saying ("improbability" in the sense of the probability calculus, of course).

The more information — just to make this point a little bit more clear — the more information a general obtains from his Intelligence Service and the more detailed information it is, the less probable it is, in the sense of the probability calculus. If a general obtains the following information: "The enemy is somewhere", then this is very little information but its probability will be equal to one. If the general obtains the information: "The enemy is retreating", then this is some more specific information. Its probability will be less than one. But if he obtains information, "The enemy is retreating towards Denver", this is still more specific information because the enemy could retreat also towards, let us say, Colorado Springs. The more specific an information is, the more does it reduce the possibilities, and by reducing possibilities it becomes more and more improbable (in the sense of the probability calculus).

I suggested about thirty years ago that the informative content of a hypothesis can be measured as some sort of inverse of its probability. So the formula

$$ct(h) = 1 - p(h)$$

says that the content of h, the content of a hypothesis, equals 1 minus its probability. I could just as well have written 1 over probability. The way we choose to write this is pretty irrelevant. The whole thing which is involved is very trivial and very simple: the informative content gets lower as the probability rises, and as the informative content rises, the probability becomes lower.

All this I think is very trivial, but it shows one thing, namely, that in science we do not simply aim at high probabilities in the sense of the probability calculus. As a consequence, we can say: *When we call a "good" hypothesis a "probable" hypothesis, then we do not use the word "probable" in the sense of the probability calculus*. This, I think, is a simple, straightforward result. There are at least two pretty opposite concepts of probability, that of a *good hypothesis* which, if you like, you can call "probable" (I don't mind the terminology), and that kind of probability which satisfies the laws of the probability calculus and for which high informative content of the hypothesis must increase with the improbability, in the sense of the probability calculus.

Now, very many philosophers have thought what I say is absurd. I could quote many passages where it is said, "Popper uses a very queer sense of probability." I don't care in the least about words. I don't mind whether you call it "probability" or "bum-bum-bum" — I don't mind this at all. I have simply shown, of the probability which satisfies the mathematical calculus of probability, that this kind of probability is not something desirable for hypotheses, because if you apply it to a hypothesis then we come to the following result: *The most probable hypothesis is the one which says nothing*.

But in science we want to say something, and we want actually to say a lot. We want to have a high informative content, and a hypothesis with a high informative content will automatically have a low probability in the sense of the probability calculus.

Now, this incredibly trivial result has not sunk in through thirty-two years. On the contary, there was a wild discussion between myself and Bar-Hillel, for example, and also some pretty violent discussions between myself and Carnap about it. I am very grateful to Hintikka who is the first philosopher, perhaps, of the very powerful Carnap school who has seen my point. I do think he has definitely seen my point.

YOURGRAU: The thing which concerns me regarding Hintikka's paper is whether he means by the term "information" what I mean by it. It is "scientific information" which Popper (and I) have in mind. Other information is logically trivial; and the question of mathematical information is better not raised in this context at all. For Polya, of course, *this* would be information. Even "tough" scientific information can become unintelligible when information theorists get hold of it. To wit, the physicist shudders when he sees what the information theorist has done to such a clear, unambiguous, physical concept — or rather quantity — as entropy. The crucial issue is: in what rigorous sense is Hintikka applying the term "information"?

TÖRNEBOHM: Perhaps one could make use of the notion of amount of information in order to express what historians do when they test or check their narratives by means of comparisons with documentary evidence. The expression

$$DC(H/E) = \frac{\inf(H) - \inf(H/E)}{\inf(H)}$$

is a measure of the degree to which the hypothesis (the historical narrative) and the evidence (as contained in the documents) share information. In other words, DC can be read as the degree of covering.

HINTIKKA: Let me repeat that what one is maximizing in my model is not probability but expected information. Of course this (semantic) information has to be defined in terms of probability, but this is only to be expected. Popper has suggested various ways of defining his own notion of corroboration in terms of probability. My sense of information is in this respect on a par with the notion of corroboration.

I agree with Popper's main point concerning the inverse relation between probability and information. The additional point I want to make is that there are many different senses of information and even different senses of probability. The inverse relation certainly obtains in the case Popper has primarily had in mind, namely, an inverse relation between absolute information and *a priori* probability (to use my own terms). As it happens, there are other senses of probability and information in which the relationship between the two is more complicated. A couple of them can be seen from my paper. (The absolute information and the *a posteriori* probability of constituents are not inversely related if one's evidence is large enough, for instance.)

I do not think that earlier theorists of semantic information have come close to exhausting the variety of different senses of information. A great deal of further work remains to be done here. (In fact, it seems to me that Törnebohm's expression gives us an interesting but largely unexplored new possibility.) My guess is the further work in this direction will bring philosophers closer to some theories of statistical decision and statistical testing than they have yet realized.

With regard to Yourgrau's concern with our respective definitions of information, surely we want to have a reasonable approximation to what passes as information in ordinary life (and ordinary science). Indeed, whatever measures of information one proposes, they have to refer to the language one is using (in the logical sense of the word). Granted, there is a great variety of different senses (or kinds) of information which include the unusual sense of information in which a mathematical or logical argument may create new information, but this variety only seems to make the subject interesting. I have a theory about this last kind of information, or at least the beginnings of a theory, and I do not see any insurmountable problems in the other distinctions between different kinds of information.

Chapter X

QUANTIFICATION AND ONTOLOGICAL COMMITMENT

Czesław Lejewski
University of Manchester

In his review of a paper by Ajdukiewicz [1], Quine makes the following comments on Leśniewski's version of the membership connective 'ϵ':

> One way of viewing Leśniewski's logic of 'ϵ' (which he called "ontology") would seem to be this: the variables stand in places appropriate to general terms, the notion of general term being construed broadly enough to include terms which are true of fewer than two objects as well as terms which are true of two or more. Then, where 'a' and 'b' are thought of as general terms, '$a \epsilon b$' is construed as true if, and only if, 'a' is true of one and only one object and 'b' is true of that object. Leśniewski, and others who have found this part of his logic useful (Kotarbiński, Ajdukiewicz), have entertained an amiable distaste for abstract entities and hence have liked to appeal to general terms, more or less as above, rather than to classes. Too little significance, however, has been attached to the fact that the variables which have been said to stand in places appropriate to general terms are subjected in Leśniewski's theory to quantification. Such quantification surely commits Leśniewski to a realm of values of his variables of quantification; and all his would-be general terms must be viewed as naming these values singly. If quantification as Leśniewski used it did not commit him squarely to a theory of classes as abstract entities, then the present reviewer is at a loss to imagine wherein such commitment even on the part of a professing Platonist can consist [2].

I quote this passage because the last sentence in it poses two problems which I want to make central to the present enquiry. First, is quantification as Leśniewski used it, and as his followers continue to use it, incompatible with the renunciation of abstract entities? Second, in what way can a professing Platonist give expression to his ontological commitment?

173

The language of Leśniewski's logic differs from the language of the traditional theory of quantification (with identity) in several respects. But as it happens, we need not go into details because we can solve our problem by first solving it within a more familiar context.

As is well known, the meaning of the quantifiers in that part of quantification theory which is described as the first-order predicate logic with identity can be determined informally, in terms of so-called expansions. We assume a universe consisting of a finite number of objects, a_1, a_2, ..., a_n, and then we construe a universally quantified expression of the form '$(x) \cdot F x$' as equivalent to the conjunction '$Fa_1 \cdot Fa_2 \cdot \ldots \cdot Fa_n$', which is said to be the expansion of the quantified expression. Analogously, an existentially quantified expression "$(\exists x) \cdot Fx$" *expands into an* equivalent alternation '$Fa_1 \vee Fa_2 \vee \ldots \vee Fa_n$'. In both expansions '$a_1$', '$a_2$', ..., '$a_n$' are singular constant names, each naming exactly one object from among the objects which are assumed to constitute the universe. The meaning we attach to the respective expansions determines our way of reading and interpreting the quantified expressions '$(x) \cdot Fx$' and '$(\exists x) \cdot Fx$.' And one thing becomes clear in the light of the expansions: expressions quantified in this manner have existential import. They imply or presuppose or are committed to what has been assumed to start with, name a universe consisting of n objects.

When we generalize the notion of quantification to apply to a non-finite universe, we replace the equivalences

$$(x) \cdot Fx \cdot \quad \equiv \quad \cdot Fa_1 \cdot Fa_2 \cdot \ldots \cdot Fa_n \tag{1}$$

$$(\exists x) \cdot Fx \cdot \quad \equiv \quad \cdot Fa_1 \vee Fa_2 \vee \ldots \vee Fa_n \tag{2}$$

by implications of the following form:

$$(x) \cdot Fx \cdot \supset \cdot Fa \tag{3}$$

$$Fa \cdot \supset \cdot (\exists x) \cdot Fx, \tag{4}$$

where 'Fa' qualifies as a conjunct, or as an alternant, in whichever non-finite expansion is involved. When we say this, or when we say that in the case of a non-finite universe a universally quantified expression is equivalent to a non-finite conjunction while an existentially quantified expression is equivalent to a non-finite alternation, we are using a metaphorical way of talking. For we are not suggesting that there are such things as non-finite conjunctions or non-finite alternations. And the metaphorical way of talking is as good a device as any if it enables us informally to determine the meaning of undefined constructions in our symbolic language. In any case what has been said so far is neither new nor controversial. There are alternative ways of interpreting quantified expressions but this does not mean that the one just described is wrong.

The constant term 'F' which occurs in our examples is to be construed as an intransitive verb. It does not name anything but it can be said to apply to objects which happen to be the values of 'x.' Thus 'F' does not add to the ontological commitment embedded in the quantifiers '(x)' and '$(\exists x)$'. But the trouble begins when besides such constant terms as 'F' we want to have corresponding variables, say 'ϕ' and 'ψ', and subject them to quantification to get, as a result, expressions such as '$(\phi): (\exists x) \cdot \phi x$' or '$(\exists \phi): (x) \cdot \phi x$' or '$(\phi): (x) \cdot (\phi x \supset \phi x)$', etc. How are we to interpret expressions of this sort?

Logicians have always felt that quantification involving predicate variables, 'ϕ', 'ψ', etc., should be understood as analogous to quantification involving name variables, 'x', 'y', etc. This having been taken for granted, the argument has then proceeded as follows. Just as the quantifiers '(x)' and '$(\exists x)$' presuppose a realm of values of the quantified variable 'x', the quantifiers '(ϕ)' and '$(\exists \phi)$' must also refer to an appropriate realm of values. The quantifiers '(x)' and '$(\exists x)$' are read 'each thing x is such that' and 'something x is such that' or 'there is a thing x such that'. Similarly the quantifier '(ϕ)' should mean something like this: 'each attribute ϕ is such that' or 'each class ϕ is such that'. And the quantifier '$(\exists \phi)$' should, analogously, be read 'there is an attribute ϕ such that' or 'there is a class ϕ such that'. The new sort of quantification carries with it a new ontological commitment just as manifestly, so it would appear, as the old one, although a different ontological category is supposed to be involved.

It is true that quantifying predicate variables is, in many respects, analogous to quantifying name variables, but to my mind the analogy suggests, before anything else, that the meaning of '(ϕ)' and the meaning of '$(\exists \phi)$' should be related to the meaning of the corresponding expansions. Now, consider the following expression: '$(\phi): (\exists x) \cdot \phi x$'. Its non-finite expansion can be visualized as

$$(\exists x) \cdot F_1 x: (\exists x) \cdot F_2 x: \ldots: (\exists x) \cdot F_n x: \ldots, \tag{5}$$

where 'F', 'F_2', $\ldots$ 'F_n' $\ldots$ are all constant predicates. Of course, we cannot have a conjunction with a non-finite number of conjuncts. But we know from what has already been said that every expression which, speaking metaphorically, qualifies as a conjunct in such a non-finite conjunction, in other words every expression of the form '$(Ex) \cdot Fx$', carries with it commitment to a universe of the values of 'x' only. No other ontological commitment can be read into such an expression. By the same token no other ontological commitment can be read into a finite conjunction of such expressions. And I fail to see how a non-finite conjunction of such expressions, if we could have one, could possibly engender an additional ontological commitment. If this argument, informal as it is, has a measure of cogency then we seem to be in the position to conclude that '$(\phi): (\exists x) \cdot \phi x$' commits us to the same realm of entities to which we are committed by admitting '$(\exists x) \cdot Fx$'.

It is fairly obvious that our interpretation of the quantifier '(ϕ)' can appropriately be extended to the quantifier '$(\exists \phi)$' with a similar result. In other words, '$(\exists \phi): (\exists x) \cdot \phi x$' commits us to the same realm of entities as that to which we are committed by '$(\exists x) \cdot Fx$'. Moreover, we can now generalize and argue, on analogous grounds, that the ontological commitment of a theory formulated in the language of the traditional logic of quantification, with no restriction on the order of quantifiable variables, is carried solely by quantification binding the variables of the first order. The nature of the commitment, that is to say the semantic relation which holds between the objects constituting the assumed universe and the variables of the first order, would have to be specified, and as a rule is specified, at a pre-systematic level, and encapsulated in the meaning of the primitive terms of the theory. Thus, for instance, we may be told, still at the pre-systematic level, that the constant terms corresponding to the variables of the first order are to be construed as names of the objects which constitute the assumed universe, and which are said to be the values of the variables. Alternatively, we may be told that the constant terms corresponding to the variables of the first order are to be construed as being true or false of the objects which constitute the assumed universe. These objects could also be described as the values of the variables of the first order, although in this case the notion of being a value has a somewhat different meaning. In fact one could say that every kind of variable, irrespective of its order, has its values in an appropriate sense of 'having a value', but the point is that in our view the values correlated in an appropriate way with the variables of a given order are the same as the values correlated in a different but appropriate way with the variables of a different order but still within the framework of the same theory.

Quantifying propositional variables within the calculus of propositions, and indeed quantifying any kind of variables available within the framework of a generalized logic of propositions such as Leśniewski's prototetic, carries with it no ontological commitment whatsoever. The meaning of quantified expressions in prototetic can be made perfectly clear by means of expansions which in prototetic are always finite. As regards quantification involving propositional variables, which in the logic of propositions are of the first order, we have

$$(p) \cdot fp \cdot \equiv \cdot f0 \cdot f1, \qquad (6)$$

where 'f' is to be construed as a constant proposition-forming functor for one argument, and '0' and '1' are the standard false and the standard true propositions, respectively. If '$(p) \cdot fp$' carries with it an ontological commitment, the same commitment should be implied by the right-hand side of the equivalence. Some logicians and philosophers have urged that the proposition '0' refers, in some sense of referring, to falsehood, while the proposition '1' refers to truth, truth and falsehood being the two truth-values with which the logic of propositions is supposed to be concerned. This kind of semantics, however, appears to be a tall story, prompted perhaps by the doctrine already criticized that to every different category of variables within the framework of the quantification theory there should correspond a different ontological category of values.

To sum up. In the traditional theory of quantification the variables of the first order are correlated with a realm of entities thought of as their values; the variables of a higher order, whether quantified or not, presuppose no additional realm of entities, and quantifying propositional variables within the logic of propositions does not commit us to entertaining the existence of any entities at all.

Now, if the view I have been advocating is accepted, how are we to read the quantifiers binding variables other than those of the first order? How, for instance, are we to read '(ϕ)' or '$(\exists\phi)$'? The noncommittal 'for all ϕ' and 'for some ϕ' suggest themselves as possible renderings, it being understood that the syntactical status of 'ϕ' as a predicate variable is somehow indicated by the context or by appropriate conventions. This way of reading the quantifiers seems to be compatible with the relevant extensions and also with the manner in which we often talk about formulae beginning with '(ϕ)' or '$(\exists \phi)$'. For, concerning such formulae, we are in the habit of saying that they hold for all interpretations of the variable 'ϕ', or for some interpretations of the variable 'ϕ' as the case may be. The quantifiers binding other variables of higher orders, and quantifiers binding variables available in the logic of propositions, could be read 'for all ...' and 'for some ...'. But there is no objection to reading the quantifiers binding variables of the first order in the traditional theory of quantification as "each entity x is such that" and "there is an entity x such that."

The doctrine that in the traditional theory of quantification every category of quantifiable variables presupposes a different ontological category of values appears to be no more than a dogma. Interestingly enough it closely resembles another dogma which used to plague logicians and philosophers. I am alluding to the once prevailing belief that, in order to be meaningful, expressions other than names must still name something. Quine has spared no pains to expose the fallacy of this doctrine, which, as he points out quite rightly, results from confusing meaning with naming. In connection with quantification theory we distinguish between substituends and values, but the distinction does not seem to be as sharp as it should be, and certain characteristics of substituends are perhaps too readily transferred onto the values. Since every kind of variable can be supplied with appropriate substituends, constant or variable, we seem to be inclined to take it for granted that every kind of variable must also have its appropriate values. In the light of the present paper this preconceived idea appears to be groundless, and the nominalist's distaste for abstract entities turns out to be compatible with quantification binding any kind of variable.

However, a new problem arises: How can a professing Platonist, to use Quine's expression, indicate his ontological commitment? It is to this problem that I propose to address myself now.

For the sake of simplicity let us assume that our Platonist countenances an ontology with concrete individual objects and abstract classes. It would appear that the commitment to such an ontology could take the form of the following expression:

$$(\exists \alpha): (\exists x) \cdot x \, e \, \alpha, \tag{7}$$

where 'α' is a class variable and 'ϵ' is the symbol for class-membership. We could read (7) as "there exists a class α such that there exists an individual object x which is a member of α."

However, it turns out that (7) is far from being unequivocal. Two different usages of 'ϵ' can be distinguished. Some logicians use it in such a way that if '$x \, \epsilon \, \alpha$' is meaningful then '$x \, \epsilon \, x$', '$\alpha \, \epsilon \, \alpha$', and '$\alpha \, e \, x$' are all meaningful. Others have maintained that if '$x \, \epsilon \, \alpha$' is meaningful then '$x \, \epsilon \, x$', '$\alpha \, \epsilon \, \alpha$', and '$\alpha \, \epsilon \, x$' are all syntactically ill-construed and utterly meaningless.

Now, if our Platonist wants us to understand (7) in the former sense then we would conclude that his ontology is in fact unicategorial. He assumes one universe consisting of entities which can be regarded as values of the variables of the first order. Constant terms substitutable for such variables are just names, each naming exactly one entity. Some of these names name concrete objects, others name abstract classes. Ontological theory like this can hardly be said to be objectionable to the nominalist, if by nominalism we understand the doctrine according to which only names can name objects, but it would come under criticism from those who hold that every entity constitutive of the universe is a concrete object extended in space and time.

In order to differentiate between concrete objects and abstract classes our Platonist would have to make use of appropriate predicates. He could have, for instance, a predicate 'P' to stand for 'is a concrete object' and a predicate 'S' to stand for 'is an abstract class'. Given these predicates the pre-systematic characterisation of his ontology could conveniently find its expression in the assertion:

$$(\exists x) \cdot Px: (\exists x) \cdot Sx: (x) \cdot Px \supset \sim Sx \cdot \tag{8}$$

As regards the syntax of (8) we are still within the framework of the language characteristic of the traditional theory of quantification. And we are still in harmony with our main thesis, which says that quantifying variables of the first order embodies the whole of our ontological commitment. We have only emphasized the role of predicates in making our ontological commitment articulate. A similar point is discussed by Church [3] in his contribution to the problem of ontological commitment.

But let us consider the other interpretation of (7). It presupposes that neither variable in (7) can take the place of the other *salva congruitate*. This would be the case if, for instance, 'ϵ' were to be understood in accordance with the following definition:

$$(\phi): \cdot (x): x \epsilon \phi \cdot \equiv \cdot \phi x \text{'}. \tag{9}$$

In the light of this definition, 'α' in (7) would be appropriate for the place of a monadic predicate, i.e. a proposition-forming functor for one nominal argument, and consequently the quantification involving 'α' in (7) would carry with it no additional ontological commitment over and above the one embodied in the quantification involving the first-order variable

'x'. The expression '$x \in a$' would thus be just a stylistic variant of the more usual 'ϕx'.

There is, however, yet another possibility to be taken into account. Although 'x' and 'α' in (7) may not change places *salva congruitate*, they may have this in common: either may be regarded as suitable to take the place of an argument in a functional expression but neither may be used, meaningfully, in the place of a functor. Thus if 'x' in (7) is a variable of the first order, 'α' too is of the first order, but in a somewhat different sense. While the values of 'x' constitute the universe of concrete objects, the values of 'α' constitute the universe of abstract classes. There is no comprehensive universe of entities to embrace the two. The constant substituends of 'x' are all names, each naming exactly one concrete object. The constant substituends of 'α' are all class-expressions, each naming, in a somewhat different sense of naming, exactly one abstract class. If our Platonist wants us to understand (7) along these lines then his ontology can be described as bicategorial. And bicategorial is his language. For in it expressions of the first order, that is to say expressions which are not propositions and which can only be used, meaningfully, as arguments, never as predicates or functors, divide into two semantic categories — I will call them fundamental — each such category collecting those and only those expressions which are substitutable for one another *salva congruitate*. We can think of a fundamental semantic category as giving rise to its own hierarchy of predicates or functors of various orders but it is only the quantifiable variables of the first order that bear reference to extra-linguistic entities.

The idea of a bicategorial language can be generalized to meet the requirements of more extravagant Platonists. And the upshot seems to be that in the case of a multicategorial Platonist the evidence of ontological commitment is to be sought in his choice of fundamental semantic categories, of which he may need several. Quantifying variables belonging to a fundamental semantic category commits him to a universe of entities of a sort. Introducing several fundamental semantic categories commits him to an ontology with several different universes of entities. But quantifying variables which, with respect to the fundamental semantic categories he makes use of, can be said to be of a higher order, seems to involve him in no additional ontological commitment.

Controversies over ontology have, as a rule, been conducted by philosophers in terms of ordinary language. It is only recently, in connection with the problem of ontological commitment, that symbolic language is beginning to play its part. It is interesting, and not at all surprising, that in ordinary usage we find tendencies which appear to favor the unicategorial conception of language, as well as tendencies which seem to point in the direction of a multicategorial discourse. Thus on the one hand grammarians regard singular names and common nouns as forming one part of speech, which can be justified on morphological grounds. On the other hand we have the view that in propositions of the form 'x is an a' singular names can be used, meaningfully, in the place of 'x' only, whereas the place of 'a' can be filled in only by a common noun, if the result is to be meaningful. This suggests that there is some categorial difference between singular names and common nouns. Since there seems to be no evidence that common nouns are ever used as functors — in this respect they are like singular names — we can think of them as constituting a fundamental semantic category, different from the one to which all singular names belong. The tendency of viewing ordinary language as a multicategorial structure in the sense of the present discussion is paralleled by such ontological, or quasi-ontological, doctrines as, for instance, Aristotle's distinction between entities which, as he puts it, cannot be predicated of other entities, and entities which can be so predicated. His doctrine of primary and secondary substances should also be remembered in the present context.

Here I propose to bring to an end my examination of the two problems arising from Quine's remarks on Leśniewski's logic of 'ϵ'. In conclusion I wish to mention a third problem, which is closely connected with the topic under discussion but exceeds the boundaries of the present paper. The problem is this: how can we give expression to what might be called a negative ontological commitment? How can we say, without contradicting ourselves, that there are no abstract entities? How can we renounce, without contradicting ourselves, the universe of classes or the universe of numbers, or the universe of any other sort of abstract entities?

REFERENCES

1. K. Ajdukiewicz, "On the Notion of Existence," *Studia Philosophica* **4** (1949/1950), published in 1951, pp. 7-22.
2. W. Quine, *J. Symbolic Logic* **XVII**, 141.
3. A. Church, "Ontological Commitment," *The Journal of Philosophy* **55**, 1008-1014 (1958).

DISCUSSION

QUINE: Concerning the quantifiers as Lejewski described them, I would say that in the case of a finite universe if we introduce the quantification by exhaustion as a conjunction of a finite number of cases or an alternation of a finite number of cases, no problem arises. Quantified expressions are simply construed as abbreviations of appropriate connunctions or alternations, and if need be, can be eliminated altogether.

Prima facie, at least, you are talking about some alleged objects when you quantify. However, you might say these objects are a fiction. Of course, that is easy; you just cross your fingers and say anything you like. But there is a sense in which we can say that they are fictions and justify ourselves, namely, say they are merely a manner of speaking and implement this by showing how to define them out, how to paraphrase. In fact, that was Bentham's word, *paraphrasis*, i.e. how to paraphrase in ontological notation. Lejewski shows us how, if you have a finite universe, you have a name for everything in the universe. Then you come to the infinite case and it doesn't work.

Now, it is true you can teach the use of the quantifier by talking of infinite conjunction or infinite alternation, but that is not an explanation, it is not a definition because we don't have infinite expressions. That does not mean there is no such thing as infinity, but there are not infinite expressions.

That would be a matter of teaching the use of a primitive notation of quantification. Unless this is paraphrased in some other way, in turn, this is, I would say, a matter of talking about those things, those individuals. They are assumed, then, in the universe, and this is all assuming them means.

I would say the same thing about predicates. Again, if you had only a finite number of predicates you would eliminate them and quantifiers would be just a manner of speaking. Of course, you can't have just a finite number of predicates because of the fact that you can make compounds.

This would be my answer to the suggestion that it is a dogma to regard the unreduced quantification as commitment to objects. This isn't a dogma as far as I can see, because you have got no further meaning of "there is." It might be a help in explaining things to somebody who is coming from a language in which the whole point of view is very different to talk about infinite conjunction as sort of a limit approached. But this would be an explanation, I would say, not of another sense of quantification but an explanation of "there is."

Now, of course, the ideal in this sort of litigation is to get the benefits of quantifying over abstract entities without being accused of assuming these. Well, being accused of what? What does it mean: assuming them? I can't see it means anything except the unreduced quantification. One needn't care whether it is Platonism or not. I don't see the point of litigation. Just go ahead and quantify — it isn't necessarily bad.

But then there is the question of whether you might paraphrase it some other way and get your quantification over predicates again as an innocent notation. And such a line is suggested again by Lejewski's point that a truth condition of such a quantification, although that is not a paraphrase, i.e. a semantical truth condition, is that the universal quantification over a ϕ is true, just in case every substitution instance with a particular intransitive verb, say, in that position, is true, and correspondingly a truth condition for quantification over an x could be based on substitution of names.

Here is a substitutional concept of quantification. Someone might say, "Well, I mean quantification in Lejewski's sense, I don't mean quantification in an ontological way; I just mean it in a substitutional way, and I haven't given a paraphrase of it, but I have given a sort of semantical description of the truth condition. This is all I mean by quantification."

This, I think, is a tenable position, too, but it should not be regarded as it stands, as eliminating the objects and having just concrete objects. I think this should be regarded, rather, as going into another idiom. Although we are teaching people from scratch how to speak this new idiom, in this case the question of reducing to universals does not arise; it is not that they don't exist: it does not arise. And if, for some reason, someone does want that question to arise, the only way he can raise it is by finding a translation of this notation into the kind of notation that does talk of quantification in an ontological way.

There may be various ways of doing this. If so, then the imputation of ontological commitment that he arrives at will be relative to that particular translation manual which he uses. It will be relative to that as a parameter and it might run in some such way as this. A Lejewskian substitutionally construed quantification can be translated into a partly semantical form of our language by saying the sentence form, schema, with a ϕ in it, is true for every general term, shall we say, or predicate or intransitive verb substituted in the position of ϕ, and here we have the word 'true'. This would have to be, in our language, explained by some sort of Tarskian construction of the truth theory. Then we would examine this whole system of translation, see how it stands in our language and what things are needed, and we would find perhaps that we were quantifying over some things in our ontological sense of quantifying, and these things were perhaps individuals and also expressions, strings of signs, which are perhaps abstract entities, or perhaps we could work the things around so the only things we had to quantify over at the expressional side were inscriptions, again concrete objects. Then we might find we were accommodating the whole Lejewskian logic in a logic of ontologically construed quantification in which, after all, there were no universals assumed; we get concrete objects in the universe. Can we in that logic, which we have nominalistically legitimated in this way, construct all of classical mathematics? I think this is very doubtful. If we could, then we would have a nominalistic solution of the problem, but not just by simply departing from the ontological notion of quantification.

There was a mention of "protothetic" where Leśniewski quantified over truth functions as values of variables. Now, this can be legitimized by the straight Benthamite fictional approach because for a given number of arguments there are only a finite number of truth functions. You can exhaust that and you can introduce quantifiers over truth functions and over propositions as fiction. So there is no problem here.

LEJEWSKI: I agree with Quine in that talking about infinite conjunctions or infinite alternations of quantified expressions does not amount to a definition of the quantifiers. I regard the latter as primitive concepts, and my reference to expansions is a mere teaching device. It is often used by logicians in connection with quantification involving variables of the first order. I tried to make use of it in an attempt at clarifying the notion of quantification involving variables of the second order. It is this sort of teaching device that helps me to understand why I am justified in construing the quantifier '($\exists X$)' as meaning the same as 'there exists an individual x such that', and, more generally, why quantifying name variables is a way of referring to individuals. But I fail to see comparable justification for construing '($\exists \phi$)' as meaning the same as 'there exists an attribute ϕ

such that'. We all seem to agree that asserting an expression of the form '*Fa*' where '*a*' is a name of an individual and '*F*' an intransitive verb or a predicate expression such as 'is-wise' or 'is-a-philosopher', commits us to accepting a universe of individuals. We also seem to agree that it does not commit us to accepting a universe of attributes or classes. Now, I construe quantification in such a way as to render it compatible with the validity of the following two inferences: (1) *Fa*, therefore $(\exists x).Fx$, and (2) *Fa*, therefore $(\exists \phi).\phi a$. Inference (1) presents us with no problem, but if '$(\exists \phi),\phi a$' presupposes a universe of attributes or classes, then inference (2) must be viewed as invalid since the premiss '*Fa*' presupposes no such universe. Instead of rejecting inferences of this latter sort I prefer to abandon the traditional reading of '$(\exists \phi)$' as 'there exists an attribute ϕ such that' and free the expression '$(\exists \phi).\phi a$' from any ontological commitment other than the one resulting from the use of the name '*a*'.

Quine refers to the ontological concept of quantification as distinct from the substitutional one. I am not quite certain whether I understand the distinction correctly. In any case, in terms of truth conditions I would describe '$(\phi).\phi a$' and '$(\exists \phi).\phi a$' as follows: (1) if '$(\phi).\phi a$' is true then so is every substitution instance with an intransitive verb of a predicate expression in the place of 'ϕ' in 'ϕa', and (2) if a substitution instance with an intransitive verb or a predicate expression in the place of 'ϕ' in 'ϕa' is true then '$(\exists \phi).\phi a$' is also true. But as Quine acknowledged, this is not a paraphrase of '$(\phi).\phi a$' or of '$(\exists \phi).\phi a$', nor is it meant to be one. However, it seems to me that a reference to truth conditions of quantified expressions may help us to understand the meaning of the quantifiers, just as truth tables help us to understand the meaning of propositional connectives.

A multicategorial or many-sorted language is certainly awkward, but it seems to be a possible answer to the problem of codifying a multicategorical ontology with its irreducible levels of being.

YOURGRAU: I don't see why Lejewski is so very anxious to avoid what we call the ontic commitment (I would prefer the term 'ontic' to 'ontological'). If one is a realist, then one has a definite viewpoint, and I think if one intends to perform quantification over lower-order variables or higher variables, one does so *before* one applies one's quantification. Hence, I submit that one should say, "This is my ontic commitment, here is where I stand," and then we have some definite rules on which to go. But the issue cannot be avoided by simply saying, "Look here, it doesn't matter whether I talk about zoom-zooms, angels, pebbles, billard balls, Queens of England—it really doesn't matter: I get very nice, neat quantification formulae out." We must have some sort of ontic commitment and we should not merely logicalize it away.

LEJEWSKI: I am not anxious to avoid ontological commitment at all. What I am anxious to do is, on the one hand, to show that by quantifying variables of various orders within the framework of the traditional theory of quantification with identity I do not necessarily commit myself to entertaining any ontology other than that which presupposes the existence of individuals. On the other hand, I want to explain what sort of language I would need if I were to commit myself to a Platonistic ontology of one kind or another.

As far as I am concerned, the problem of avoiding ontological commitment does not arise. For my own purposes as an ontologist, the traditional theory of quantification with identity will do. I only argued that notwithstanding my quantifying variables of any order I commit myself to a universe of individuals and to nothing else. Such an austere ontology, however, is objectionable to Platonists of various persuasions. Hence, for me as a logician, the question arises of the nature of a language which commits one to a Platonist ontology.

Or indeed, to none. This last is closely connected with the problem I mentioned in the last paragraph of my paper, and which takes the form of the following questions: how can one effect a 'negative' ontological commitment? How can one say, without contradicting oneself, that there are no attributes, no classes, no numbers? This seems to be an important problem, and it is because it has not yet been solved that we talk about avoiding ontological commitment, say, to a universe of attributes rather than talking about denying the existence of attributes. For if I concede, for the sake of argument, that '$(\exists\phi).\phi a$' does commit me to accepting a universe of attributes, and if I now try to register my negative commitment by asserting '$\sim(\exists\phi).\phi a$' then I find that my assertion implies '$(\phi).\sim\phi a$', which in turn implies '$(\exists\phi).\sim\phi a$', which again commits me to a universe of attributes against my original intention. Thus it appears that on the Quinean interpretation of the quantifiers we either commit ourselves to an ontology or avoid the commitment by restricting our use of quantification. The denial of an ontology does not seem to be possible. The sentence 'there are no attributes' cannot be translated, so it seems, into the language of the traditional theory of quantification. And if this is in fact the position then, I would say, the customary way of interpreting the quantifiers calls for re-examination.

KAPLAN: I think there probably is agreement between Lejewski and Quine; there are two different interpretations of quantification, neither of them formally inadequate. But I think there may be some disagreement as to the extent to which results that can be obtained using one style of quantification can be duplicated by using the other style.

Rather than going directly to the case of quantification on predicates, I want to discuss the question of a similar kind of quantification on individual variables. I wish to consider this kind of quantification, now, on individual variables: $(\exists x)Fx$.

Before considering that, let's consider these formulas with the individual constants in them: Fa. How are we to treat them? Are we to take the individual constant as making an ontological commitment? Well, I am not sure. People are now exploring languages in which the individual constants don't make any ontological commitment. It seems that there are lots of names in English, for example, that don't contain any ontological commitment, names for which just this particular law $Fa \supset (\exists x)Fx$, would fail. Now, using the superscript 's' for the substitution sense of the quantifier, the law $Fa \supset (\exists^s x)Fx$ will certainly be true. In fact, I think the best way of understanding the substitution sense of the quantifier is as follows: We should here adopt Quine's device of semantic ascent and say that the sentence '$(\exists^s x)Fx$' should be understood in this way: not that 'There is some individual x which runs' is true, but rather, there is some individual constant α such that the result of writing α before the verb 'runs' is true. I am using referential quantifiers in the metalanguage here, of course.

So in the substitutional sense it is clear that if the sentence 'Fa' is true, then there is some individual expression, namely 'a'; so '$Fa \supset (\exists^s x)Fx$' is always true.

Now, what about '$Fa \supset (\exists^r x)Fx$'? Well, what about 'If Medusa had snakes for hair, then there exists (in the referential sense) some individual which had snakes for hair'. Clearly this is false, because the antecedent is true and the consequent is false. It depends, of course, on allowing individual expressions or names in our language which don't carry any ontic commitment with them.

Yet Lejewski said clearly that there is no ontic commitment in '$(\exists\phi)(\exists x)\phi x$' because it follows from '$(\exists x)Fx$'. Now, '$(\exists^s\phi)(\exists x)\phi x$' follows from '$(\exists x)Fx$', of course, in the substitution sense; that is quite clear. But whether it follows using the referential quantifier '$(\exists^r\phi)$' depends on whether or not there exists something that is denoted, designated, named (I am not sure what semantic relation to use) that stands in some relation to the predicate 'F'.

So I think there are these two different senses of quantification. Both I think are interesting and worthy of investigation.

Now, I want to add this point. If we try to avoid ontic commitment by sticking to the substitution sense of quantification (I want to emphasize the fact that as far as I can see, it doesn't matter whether we are talking about quantification on predicates or individual constants; it is the quantifier, not really the style of variable that makes the difference) then we lose something, and that, it seems to me, is the critical issue. Can we get everything using the substitution sense of quantification that we can get by using the referential sense of quantification? I think not, and this becomes clear if we consider those entities that we may want to quantify over, which are such that we do not have names for all of them.

Take the real numbers; take the sets Lejewski suggested. The difficulty is that there may be an infinite number of them and so there are an infinite number of names; thus the quantifier '$\exists^s$' could not be introduced by a definition within the language. I don't think it makes an important difference whether or not we can write out a sentence within the language which defines this quantifier. The crucial thing is the interpretation of the quantifier.

The case that we have been talking about is the case of quantification over predicates. Quine has referred to the definition of 'ancestral'. Let us then take the substitution interpretation of the predicate quantifier so that the sentence which defines the ancestral of the relation 'parent of' is understood as saying something like this: 'x is an ancestor of y, if, and only if, it is not the case that $(\exists^s\phi)$ (ϕ is heredity, ϕx, and not ϕy)'.

Now, the situation is just this. The crucial predicate, the crucial relation that we need to refer to in order to get this definition to work is not a relation which has a name in this language. So, if we adopt the substitution interpretation (which simply says that 'ϕ is hereditary, ϕx, and not ϕy' will be false for all substitutions of 'ϕ', where we substitute some relational expression which lies within the language), we are not going to get defined what we want defined.

LEJEWSKI: Concerning the extent of the "agreement in essentials" between Quine and myself, the positions seem to me to be this. There is no difference in interpreting quantification which involves variables of the first order in the traditional theory of quantification with identity. We also seem to agree that an expression of the form 'Fa', where 'a' is a constant name designating an individual and 'F' a constant intransitive verb or a predicate expression, presupposes a universe of individual objects only and commits us to no universe of abstract entities. But there is no agreement as regards the import of quantifying variables of higher orders. It is not simply the case that I offer an alternative interpretation of '$(\phi).\phi a$' or '$(\exists\phi).\phi a$'. I go in fact a little further, so it seems to me, and question the rationale of the view that these two formulae mean, respectively, the same as 'each attribute ϕ is such that. . .' and 'there is an attribute ϕ such that . . .'. In other words, I question the rationale of the view that these formulae commit us necessarily to a universe ot attributes.

I think that I am now a little clearer in my mind about the distinction Kaplan and Quine make between the ontological concept of quantification and the substitutional concept of quantification. Kaplan seems to suggest that an expression such as '$(\exists x).x$ runs', for instance, can be understood in two different ways, namely, it can be understood to mean the same as (1) 'there exists an individual object x which runs' or the same as (2) 'there exists a constant name which, followed by an occurrence of "runs" forms a true sentence'. Kaplan also seems to hold that I favor the latter interpretation rather than the former. Well, to begin with, I do not regard (2) as an acceptable rendering of '$(\exists x).x$ runs'. It is, if you like, a metalogical analogue of (1), and can be said to give truth conditions of (1). I would say that the concept of quantification as advocated in my paper is ontological. I use quantification, without restricting it to variables of the first order, for the purpose of saying something about individual objects in the widest sense of the term and not for the purpose of making meta-statements about an object language.

However, Kaplan's reference to Medusa seems to indicate that he may have in mind an alternative ontological concept of quantification, which by having recourse to the idea of expansions could be characterized as follows: we assume that the expansion of '$(x).Fx$', and that of '$(\exists x).Fx$', contains a conjunct, or an alternant, of the form 'Fa', where 'a' is a name-like expression which does not designate anything. On this assumption quantifying variables of the first order loses its existential import, and we can no longer read '(x)' and '$(\exists x)$' as 'each individual object x is such that' and 'there exists an individual object x such that', respectively. Moreover, if we interpret quantification on these lines, we have to abandon the law of absolute reflexivity for singular identity as it is normally understood. We can, however, define, with the aid of singular identity, the concept of weak identity, which is reflexive but too weak for the purpose of defining singular identity in turn. In the traditional theory of quantification, '$(x).x$ exists' is a tautology and '$(\exists x).x$ does not exist' is a contradiction, the statement '(x): exists. $\equiv .x = x$' serving as the definition of the concept of existance. It is the other way round if the alternative concept of quantification, as I understand it, is adopted. Personally, I am in favor of this concept of quantification. It gives rise to a different type of language, which has its advantages. Both types of language, that of the traditional theory of quantification and that which accommodates empty names, seem to me to be adequate for the purpose of an ontology with individual objects only.

Now, to return to Kaplan's main question, which I understand to be this: does the metalogical analogue of the axiom of induction in arithmetic yield as much as the axiom itself? I doubt that it does, unless we are allowed to assume a nonfinite supply of numerals and predicate expressions. It is perhaps a little risky to try and give a decisive answer without further investigation. In any case, the problem seems to be outside the scope of my paper, since in it I do not interpret the quantifiers as presupposing an object language. Nor am I prepared to accept that '$(\exists \phi)$ . . .' means the same as '$(\exists \phi).\phi$ is-a-predicate-expression. . .'.

MERCIER: The physicist is looking for a's and for ϕ's. Perhaps he is also doing something else. For example, he is doing experimental research.

Then we may ask: are there a's, are there objects, and if yes, which ones are there? Further, are there ϕ's, and ones that make sense in some orderings that I can call physical theories? Around these questions, some difficulties arise.

First, are there a finite number of a's, and are there a finite number of ϕ's? Probably not — I say probably not, though I am not sure how to interpret the word 'probably' — because I cannot think of an end to the possibility of constructing physical entities.

Second, are there infinitely many? I doubt that, insofar as physics aims at dealing with concrete objects, though not strictly concrete *individual* ones, but general models of such. Wolfgang Pauli said once that there are probably not infinitely many things in the world, and he was a very good physicist. I dare say he meant concrete individual objects. However, are the objects of physics concrete individual objects, or are they constructs of our mind to which we attach at least the meaning of being models of such concrete objects?

Take for instance the case of statistical mechanics, which considers ensembles of elements as models of real bodies. There, the question of finiteness or of infiniteness is 'neither-nor', because it posits a problem of innumerability, which is something different from, or between, finiteness and infinity. Can that be included in some kind of scheme that would make sense from the logical point of view?

Now, certainly the objects of physics proper are a's that we might name, but which are never the final objects existing in the real world because they undergo epistemological changes in the course of the history of physics; they are objects which we initiate and eventually reduce to other ones. It can therefore happen that the same name is applied to objects of various or of variable meaning. Hence, physical objects are abstract entities.

But even if we talk about the same invariable physical objects and name them in the same way, the properties ϕ we apply to them change when a new theory is substituted for an old one concerning these objects. For example, space and time remain space and time when passing from the Galilean to the Lorentz group, and these groups are ϕ's on or about space and time.

Anyway, suppose physics would at least consist of naming these a's and the ϕ's and this procedure would not lead to ambiguities, my further questions would be — and there I can but make a guess for myself, for I am not a historian — "What is history doing?" "Does history look for a's and/or for ϕ's (and what is the concreteness of such a's) or does history do something else?"

It seems to me, at least, that the fundamental difference between physics and history is to be found in the fact that whereas history is looking for ever so many particulars as can be found, physics does look for universals that appear to be verified (or not to be falsified — in accord with Popper's view) by classes of particulars. Therefore physics is a science, whereas history is not.

HINTIKKA: I don't think I have seen the substitutional sense ever explained in a philosophically satisfactory way, that is to say, in a way that would enable me to understand what it would mean to use in actual practice a language in which quantifiers are only interpreted in terms of substitution.

In addition, I haven't seen iterated quantification explained in terms of substitution in a satisfactory manner. This might be forthcoming, but there seem to be reasons to be sceptical about that.

Further, I believe that the substitutional interpretation of quantification simply misses everything interesting that can be said about second-order logic. The possibility of formulating a complete system of arithmetic in the second-order logic is just one of many things that we can do if the ordinary nonsubstitutivity sense of quantifiers is presupposed. Assuming a standard semantics, we can formulate in a second-order language, I would say, most of the interesting things in the foundational studies, e.g. the axiom of choice, the axiom of reducibility, etc. Surely we want to be able to approach these assumptions and to discuss them, and second-order logic might even be the most interesting way of doing so.

Finally, I think the problem of the different kinds of quantifiers and their ontic commitments is a highly complicated one. In studying the logic of our perceptual terms, I have discovered differences between different kinds of quantifiers, differences which I would argue are even imbedded in our ordinary conceptual system and which obtain although the quantifiers in question in some sense involve the same ontic commitments. The obtain because the quantifiers themselves are logically not on a par.

POPPER: Logicians have found that the ordinary languages we use lead into trouble: They imply (though we are rarely aware of this) the existence of everything we speak of; whether it is God or the Devil, or whether it is Achilles or Pallas Athene. This is one of the reasons why logicians have tried to reform language; but they are still quarreling about the best way to reform them, and to reform them in such a way that, on the one hand, one can say that I assume such and such an ontology, and that, on the other hand, one can also *discuss* ontologies. Thus, if Adam has the ontology *alpha*, and Benjamin has the ontology *beta*, there ought to be a common (and ontologically neutral) language in which the quarrel between Adam and Benjamin can be properly and rationally thrashed out.

Such a neutral language has not been quite satisfactorily established; at least not to the satisfaction of some logicians. I think that is the main point of the present problem.

LEJEWSKI: The problem raised by Mercier appears to be that of ontology. As Quine has pointed out in one of his papers, it reduced to the question "What is there?" Now, the problem of ontological commitment is different. In ordinary language we commit ourselves to an ontology whenever we assert or imply existential statements such as the following ones: there are individuals, there are attributes, there are classes, there are relations, there are numbers, etc. Can these statements be adequately translated into the standardized language of the theory of quantification with identity? If so, how? If not, what sort of standardized language would be sufficient for the purpose? It is questions like these that constitute the problem of ontological commitment. By solving it we may be able to contribute to the clarity of ontological theories, but there is no claim that, by the same token, solutions will be found to ontological problems.

I would like to reiterate my interpretation of the quantifiers. First of all, my interpretation of the quantifiers binding variables of the first order is ontological. I interpret '$(x) \cdot fx$' as meaning the same as 'each individual object x is such that Fx'. I justify this interpretation by referring to the expansion of the formula, and I characterize the meaning of the formula in metalinguistic terms, by saying that every substitution instance of 'Fx' is true provided '$(x) \cdot Fx$' is true.

Concerning '$(\phi) \cdot \phi a$' I also say, at the metalinguistic level, that every substitution instance of 'ϕa' is true provided '$(\phi) \cdot \phi a$' is true; I give the expansion of the formula, and I find that the expansion does not justify the usual interpretation of '$(\phi) \cdot \phi a$' as meaning the same as 'each attribute ϕ is such that ϕa'. What appears to be warranted by the expansion is an ontologically weaker interpretation, an interpretation which, without ceasing to be ontological, does not commit us to an ontology with attributes. I have, therefore, suggested that until a better rendering is found, '$(\phi) \cdot \phi a$' should be read as 'for all $\phi, \phi a$'.

As far as I can see, there is no difficulty in explaining iterated quantification in terms of appropriate expansions.

Chapter XI

POSSIBLE INTERNAL SUBQUANTUM
MOTIONS OF ELEMENTARY PARTICLES

Jean-Pierre Vigier
Institut Henri Poincaré

Most of the past discussions of quantum mechanics have concentrated on the behavior of matter under the influence of the so-called long-range fields (with singularities of the Coulomb type), namely the gravitational and electromagnetic fields. Indeed, Einstein's last attempts were devoted to the construction of a unified geometrical description of their effects at the classical level.

In this paper, I will make an effort to present the essential results of the most modern type of high-energy physics: the theory of elementary particles. Such a step is indeed indispensable for two fundamental reasons. First, it is clear that only within this field can we look for experimental confirmations (or infirmations) of any assumption we can make on the deep nature of matter itself (see, for example, de Broglie [1]). Second, because all discussions of the epistemological problems raised by the development of physical theories within the last fifty years (they have been more important and deeper than ever before in the history of human knowledge) must, in my opinion, keep abreast in principle with the progress of the most advanced type of physical theories. If that were not the case, our discussions would soon resemble the bar talk of retired war veterans. They would turn into a Byzantine exchange of arguments on dead problems foreign to the preoccupations of the 'plumbers'* – the men who really raise (and solve) the most advanced types of philosophical problems of our world: the scientists themselves.

In my opinion, the most striking fact in the recent period has been the appearance of new quantum numbers associated with the new short-range types of fields (strong and weak interactions), introduced to describe the behavior of matter in the high energy domain. Indeed, along with the "old" quantum numbers such as electromagnetic charge e, mass (defined

---

*Yourgrau, who coined this label, seems to enjoy not only rational controversy, but even polemical, plainly crude, terminology. Well, this is perfectly alright with me!

by $P_\mu P^\mu = -\mu_0^2$, P_μ being the 4-momentum), and spin J, we have been obligated to introduce "new" quantum numbers such as T (isobaric spin), Y (hypercharge), and B (baryon number), all related to new conservation laws which have been discovered in the fields of strong interactions.

This immediately raises new fundamental problems:

A. What is the physical nature of the new fields? Is it possible to "geometrize" them as Einstein had done for the old long-range fields?

B. What is the meaning of the new quantum numbers? Can they be associated with the quantization of some dynamical behavior? Or should we interpret them in a completely new way?

Note here that (still in my opinion) one of the most mysterious aspects of modern physics is encountered in the existence of empirical relations which connect the old with the new quantum numbers. Indeed, we have

$$Q = T_3 + \frac{Y}{2},$$

the Nishijima-Gell-Mann relations, and

$$m = a + bY + c\left[T(T+1) - \frac{Y^2}{4} \right],$$

the Gell-Mann-Okubo formula; though the second (as is well-known to physicists) is not as satisfactory as the first.

This suggests that there must be something in common between them, and one of the essential theoretical problems of our time is precisely to justify and understand this connection.

C. If one accepts the idea that quantum numbers are associated with dynamical behavior (so that the new quantum numbers are somehow connected with a dynamical behavior at a deeper level, that is, with strong interaction regions with radii $r \approx 10^{-13}$ cm), then we must raise the problem of the connection of the old external invariance groups (associated with the long-range forces), such as P (Poincaré) or de Sitter SO(4,1) or U(1) (electromagnetic gauge invariance), with the new strong invariance groups such as SU(3).

This raises two subquestions:

C_1. Until now, one always assumed that the global symmetry group could be written in the typical form P × SU(3) (× denoting the direct product) which implies that the operator $P_\mu P^\mu$ commutes with the SU(3) multiplets which are thus degenerate in mass. This entails that, if one wants to remove this degeneracy, one should break this symmetry — a very disturbing fact since conservation laws result generally from the existence of exact symmetries.

Question C, however, raises the necessity of a possible unification [2] (at least at the level of the Lie algebras) of external [P or SO(4,1)] with a non-compact internal symmetry group [such as SU(2,1)] within the usual conform group SU(2,2) to obtain a mass splitting without symmetry breaking.

C_2. If we want to construct a new internal dynamic of such a unifying group [SU(2,2), for instance], one should try to consider it as an isometry group of motion of some Riemannian manifold which will be considered as a geometrical description of the internal space-time of elementary particles.

This means that we introduce the following assumptions:

a. External and internal motions of elementary particles are isometric groups of motions on different external and internal 5-dimensional curved Riemannian manifolds (Lichnerowicz's domains of isometry[3]).

b. These domains are connected through an isometry boundary (time-like hypertube built with trajectories of the internal motions) and geodesics of the external domain. On the boundary, we unify (in the sense of M. Flato *et al.* [2]) the external Weyl group SO(4,1) $_\times$ U(1) with the internal motion group SU(2,2), which contains our new strong group SU(2,1).

This extends Einstein's general relativity ideas within elementary particle theory. Particles are now considered as extended regions with very strong internal curvature connected with 5-dimensional external-event space-time of the Einstein-Kaluza-Klein type [3].

Let us now come back to the question of the isometry groups of motions. As is known, these groups (defined on Riemannian manifolds), due to the fact that the Lie derivatives of the $g_{\mu\nu}$ vanish ($g_{\mu\nu/\sigma} = 0$) along the corresponding trajectories, have aroused a wide interest for three main reasons:

1. In any curved space-time, the invariance of natural laws under the change from one observer frame to another implies that such changes correspond to isometric transformations.

2. The choice of a given group of motion on a given manifold entails the choice of a well-defined dynamical group, which in turn (when suitably quantized) implies the conservation of specific quantum numbers associated with constants of the said motions.

3. This choice also determines the field equations on the $g_{\mu\nu}$ and the form of the metric, so that the determination of an empirical dynamical group of motions appears closely related to the structure of space and time.

I shall now develop some possible applications of these ideas. Before I do this, however, I want to recall that the general mathematical theory of groups of motions on Riemannian manifolds has been extensively studied by mathematicians (such as Fubini [4], Cartan [5], Vranceanu, Egorov [6], Ishihara [7], Lichnerowicz, and Yano), so that the physicist's work in this field is mainly a problem of physical interpretation of known mathematical results.

For example, on a V_4, depending on the choice of the stability subgroup, we get:

(1°) 10-parameter groups of motions G_{10} (that is, G_r of order $r = 10$), namely, SO(5), SO(4,1), SO(3,2), ($R_4 \times T_4 = P$). The corresponding Riemannian spaces have a constant curvature. This results immediately from the fact that in a V_n the maximum order of a possible group of motion G_r is $r = n(n + 1)/2$, which is only reached when V_n is of constant curvature.

(2°) Since no G_r with $r = 9$ exists, the next set is a set of G_8's, namely, SU_3, SU(2,1), $T_{2c} \times U_2$, where T_{2c} denotes translations in a 2-dimensional complex space.

Both sets (1°) and (2°) correspond to Einstein-spaces ($R_{\mu\nu} = \Lambda g_{\mu\nu}$) with and without constant curvature [8].

Of course, similar classifications can be performed for $n > 4$. For $n = 5$, for example, we discover, among others, in categories (1°) and (2°) the groups $G_{15} = SU(2,2) \approx SO(4,2)$, the Weyl groups $G_{11} = P \times A_0$, and $G_{11} = SO(4,1) \times A_0$ (which can be used in Kaluza-Klein or Jordan-Thiry types of theories), the corresponding manifolds being also Einstein-spaces with or without constant curvature.

I shall now summarize two concrete examples of application of the preceding considerations to elementary particle theory.

The first has been proposed by Raczka [9].

1. If we observe that the introduction of gravitational masses in an empty space gives rise to gravitational interaction which changes the geometry of the space itself, then we may expect that the strong interaction of elementary particles also changes the geometry of an interaction region.

In order to apply this idea to elementary-particle theory, Raczka then chooses for this interaction region a V_4 manifold invariant under the group $G_4 = T_{x_0} \times T' \times R_3'$ in order to have the usual time-like displacement, the V_4 being compact with respect to three space variables. In that case, V_4 is homeomorphic to the direct product of a straight line (time) L, 1-dimensional torus S_1, and 2-dimensional sphere S_2. The ds^2 of this space is:

$$ds^2 = -dx_0^2 + R_T^2 \, d\alpha^2 - R_s^2 \, (d\beta^2 + \sin^2\beta \, d\gamma^2).$$

On this V_4, the Laplace-Beltrami equation,

$$\left[\frac{1}{\sqrt{|g|}} \partial_\mu \, g^{\mu\nu} \sqrt{|g|} \, \partial_\nu \right] \Psi = \chi^2 \Psi$$

where $g_{\mu\nu}$, g, and $\partial_\nu = \partial/\partial\nu$ represent the usual symbols, becomes:

$$\left[-\frac{\partial}{\partial x^2} + \frac{1}{R_T^2} \frac{\partial^2}{\partial\alpha^2} - \frac{1}{R_s^2 \sin\beta} \left(\frac{\partial}{\partial\beta} \sin\beta \frac{\partial}{\partial\beta} + \frac{1}{\sin^2\beta} \frac{\partial^2}{\partial\gamma^2} \right) \right] \varphi = \chi^2 \phi.$$

Its solutions can be developed on a set of orthonormal functions:

$$\Psi = e^{iEx_0} \, e^{iYa} \, Y_{TT_3} \, (\beta,\gamma),$$

subject to the condition that E (energy), χ, Y (hypercharge), and T (isobaric spin) satisfy the relation:

$$E^2 = Y^2 - \left(\frac{1}{R_s^2} \right) \left[T(T+1) - \left(\frac{R_s}{R_T} \right)^2 Y^2 \right]$$

which is just Okubo's mass formula for bosons if we take $R_s/R_T = \frac{1}{2}$ and $R_s \simeq 2/5m_\pi$. This very simple calculation is interesting because we obtain a mass formula without any breaking of symmetry. Its main defect is, of course, that we only obtain integer values of T on S_2. However, one could remedy this defect by passing to a V_5 and a G_5 such that $G_7 = T_{x0} \times T'_1 \times SU(2)$ with $V_4 = R \times S_1 \times S_3$.

The second, more developed example, which I want to discuss at more length here, is an attempt to represent geometrically the unification procedure. The external dynamical group $SO(4,1)$ and internal symmetry group $SU(2,1)$ are both considered to be isometry groups of motions on a V_5 which represents an internal space-time denoted by Σ_5. We "fit" Σ_5 on an external space-time V_5 invariant under the Weyl group $SO(4,1) \times U(1)$, which corresponds to the combined effects of gravitation and electromagnetism.

If we recall a well-known result of Kobayashi, we see that the 5-dimensional hyperbolic sphere Σ_5 can carry isometric transformations which yield the Lie algebra of the locally isomorphic groups $SO(4,2)$ $SU(2,2)$. That is, a G_{15}.

This Σ_5 (our new internal space-time) can be considered as a curved 5-dimensional surface imbedded in a 6-dimensional space $E(4,2)$ with a metric $ds^2 = g_{\alpha\beta} dx^\alpha dx^\beta$ ($\alpha, \beta = 1, \ldots, 6$), that is, $ds^2 = dx_1^2 = dx_2^2 - dx_3^2 - dx_4^2 + dx_5^2 - dx_6^2$. Indeed, we can parametrize it with

$$\begin{aligned}
x_1 &= \text{ch } \theta \cos \zeta \sin \Phi_2 & x_2 &= \text{ch } \theta \cos \zeta \cos \Phi_2 \\
x_3 &= \text{ch } \theta \sin \zeta \cos \Phi_3 & x_5 &= \text{ch } \theta \sin \zeta \sin \Phi_3 \\
x_0 &= \text{sh } \theta \cos \Phi_1 & x_6 &= \text{sh } \theta \sin \Phi_1
\end{aligned}$$

with $-\infty \leqslant \theta < \infty$, $0 \leqslant \zeta < \frac{\pi}{2}$, $0 \leqslant \Phi_i \leqslant 2\pi$ $(i \neq 1,2,3)$, the surface being defined by the relation

$$x_1{}^2 + x_2{}^2 + x_3{}^2 - x_0{}^2 + x_5{}^2 - x_6{}^2 = 1.$$

We can obtain the following results:

(a) The ds^2 of the Σ_5, which can be written

$$ds^2 = -d\theta^2 - \mathrm{sh}^2\,\theta\,d\phi_1{}^2 + \mathrm{ch}^2\,\theta\,(d\zeta^2 + \cos^2\zeta\,d\phi_2{}^2 + \sin^2\zeta\,d\phi_3{}^2).$$

Σ_5 is an Einstein-space with $R_{\mu\nu} = (4/R^2)g_{\mu\nu}$.

(b) The exact form of the second-order wave equation, namely, the Laplace-Beltrami operator

$$\Delta_5 = -\frac{1}{R^2\,\mathrm{ch}^3\,\theta\,\mathrm{sh}\,\theta}\frac{\partial}{\partial\theta}\left(\mathrm{ch}^3\,\theta\,\mathrm{sh}\,\theta\frac{\partial}{\partial\theta}\right) - \frac{1}{R^2\,\mathrm{sh}^2\,\theta}\frac{\partial^2}{\partial\phi_1{}^2}$$

$$+\frac{1}{R^2\,\mathrm{ch}^2\,\theta}\left[\frac{1}{\sin\zeta\cos\zeta}\frac{\partial}{\partial\zeta}(\sin\zeta\cos\zeta)\frac{\partial}{\partial\zeta} + \frac{1}{\cos^2\zeta}\frac{\partial}{\partial\phi_2{}^2} - \frac{1}{\sin^2\zeta}\frac{\partial^2}{\partial\phi_3{}^2}\right],$$

which can be solved explicitly. Its solutions correspond to internal quantized states of motions labelled, as we shall see, by the observed quantum numbers of strong interactions.

(c) The explicit for $M_{\alpha\beta}$ of the fifteen generators $M_{\alpha\beta} = x_\alpha\partial_\beta - x_\beta\partial_\alpha$ $(\alpha, \beta = 1,...,6)$ of the isometry group of motion $G_{15} = SO(4,2) \cong SU(2,2)$.

Among its subgroups evidently appear its strong interaction subgroup $SU(2,1)$ as well as $SO(4,1) \times U(1)$.

Its maximal compact subgroup $SU(2) \times SU(2) \times U(1)$ contains two $SU(2)$ groups, which one can associate with spin J and T spin, the $U(1)$ corresponding to hypercharge Y.

(d) The explicit form of the generators of $SO(4,1)$ and $SU(2,1)$ and the linearized internal Dirac-like wave equation $\Gamma_\alpha\Gamma_\beta M^{\alpha\beta}\Psi = \chi\Psi$ [with $(\Gamma_\alpha, \Gamma_\beta)_+ = 2g_{\alpha\beta}$]. Evidently the mass^2 operator [casimir of $SO(4,1)$] splits the $SU(2,1)$ multiplets and one obtains a mass formula for baryons of the type

$$m^2 = a + bY + cT(T+1), \quad (a,b,c \text{ constants})$$

(e) The explicit form of the isometry boundary Σ_5 (a de Sitter space) on which the $g_{\mu\nu}$ of constant curvature internal space-time (satisfying $R_{\alpha\beta} = \Lambda g_{\alpha\beta}$) are homothetic to those of the external space-time (satisfying $R_{\alpha\beta} = \lambda g_{\alpha\beta}$ but with non-constant curvature; the ratio Λ/λ is related to the fundamental length l).

(f) The Σ_4 can be considered as the wall of a $g_{\mu\nu}$ well which contains discrete levels of energy (mass) corresponding to different types of elementary particles. In this picture bosons appear as massive quanta when one jumps from one internal level of energy to another.

(g) Finally, the "quarks" of such a picture appear as independent massive (with a mass $m_q \approx M_n/3$, M_n = nucleon mass) particles combining in this well to form the observed baryons or resonances.

Two remarks to conclude. The first (if such a line of research should be confirmed) is that the new subquantum level of matter must contain new parameters, associated with the new quantum numbers, so that the "old" quantum description would not be "complete" as assumed by its promoters.

The second remark is that the existence of such internal structures would corroborate the "level" picture of matter proposed by de Broglie, Bohm, and others. If the point-like character of a material particle vanishes, we are forced back to Pascal's profound view. Pascal had, indeed, foreseen the discovery of an inexhaustible set of new properties of matter as we descend step by step the ladder of distances which propel us upward into higher and higher energy domains.

REFERENCES

1. L. de Broglie, *Une tentative d'interprétation de la mécanique ondulatoire*, Gauthier Villars, Paris, 1956.
2. M. Flato, D. Sternheimer, and J.P. Vigier, *Compt. Rendus* **260**, 3869 (1965).
3. A. Lichnerowicz, *Théorie globale des groupes d'holonomie*, Cremonese, Roma, 1962.
4. G. Fubini, *Annali. di Mat.* **8** (3), 39 (1903).
5. E. Cartan, *Bull. Soc. Math. France* **54**, 215(1926).
6. I.Egorov, *Dokl. Akad. Nauk SSSR* **9**, 103(1955).
7. S. Ishihara, *Jour. Math. Sci. Japan* **7**, 345 (1955).
8. A. Z. Petrov, *Einstein Spaces*, Phys. Math. Editions, Moscow, 1961.
9. R. Raczka, Preprint, Trieste, I.C./65/32.

DISCUSSION

YOURGRAU: "Stupid people never listen and clever people never talk." Well, I do not wish to be stupid, but I have to talk because I think the arguments presented by Vigier have one tremendous advantage: They are elegant — they are also highly dubious, not merely controversial.

The way he treats theoretical physics reminds me of a guild in the Middle Ages. He says, "we physicists"; one is reminded of "we barbers" or "we plumbers."

VIGIER: But physics *is* a trade!

YOURGRAU: I thought Vigier adopted rather the attitude of a magician. He admits that for the new physics our household, everyday language is not rich enough. We must develop a new vernacular, and the trouble is that we create formalisms, we talk 'technicalese', we use jargon. But when we try to explain our formalism in ordinary language, we fail and resort once more to a formalism. In other words, we escape into the abstract, imaginative realm of the mathematician.

I am afraid our friend Vigier is bitten by the bug of invariance principles, of symmetries and groups. Now, although these techniques have produced some fantastic results, they are nevertheless no panacea for all the ills in physics. For instance, I could demonstrate how in 3 or 4 very serious cases the group approach to physics breaks down — a fact admitted even by Wigner himself. You mention the structure of space and time. In what time?

VIGIER: Your own time, proper time.

YOURGRAU: You mean the time of my watch?

VIGIER: When your watch moves, it has its own time.

YOURGRAU: Of course. But now a nasty question: are you with Fock or with Einstein?

VIGIER: Einstein.

YOURGRAU: Needless to say, I am not satisfied with such an unqualified answer, but then we have had that kind of feud for more than 10 years. It looks as if we shall remain in opposite, or at least different, camps for the next 20 years — our time!

MERCIER: I agree with the procedure Vigier is trying to follow; some of my collaborators tried similar models and may arrive at similar conclusions.

I want to draw a historical comparison which is well illustrated by Vigier's paper. At the time when Newton founded mechanics, his only purpose was to explain gravitation as it is manifested in the motion of planets around the sun. It was discovered very quickly that his fundamental postulate was applicable to many other laws of forces like friction and others, and this endowed Newtonian mechanics with so great an epistemological power that scientists and philosophers soon believed that it would give the solution to all problems of nature. Some time later, even Kant in his *Critique of Pure Reason* tried to ground Newtonian mechanics on apodictic reasoning, which, of course, was an impossible task.

Now, the same can be said in a certain way about quantum theory. Quantum theory was at the start only meant to explain electromagnetic interaction within atomic distances, just as mechanics at the beginning was meant to explain gravitation within planetary distances. But we know today that there are several fundamental interactions: gravitational, electromagnetic, so-called weak and strong interactions and a few more that, as far as we know, will not reduce to each other or be subsumed under one single heading.

Well, quantum theory was so successful that very soon everybody began to believe that all that is happening in nature would also be taken care of by quantum theory: Quantum field theory is often assumed to be valid in the fields of all these interactions. This is a dangerous assumption, and I am glad to see that physi-cists like Vigier and so many others now try to further efforts towards develop-ing a kind of super-theory that, of course, will include the former theories like present quantum theory as some kinds of approximations.

KEYES: The question I want to raise is perhaps peripheral to the main concern of Vigier's paper, but it is a question which I feel is important if we are to ask about the relation between the sciences and the philosophy of history — a question suggested by Vigier's program for the relation of science and philosophy, a pro-gram which I understand to involve the reciprocal interrelation of praxis and theory.

Now, I don't propose to raise any criticism at all of this program, as far as it is applied to the philosophy of science as such. I rather want to ask whether it can be applied univocally to the philosphy of history, because there is a different kind of subject matter with which the philosophy of history is concerned, namely, a subject which has a unique kind of identity with the praxis about which he is theorizing.

The real basis of my question is a constructive one, namely, to ask whether there is a philosophical schema which can do justice both to the program for the relation of praxis and theory, which Vigier has outlined, and also to the unique kind of theorizing which is involved in the philosophy of history. I want to suggest that such a possibility could perhaps lie along the basic line of trying to discover what *a priori* conditions render it possible for there to be any kind of praxis, what *a priori* conditions render it possible for there to be any kind of induction. And I would suggest that perhaps if we directed our attention to the question of the subject's apriority, we would already have moved in the direction of understand-ing the apriority not only of scientific praxis but also the kind of praxis involved in historical constructs.

BONDI: I would like to raise the question of the scale of the object one is dis-cussing. In general relativity we have a scale-free theory. There is no length in it; whatever applies to bodies of a certain size will apply to bodies ten times that size. Similarly with groups of motion. I can turn around even if I am very fat, and before we can identify the structures that have all these very appealing and persuasive properties with models of elementary particles, somewhere, somehow, the characteristic length has to come in to define just how big they are.

POPPER: I would like to relate something Vigier noted to the effect that the "old" quantum description was not "complete," as assumed by its promoters, to Mercier's remark that quantum theory is a theory of electromagnetic interaction. Now, between 1927 and about 1935 or 1940, quantum mechanics was regarded as quite a different thing: as the final or almost final form of *the electromagnetic theory of the structure of matter*. This theory of matter had sprung up about 1900 and was, for example, recommended as *the* theory of matter by Becquerel in 1907. It won a tremendous victory with Bohr's model of 1913, and it was re-garded by Einstein as *the* theory of matter at least from 1914 on. This once dominant theory of matter is dead. It died without an obituary notice — without anybody speaking about its demise.

It cannot be denied that the so-called "new" quantum theory, or "quantum mechanics," of 1925-1926, was regarded by its promoters not merely as a theory of electromagnetic interaction or as a theory of electronic shells but, I repeat, as *the* most satisfactory or even *the* final form of the electromagnetic theory of matter; nor can it be denied that this electromagnetic theory of matter has been abandoned. Quantum mechanics, called today by Vigier the "old" quantum description, has now become the theory of the electronic shells of atoms, and in addition, an important general method of quantization. But as a theory of the structure of matter, quantum mechanics has clearly turned out to be *incomplete*.

Thus a dominant theory of physics was refuted without anybody speaking about its refutation. Since its breakdown, it is customary to say that quantum theory is the theory of electromagnetic interactions, which form some part, probably some small part, of the theory of matter which physicists like Vigier now try to build up, incidentally without expecting to complete it.

This is not only interesting in itself; but it is interesting also in view of the fact that in the famous controversy between Bohr and Einstein, which culminated in 1935, the main question was whether quantum mechanics was *complete*. Bohr said it was complete; Einstein argued against this assertion which he felt to be dogmatic. At the time, it was a little vague what was meant by completeness. But there is no doubt that it involved not merely the completeness of quantum mechanics as a theory of electromagnetic interaction, but the completeness of quantum mechanics as a theory of matter. Now the upholders of quantum theory believe even today that it was Bohr who won in that famous battle with Einstein: Even some great admirers of Einstein accept this as a historical fact. I think it is a relevant comment on the history of physics that they are mistaken.

I may perhaps add just one thing to what Vigier has said. His attempts to carry ideas deriving from relativity into the inner structure of particles seem to me a marvelous and bold attempt at a new theory of matter. Not the least marvelous and bold of its aspects is the fact that from the very start Vigier does not expect to obtain a complete theory. For from the very start he operates with the idea that there may be deeper levels still. It was very interesting to hear from him him that this idea, shared by de Broglie and Bohr, can be traced back to Blaise Pascal.

KAPLAN: Vigier indicates at the end of his talk that physicists will always be in business because the job will never be done, that he will find particles within particles, and particles within those particles, and so on. People sometimes in a rather humble, democratic spirit, noticing that we have been wrong so many times in the past, say, "Well, we are probably wrong now, but...." I wonder if that was the basis for his remark or whether he had some theoretical reason for thinking this.

TREDER: The synthesis of general relativity and quantum physics requires a fundamental unit of length. We have such a unit, namely, Planck's length,

$$l = \sqrt{\frac{hf}{c^3}} \approx 10^{-33} \text{ cm,}$$

which is the geometrical mean of Compton's wavelength and Einstein's gravitational radius of a particle. But now I should like to ask a technical question.

In the interior of Vigier's particle-model, the space-time has a very strong curvature. Because of continuity governing the exterior of the particle—over small distances—the gravitational field must be very strong too. What is the physical meaning of this very strong gravitational field? We have good experimental reasons to assume that the gravitational fields of the elementary particles are very weak!

VIGIER: With respect to Yourgrau's question, I am more Wignerian than Wigner himself in the sense that all we know of the physical law is the dynamical mapping of phenomena on space and time, in the Einstein sense of the expression, and this can only be done through a series of invariants, and a series of group motions on curved manifolds, of course.

With respect to the question raised by Mercier, I wish to remark that usually all philosophical predictions based on analysis of data which fall within the realm of scientific practice turn out to be wrong. For example, if you want to discuss time in philosophy, you must bear in mind that the conception of time was founded for two centuries on the experimental data of classical mechanics, and no progress was made beyond physical knowledge for a long time. A true advance in the theory of time was finally accomplished not by general discussions on the problem of time but by Einstein himself. How? Through his analysis of the methods adopted by the experimentalists of his age.

Of course, one could have made a wild guess and said that one should have a proper time attached to every piece of matter. However, that was not done. When Einstein's theory appeared, Bergson published a fantastic paper (which he finally had to delete from his complete works) where he tried to make a distinction (which turned out to be completely ridiculous) between psychological time and the physical time of Einstein. He criticized Einstein from that point of view, enlarging upon the distinction between the so-called philosophical conception of time and the time of the physicist, but the discussion finally turned out in favor of the physicists.

Now, Keyes discussed the causal laws of nature. It would be, I think, completely wrong to state that we desire to have a unique set of laws which is going to govern the totality of nature. What is really happening here is that we deal with levels of structure and organization where new types of laws appear.

I want to cite an example with respect to existentialism and some discussions Sartre and I have been having on those questions. Sartre has nothing to do with Heidegger. What he believes is fundamental in his contributions to these questions is the fact that when you want to discuss psychology or when you want to discuss history you must introduce a new dimension: the permanent feedback of consciousness at the levels of the individual and of scientific consciousness. He has stressed the impact of the effect of consciousness concerning the laws of history on the notion of history itself. Thus when structures become more complex at higher organizational levels, new situations (events, states of affairs) appear. Indeed, the formation of macromolecules is accompanied by the sudden appearance of qualitatively new properties. DNA is not just the simple superposition of its chemical components but it also carries coded information related to its internal spatial structure. This structure itself introduces a new property, which is precisely the fundamental fact which explains the behavior of the living cell. I don't think one should look for a unique and fantastic synthesis of all *scientific* knowledge, for such a synthesis is of concern to the *whole* system of *human* knowledge. This does not mean that one should throw away deterministic laws. It simply means that there is a specific deterministic behavior at different structural levels of nature. I believe in laws of history even if we are not able at a specific time to give a complete description of all historical processes. Nor do I believe that the higher you go the more freedom or indeterminacy appears. I just think new structural laws appear and new dimensions must be introduced in our analysis, which becomes more and more complex. The use of Sartre as a weapon against materialism or against science would, I think, be disavowed in much stronger terms than I am doing here by Sartre himself.

Now, with respect to the objection of Bondi. He has raised a very interesting issue. When one invokes five dimensions in external space-time, one has four space dimensions and one time dimension, and the fifth dimension as proposed by Dirac closes on itself. The question of scale becomes then fundamental. It all depends, of course, on what field we intend to consider. If one chooses, say, electromagnetic interaction, then it turns out that the radius of this ring must be the radius of the particle itself. If we want to recover electromagnetism, because it is one of the properties we can deduce from that model, then the radius turns out to be fixed and introduces a fundamental length in the series, of the order 10^{-31} centimeters. That is the scale one has to use within our model. I think this answers both the question of Treder and Bondi. Of course, this is not *the* ultimate solution, since we don't know yet if all this fits the data. I just want to add that if particles indeed have a radius, then the de Broglie wavelengths must be bigger than this radius, and there is a natural cutoff because $E = h\nu$. So the only way to test such an assumption is to compute the self-energy of this cutoff. Unfortunately, the calculations are difficult, so I am not in a position to give any definite answer.

With regard to Treder's point, that there is no experimental evidence in favor of any type of very strong gravitational field in the neighborhood of particles, I am going to answer in two steps:

A. Following a proposition of quantum mechanics, all fields have to be quantized, for experimental reasons. True, we haven't tested yet the gravitational field, but there is some reason to believe it is all right in that case too. If you accept this claim, then there is no possible qualitative physical distinction between long-range and short-range fields. What is the basic physical difference between a Yukawa field and the Coulomb solution, for example? The only difference, of course, is mass. But even in relation to this quantity, one must be very careful. For instance, workers at Los Angeles have recently found solutions of the Yang-Mills equations where you have no mass at all and you still have short-range bunch-like regions of the field. It is thus very tempting to treat all fields on the same footing, despite the fact that the short-range forces with which we experiment in nuclear or in elementary particle domains are far more important than anything we have observed before in the microscopic domain.

B. If long-range forces can be "geometrized" (that is, represented by curvature in five-dimensional space-time), why not short-range forces too?

Thus the intensity of these forces would prove the existence of very strong curvature in the particles' neighborhood. Of course, this does not at all prove my point that particles are gravitational bunched solutions, but it does prove that there remains at least a hope to use the arguments of Einstein.

Chapter XII

THE THREE KINGS OF PHYSICS

George Gamow
University of Colorado

The measurement system in physics and other related sciences is based on three fundamental units, arbitrarily chosen as those of length, mass, and time, through which all other quantities — viscosity, the electric potential, the interaction constants between the elementary particles, and so on — can be expressed. Somewhat apart stands the temperature unit, which figures in the definitions of thermal concepts such as entropy, chemical potentials, etc. This fourth unit is not actually basic since, using statistical thermodynamics, we can define temperature in terms of energy.

The choice of these three 'kingly' units is purely arbitrary and historical, and only the English unit of length may have a claim to royal origin, being originally defined as the length of the foot of an English king whose name I do not remember. It is of course possible for us to elect three other kings, provided that they are dimensionally independent and, solving the system of dimensional equations, to dethrone the old kings and make them vassals of the newly elected triumvirate. In doing so, it is rational to choose for the new kings the physical quantities which have naturally defined and indisputable values, such as the velocity of light in vacuum in the absence of a gravitational field, or the mechanical quantity known as action. Personally, I throw my votes for His Royal Majesty *Cee*, and His Royal Majesty *Aich*. But who should be the third candidate? Of course, a valuable candidate would be the elementary electric charge, but unfortunately he would not go along with the first two, because he is connected with them by the formula:

$$\frac{c\hbar}{e^2} = 137,$$

(1)

the figure 137 being a dimensionless number. Now what about the Newtonian constant γ? Since $m^2\gamma = \text{const.}e^2$, we may write for the mass

$$m = \beta \sqrt{\frac{c\hbar}{\gamma}}, \tag{2}$$

where β is the dimensionless constant which must be equal to 10^{-20} in order to get the mass of a nucleon. It is a deep belief of theoretical physicists that pure numbers which appear in formulae derived by dimensional analysis must be obtained in a purely mathematical way from some not yet existing theory. Otherwise what fun would there be in theoretical physics?! If Galileo had thought of taking his pendulum to the top of a mountain and measuring its period by using his pulse, he would have derived the formula

$$T = 6.283 \sqrt{\frac{l}{g}}$$

and could have noticed that 6.283 is simply 2π. But it took about a century before Newton proved that point by using his infinitesimal calculus.

Eddington spent the last part of his life without much success in an attempt to derive the number 137, but we still believe that somebody, someday, will do it.

But in the case of gravity we deal with the terrifically large pure number 10^{20}, or rather its square, 10^{40}, which represents the ratio of electric and gravity forces between two nucleons. How on earth can a mathematician get such a big number? In 1937 Dirac got a brilliant idea [1]. "It may not be a mathematical number," said he, "but just the present value of some universal variable parameter."

Let us express the time not in seconds but in the duration necessary for light to cover the distance equal to the radius of an electron. This unit of time, occasionally called by Frenchmen a *"tempon"*, is

$$\tau \approx \frac{3 \times 10^{-13}\,\text{cm}}{3 \times 10^{10}\,\text{cm/sec}} = 10^{-23}\,\text{sec} \cdot$$

Now let us use this unit to express the age of the universe, which is at present estimated to be about ten billion years or 3×10^{17} sec. The result will be of the order 10^{40} tempons, i.e. just the (present) ratio of electric and gravitational forces between two elementary material particles. Thus Dirac concluded that the gravitational constant is not a constant at all and decreases in inverse proportion to the age of the universe.

Ten years later Teller published an article [2] in which he proved that a decrease of γ with time would contradict the paleontological evidence. On the basis of the thermonuclear theory of energy production in the sun, one can show that the larger value of γ would result in an increase in the sun's brightness because of higher compression and a rise of temperature in its interior. Teller calculated that if $\gamma = \text{const.}/t$ (t counted from the origin of the universe), the luminosity of the sun would have decreased as t^{-7}, so that it would have been much brighter in the past geological era. On the other hand, if γ was larger in the past, the radius of the earth's orbit must have been smaller. Putting the two things together, Teller showed that if Dirac was right, the oceans would have been boiling during the Cambrian era. Since Teller's original paper was published, the astronomically-estimated age of the universe has more than tripled. This would permit the oceans to have existed, though in a very hot state, for over one billion years before the present time. Thus the paleontological objection was somewhat eased, since very little is known about organic evolution in such a distant past. The hope was expressed by many people, including myself, that this long-lasting hot bath could be helpful for explaining the origin of life and its rapid evolution during the early stages, because of the increased mutation rates.

Amusingly, in his paper Teller overlooked one important point which had passed unnoticed for a number of years until it was stumbled on by me quite recently [3]. It has nothing to do with the temperature of the earth or the earth itself, but with the sun. If the sun had been using its nuclear energy supply much more liberally in the past, when did the sun start shining in order not to have run out of fuel by today?

Let us calculate the total amount of energy which must have been emitted by the sun between some distant era t_1 in the past and the present era t_2, on the assumption that its luminosity $L(t)$ was decreasing in inverse proportion to the 7th power of the time. If the sun's luminosity today is $L(t_2)$, its luminosity at the time t must have been

$$L(t) = L(t_2) \left(\frac{t_2}{t} \right)^7 \tag{3}$$

and the amount of energy emitted between $t = t_1$ and $t = t_2$ must have been

$$E_{12} = \int_{t_1}^{t_2} L(t)\, dt = L(t_2)\, t_2^{\,7} \int_{t_1}^{t_2} t^{-7} dt$$

$$= \frac{1}{6} L(t_2)\, t_2^7 \left[\frac{1}{t_1^6} - \frac{1}{t_2^6} \right]$$

$$= \frac{1}{6} L(t_2)\, t_2 \left[\left(\frac{t_2}{t_1} \right)^6 - 1 \right]. \tag{4}$$

This amount should not exceed the total amount of nuclear energy stored in the sun's interior at its birth. Let us suppose that the original amount of hydrogen in the interstellar material from which the sun was formed was of the order of 0.75 gm per gm of material. Then the total number N of hydrogen atoms in the primeval sun was $\sim (3/4) M_\odot / m_p$, where $M_\odot$ is the sun's total mass and m_p the mass of the hydrogen atom. Let us further assume that half of the hydrogen could thus far have participated in energy production, so that the total number of hydrogen atoms which were available for thermonuclear reactions in the past could not have exceeded $(3/8) M_\odot / m_p$. Writing ϵ for the amount of energy per proton, one finds that the total energy so far is

$$E_{max} = \frac{3 M_\odot \epsilon}{8 m_p} \tag{5}$$

Since its origin the sun could not have spent more energy than it had to start with; hence we write

$$\frac{1}{6} L(t_2) t_2 \left[\left(\frac{t_2}{t_1} \right)^6 - 1 \right] < \frac{3 M_\odot \epsilon}{8 m_p} \tag{6}$$

from which it follows that

$$\frac{t_2}{t_1} < \left(\frac{18 M_\odot \epsilon}{8 m_p L(t_2) t_2} + 1 \right)^{\frac{1}{6}} \approx 1.22. \tag{7}$$

If we accept for t_2 (the present age of the universe) 10^{10} years $= 3 \times 10^{17}$ sec, then $t_2 - t_1 \approx 2 \times 10^9$ years, which is much smaller than the estimated age (5×10^9 years) of the solar system. Thus apparently Dirac's assumption does not work. I must admit that I am very sorry about this conclusion, since I liked his hypothesis for its elegance. I still do believe that 10^{40} is some kind of parameter and not a "number of the beast" [4] invented by the Almighty, but is must be a different sort of parameter.

Now what about another candidate? The private Gallup polls seem to favor the 'elementary length' $\lambda \approx 10^{-13}$ cm, proposed by Heisenberg in the late twenties.[1] It has appeared twice quite independently in physics; theoretically as Abraham's classical radius of the electron e^2/mc^2 and experimentally as the interaction-range between various elementary particles. If crowned as Her Royal Highness, Queen Lambda,[2] this fundamental constant is bound to do a lot of good. Unfortunately, very little progress has been made to date in her relativistically invariant election to the third disputable throne of the Great Potentates.

Editors' Note

Ralph A. Alpher and Wolfgang Yourgrau found it necessary to make several corrections in Gamow's manuscript. These changes, however, have not affected the spirit of the original paper.

REFERENCES

1. P. A. M. Dirac, *Nature* **139**, 323 (1937); *Proc. Roy. Soc. A.* **165**, 198 (1938).
2. E. Teller, *Phys. Rev.* **73**, 801 (1948).
3. G. Gamow, *Proc. Natl. Acad. Sci. U. S.* **57**, 187 (1967).
4. *Revelations* 15:2.

DISCUSSION

MERCIER: The question of the fundamental constants has attracted the attention of physicists since the time of Planck. Now, Gamow says there are three kings. In my opinion, there are more than three. This, it is true, depends on whether you consider the world as a reversible system or as an irreversible one. But as soon as you include statistical mechanics, then one more constant has to be introduced, and that is Boltzmann's constant.

At the end of his talk, Gamow calls attention to the importance which still unused or nearly unused chapters of mathematics might assume in future physics. I also believe, as do some other theoreticians, that we may possibly explain much of what is going on in nature by means of topology.

YOURGRAU: I agree with Gamow's questioning of the plausibility of Dirac's aesthetically satisfying expression where γ varies in inverse proportion to time. I have argued that this is entirely implausible because it transforms γ from real constant into a parameter. Once it is a parameter we cannot speak of fundamental constants in nature any longer. This is a contradiction in terms. Besides, the crucial equations of general relativity theory would have to be rewritten.

There is a paper by Wallace Kantor which proves very convincingly — *if* his experiment is valid — that the second postulate of Einstein's special theory of relativity (i.e. the speed of light is independent of the uniform motion of its source) is untenable. If Kantor's results are correct, then we have to forget about c as a basic unit.

[1] Private communication.
[2] In Russian, the length (*dlina*) is feminine gender.

Finally, it is also my conviction that in topology lies one of the possible answers to some of the very important problems physicists have to face.

BONDI: I feel that Gamow's kings don't enjoy much independence in sovereignty, and I think there are emperors ruling over them. His kings depend on our choice of measure. If I can dethrone his first king straightaway by using radar, that is, if I measure my distances in light years and my time in years, then I know that the velocity of light is one. Light takes a year to cover a light year. And I think we could, though with greater difficulty, deal with his other kings that way, and therefore show that they are of our own making. The emperors above them are the pure numbers. These are the real rulers in history and there is no known way of dethroning them.

TREDER: With regard to the experiment of Kantor, which was supposed to disprove Einstein's second postulate, it has, indeed, been shown to be invalid. But I agree that had this experiment turned out to be reliable, its consequences would have affected the special theory of relativity in a most dramatic manner.

Now, in geophysics we have many new good arguments for Dirac's hypothesis in connection with the theory of an expanding universe and the rehabilitation of Wegener's continental-drift theory. I think that Teller's arguments against Dirac are founded on the old estimate of the age of the universe ($\sim 2 \times 10^9$ years). With respect to the new estimate of the age of the universe ($> 10^{10}$ years), Teller's arguments are no longer valid.

GAMOW: I do not agree with the proposal to consider Boltzmann's constant as the fourth king. In statistical physics one can do very well without k if one measures temperature in ergs. But one cannot reduce the number of basic constants to less than three by expressing, for example, the masses of elementary particles through only two other constants.

I am sure that Kantor's experiment is wrong. Indeed, if the velocity of light *in vacuo* were to depend on the velocity of the source, astronomers would obtain bizarre results in their observation of binary stars, since the light emitted by the approaching component would travel faster than light from the receding one. This would cause delays of hours, days, or even years in the arrival of light waves emitted simultaneously by the two components.

Dimensionless numbers appear whenever various universal constants are expressed through the three chosen to be kings. Thus, if I choose H. M. Velocity of Light c, H. M. Quantum Constant h, and H. M. Elementary Length λ, the charge and mass of an electron will be defined by the formulae $e = 1/\sqrt{137}$ and $m = 137\sqrt{\hbar c}$.

Chapter XIII

SCEPTICISM AND THE STUDY OF HISTORY

Richard H. Popkin
University of California, San Diego

During the two centuries prior to the publication of David Hume's *History of England*, the attitude of the sceptical thinkers regarding the study of history had changed greatly. The ancient Greek sceptics and the Renaissance sceptics had limited interest in collections of historical data, and saw the work of historians as mostly just fables, poetry or lies, not really contributing to the search for truth. However, in the period 1560-1760, sceptics became more and more concerned with historical studies, in both a positive and negative way. They questioned the reliability of historical data and the kinds of inferences employed in reasoning about the past. On the other hand, they used historical studies as a basic form of sceptical argumentation. Issues raised in the course of this transition in sceptical concern with history, and factors involved in bringing this about, played a major role in the development of the theory and methodology of what is now considered historical research. The great achievements of Pierre Bayle (1647-1706) and David Hume (1711-1776), both philosophical sceptics and practicing historians, can, perhaps, be better appreciated when seen in terms of this phase of the history of modern scepticism.

During the Renaissance revival of scepticism, two attitudes toward historical study occur. One is the rather dim evaluation of history as a form of knowledge. The other is the employment of historical data, often completely uncritically, as a means of undermining confidence in the reliability of theories on any possible subject. The historical fact that humans disagree about almost all subjects and the fact that theories accepted by the experts were later rejected, were offered as evidence that any given theory was dubious. Montaigne used some of these kinds of historical materials to cast doubt on the new scientific achievements. Data about the personal and social factors that led persons to adopt certain theories (vanity, fear, hopes for reward or advancement, and so on) were supposed to lead to distrust of the theories themselves. Also, the Renaissance interest in the past led to the use of a kind of sceptical genetic argument intended to dispel confidence in allegedly new discoveries of the time. By showing that Copernicus, Galileo, Descartes, and others were actually not original in terms of the Greco-Roman and the Judaeo-Islamic theories, some attempted to make contemporaries dubious about the presumptuous claims of these innovators.

The employment of historical materials for sceptical purposes was both intensified and completely changed by applying sceptical arguments to the religious controversies of the time. As historical claims about the accurate text of the Bible, the original views of the Church, the development of the religious institutions, the nature of Judaism in the first century, etc., became crucial issues in the struggle between the Protestants and the Catholics, a kind of scepticism with regard to historical data arose. This scepticism posed basic epistemological problems about both historical materials and the evaluation of them. In addition, another kind of scepticism developed, a questioning of both the uniqueness and the authenticity of the Judaeo-Christian tradition and its philosophy of history. These two kinds of scepticism led to the view called *"le pyrrhonisme historique"* — historical Pyrrhonism, which involved doubts about the reliability of all past information, and about the possibility of learning anything from history.

As a result of the sixteenth-century debates about what is the accurate text of Scripture and how it is to be understood, certain Catholic arguers opposed Protestant historicism by introducing "a new machine of war" to destroy the Reformers intellectually. Starting with Juan Maldonado, the first Jesuit professor in Paris, and Gentian Hervet, the Cardinal of Lorraine's secretary, a kind of sceptical argument was set forth to show that the Protestants would be reduced to complete doubt about religion as long as they based their case on their personal reading of Scripture. Hervet prefaced his edition of Sextus Empiricus with the claim that Pyrrhonism would destroy Calvinism by showing that since nothing could be known, Calvin's claims were unknowable. Maldonado argued that without the aid of the teachings and traditions of the Church, the Reformers would be unable to make sense of Scripture, and would fall into complete doubt and irreligion. St. François de Sales and the Cardinals Bellarmine and du Perron developed these attacks further, and finally, in the hands of two Jesuits, Jean Gontery and François Veron, it became the complete "machine of war" against Reformation historicism. These Jesuits contended (1) that the Reformers could not really tell what book actually is the Bible, (2) that if they could, they could not *really* tell what it said, and (3) that if this were actually determined, they could not tell what to do about it. Traditional sceptical arguments about sense data were applied to raise problems. A book has the title, *The Bible*, but is it "The Bible"? What standards can be used to answer this besides just appeals to personal convictions? If the right book were found, how does one tell what it says? The book just contains ink marks on paper, perceived by variable and fallible human sense organs. How do we know these are words, and if they are, what they might mean? Here, the fallible human faculty of reason comes into play, and we may judge incorrectly unless we have some infallible criterion. And if we find out what the words mean, how do we tell what conclusions to draw and what to do about it?

Scripture reveals no logical rules. Should we apply the pagan logical views of Aristotle, Zeno, etc., to the Holy Writ? If we do, are we sure we have applied them correctly? Appealing to what the Church Fathers did doesn't help, since another level of sceptical problems arises about who is a Church Father, what did he say, what did he mean, etc. Due to our fallibility, we may come to the wrong conclusion at any point in this search for religious truth in the Bible. Therefore, these Jesuit arguers claimed, man should accept the Church as the infallible judge to avoid this "sink of uncertainty and error".

The next generation of Catholic arguers embellished this machine even more. Nicole and Pellison pointed out that if fallible man searches for the Truth in the Bible, he falls into an infinite series of problems. Finding the right book and interpreting it would involve surveying all likely books, knowing all relevant languages, all possibly relevant historical data, etc. Without any absolute standards, this examination would go on forever, since it could not be determined if one had, per-chance, found the Truth, and not just something that seemed true to a fallible man.

Protestants saw "the new machine of war" as a monumental menace to all inquiry, especially that which relied upon some documentary historical material. Jean Daillé sought to show the far-reaching effects of this scepticism with regard to historical data. Developing what he took to be a *reductio ad absurdum* of the Jesuits' arguments, he showed that the same difficulties could be raised against Catholicism, since it also relied on documents and testimonies — namely Conciliar and Papal proclamations, and so on. How could one tell what was actually said at a Council, or what was meant? How could one tell what the Jesuits said, and what they meant, and whether these views represented senility or youthful enthusiasm? The high point in this counter-attack was the Calvinist La Placette's book, *Of the Incurable Scepticism of the Church of Rome* (1688). He argued that since all are fallible except the Pope, then only the Pope is in a position to ascertain who is the Pope, and only the Pope can ascertain when an infallibly true pronouncement is made. Everyone else has a historical problem. His data about who the Pope is and what he said are based upon unreliable sense information, dubious inferences, questionable reading of documents, etc. Therefore, La Placette contended, the Catholic Church can at most have one member whose faith is certain, namely the Pope.

Throughout the seventeenth century, polemics appeared applying scepticism to the quest for religious truth. All sides, including the Jews, joined in using sceptical arguments to show how unreliable were the religious claims of other groups. These arguments revealed the epistemological problems involved in ascertaining the reliability of historical data based on documents or testimonies. Some could envisage that these arguments would lead to historical suicide. One strange figure, Father Hardouin, went so far as to challenge the authenticity of all documents (except Tacitus) and to claim they were all medieval forgeries. Others sought some solid data that could not be doubted, and argued that coins, monuments, and inscriptions were the "hard data" of history that could not be swept aside by scepticism. Some sought to find some "reasonable historical standards" that could not be doubted by sane and "reasonable" men. Liberal theologians like Grotius, Chillingworth, the early leaders of the Royal Society, and John Locke, developed a type of historical probabilism based on employing the criteria that 'unprejudiced', 'reasonable' men accept when judging conflicting or questioned data. Along with the English jurists of the time, they sought to set up acceptable rules of evidence, and standards for adjudging documents and witnesses. They formulated the criterion of 'reasonable doubt', namely, that data and testimony would be accepted if there were no "reasonable" cause for doubting it. There might be theoretical reasons for doubt, but since no sane person could or did take them seriously, they could be set aside as long as there was no reasonable basis for doubting.

While some sought ways of resolving the sceptical attacks on historical data, others developed another kind of sceptical attack, an attack on the historical framework of the Judaeo-Christian tradition. Instead of challenging the data, some attacked the assumptions involved concerning man's past. In so doing, the role of historical research became transformed.

In the sixteenth and seventeenth centuries, much historical research resulted from the religious struggles. Each side was trying to show the errors of the other side, and to justify its own claims. Some historical research was due to political-dynastic controversies, seeking to establish or deny the legitimacy of certain political developments. By and large, the general interpretative framework accepted by scholars was that the human historical record is part of a theodicy and constitutes a Providential history, the development of man's relations with God. Human history began with the events in *Genesis*, and developed afterwards to its present diverse stages. Its most important events are those that are crucial in the journey from Creation to Redemption and the Last Judgment. In human history the turning points are the Fall, the Flood, the Exodus and the Covenant of Sinai, and the Incarnation, Crucifixion and Resurrection, which is the end of man's developmental religious history. After Jesus, on this reading, human history is in a state of watchful anticipation, preparing for the Second Coming and eternal salvation. In the Christian theodicy, historical events after the Crucifixion up to the Second Coming are merely the "Divine Comedy". Human beings may be damned or saved in this period, but no crucial Divine historical events occur. The present human drama has no ultimate meaning except for the problem of salvation and for those events that foretell the Second Coming. 'Who is king of what?', is *really* unimportant unless this aids in preparing for the next stage of Providential history. Hence, wars against the infidels, conversions of Jews, or such events can be significant, but everything else is basically just a chronicle of man's struggles with temptation.

However, for seventeenth-century Jews and Millenarian Christians all history remained Providential. All events were seen as parts of a Divine Drama, signifying man's relations with God and possibly suggesting when the Messiah would turn up. While others wrote court chronicles and annals, Jewish historians interpreted the dynastic struggles, the voyages of exploration, the economic developments in the same manner that the Biblical authors discussed happenings in Palestine, as part of man's relations to God, and man's struggles to live in a Divinely governed world. The expulsion of the Jews from Spain, the flourishing of Jews in France, Italy, Turkey, Holland and America, the misfortunes of the persecutors, were all taken as signs of the coming of the Messiah. Judah Halevi, in the Middle Ages, enunciated the thesis that Jewish history is the heart of world history. Rabbi Menasseh ben Israel wrote the last great work in this tradition of Providential history, *The Hope of Israel*, in 1650. The drama of modern Iberian history was seen as part of a Divine Scheme involving the Chosen People. All this was leading to the Messianic Age. A pre-condition was the location of Jews in all the four corners of the world. Menasseh ben Israel's greatest contribution (besides trying to reintroduce Jews into England to get that corner covered) was his "discovery" that the Indians in America were Jews, and that thus New World history was also Providential. Millenarian Christians and Jews saw the culmination of history as imminent, and set the stage for the tragic Messianic drama of Sabbatai Zevi in 1666. Some Protestants saw the Glorious Revolution as the fulfillment of Divine Prophecy. And Archbishop Ussher tried to portray all present history as a development from the Biblical Creation of 4004 B.C.

An alternative to this view of history as Providential was set forth by Machiavelli, in which human events were considered apart from Divine Providence, and discussed and analyzed in purely naturalistic terms. Machiavelli's approach led some seventeenth-century thinkers to interpret even religious history in political terms. Uriel da Costa proposed that all existing religions are man-made, and should be interpreted solely in terms of human developments.

A more basic challenge to the Judaeo-Christian historical view came from the attack on the uniqueness or authenticity of the Bible. Much had been found both in ancient history and by explorations that did not seem to fit into the framework of Biblical history. Menasseh ben Israel and Archbishop Ussher travailled to put all the pieces into this framework. At the same time a brilliant Frenchman, Isaac La Peyrère (Pereira; 1594-1676), set forth the shattering hypothesis that all human history did not begin with the events in *Genesis*.[1] In 1655, he advanced his pre-Adamite theory, namely that there were men before Adam, and that Biblical history is the story of the origins of the Jews, and not of all mankind. Pereira burst out of the whole Judaeo-Christian framework by introducing evidence that was to overwhelm any attempt to keep all the pieces together. He pointed to the facts of Chinese, Eskimo, Mexican and other histories that indicated that these cultures began long before 4004 B.C., and that they had developed apart from the events described in *Genesis*.

The Flood, he claimed, was a crucial local event in Jewish history, but had no effect on other independent histories. The Jews descended from Noah and his family, but the rest of mankind had separate and independent beginnings.

[1]On Pereira, see D.C. Allen's *Legend of Noah*. His pre-Adamite theory was published in English in 1656, under the title *Men before Adam*. His Messianic views appear in his early work, *Rappel des Juifs* (1643). The British Museum has a copy of his suppressed translation and commentary on the Bible, from 1670, in which Pereira was still gathering evidence to support his pre-Adamite views. His friend and associate, Father Richard Simon, reported he was working on a new version of his *Rappel des Juifs* at the time of his death. This work exists in manuscript at Chantilly.

Pereira's monumental hypothesis was seen at the time as most dreadful and blasphemous, and was to be immediately repressed. His book was burned. He was imprisoned, and only released when he apologized to the Pope. While recanting and retiring to the devout Oratory, Pereira continued accumulating historical, anthropological and geographical evidence to show that Judaeo-Christian history was not the same as world history. However, he himself propounded a bizarre revision of Providential history, claiming that there were two Messiahs, one for the Gentiles, who had already come, and one for the Jews, who was soon to arrive. Therefore, the culmination of local Judaeo-Christian history was imminent, and he argued that Christians and Jews should unite under the King of France to liberate Jerusalem, and to set the stage for the private Messianic Age and Millenium of Judaeo-Christianity, while the rest of the world continued in its independent way.

Right after Pereira's bombshell, Spinoza set forth a more far-reaching thesis, that the Bible is not genuine history, and that all religion should be interpreted as features of human natural history. In the *Tractaus Theologico-Politicus*, Spinoza challenged many particular points (such as whether Moses wrote the Pentateuch) as well as challenging the entire Providential framework set forth in the Bible. With reason as the judge of all matters, religious or scientific, Spinoza proposed his theory of the world apart from the entire Judaeo-Christian conception of God and his role — a world, seen from the aspect of eternity, as a set of unchanging physical and psychological laws. In such a world, human history was a somewhat illusory picture of one aspect of the essentially immutable God or Nature. Historical religion could be understood as an effect of human fears and superstitions. It is just a short step from Spinoza's naturalistic analysis of religion to Hume's *Natural History of Religion*. Spinoza's cosmology did not see religion as providing a framework for interpreting human history. Rather, religion was just one more item to be studied as part of natural history.

These two kinds of scepticism with regard to history, that of an epistemological critique of historical data, and that of a critique of the Judaeo-Christian assumptions about history, come together in the work of Father Richard Simon and of Pierre Bayle. Simon, the founder of modern Bible scholarship, was a friend of Pereira's and a fellow Oratorian. He was the foremost scholar of Jewish and early Church history, of Near Eastern languages, and of the history of the Biblical texts. His (suppressed) *Critical History of the Old Testament* (1678) was part of the sceptical attack on Protestantism. Simon tried to undermine the Reformers' claim of a Biblical basis for their religion by raising a host of scholarly

problems. To tell what *the* text of the Bible is, one would have to study
the present texts, their immediate sources,* and the earliest sources
as well. One would have to decide which were the most reliable ones, and
would have to correct or interpret these in terms of historical data from
Jewish and early Christian history. Then would one have *the* text, or just
a guess based on copies of copies of copies, each stage of which might
contain errors and variants? Our "best" texts are based on *our* data,
our interpretations, and *our* evaluations. If we decide on such a text,
how do we tell what it means? How do we tell what ancient Hebrew and
Greek terms *really* mean at this day and age? Every attempt is just
based on our best critical historical investigations and may well be
mistaken, since it involves human judgment.

Simon built upon most of Spinoza's critical points about the history
of the Bible. But, he contended, Spinoza drew the wrong conclusion. The
present Bible is the result of tens of centuries of human effort to record,
preserve, and interpret the Divine Revelation. Today's documents are not
the words of Moses or of Jesus, but copies of copies of copies by human
scribes trying to record, within their fallible limits, what they thought
were the Words of God. (Simon pointed out that the problem of the New
Testament text is more complicated, since Jesus did not ask the Apostles
to sit down and write, but instead to go out and preach. As a result,
the written record came much later, in a language different from that
actually spoken by Jesus or the Apostles.)

On the basis of this, Simon claimed that the Protestants can hardly
stand very securely on *the* text of *the* Bible. He contended it is doubtful
that there is today an accurate text of the Bible, or that anyone can tell
precisely what the text originally meant when revealed by God. And so,
fallible men need some infallible authority to guide them, and the Church
and its traditions provide this.

Catholic leaders immediately denounced Simon's scholarly efforts as
the most dangerous kind of historical Pyrrhonism. Without an accurate
text of the Bible, what could be the basis for knowledge of revealed
religious truth? Simon's kind of historical investigation would only
result in probable views, not doctrinal certainties. Each conclusion
would only be tentative, to be corrected by ever on-going scholarship.
Simon contended that this, in fact, was what the work of St. Jerome and
the decision of the Council of Trent amounted to. They were each the
best decisions of their time about what the true text of the Bible is,
based on the best available scholarly evidence at that time. Since more
data had been uncovered by the seventeenth-century scholars, new
hypotheses had to be offered as to what could actually be God's message

to man. As critical scholarship developed from Erasmus to Simon, there was a constant need for reevaluation of the documents and their interpretation, and for more historical scholarship. And futher historical investigations revealed more problems rather than *the* answers. Simon's incredible erudition and his scholarly techniques revealed on the historical rather than the philosophical plane the endless nature of the search for historical truth. Simon appears to have believed that he was approaching a definite characterization of the Spirit of the Message, which had been expressed diversely in various sources and traditions. He sought the common core of Judaeo-Christianity which appeared differently in each of the historical expressions. The opposition saw him as forcing everyone into an endless quest for historical truth. They could not see how his method could lead to anything but historical Pyrrhonism and a Spinozistic interpretation of Judaeo-Christianity — that it is only a human way of discussing religion and can thus only be adjudged in human terms.

The finale of this seventeenth-century sceptical concern with history appears in the work of the sceptical historian, Pierre Bayle. He was constantly involved in the religious controversies, having converted from Calvinism to Catholicism and back again. The end result was that he became super-tolerant, fighting the fanatics, orthodox and liberal, of all sides. Bayle's famous *Historical and Critical Dictionary* was a sceptical critique on all kinds of subjects: theological, philosophical, scientific, and historical. He employed all the standard sceptical gambits and advanced new ones, such as his attack on the criterion of truth, that of self-evidence, even arguing that a self-evident proposition could be demonstratively false. Bayle challenged all sorts of contentions in the *Dictionary*, trying to show that man's most rational achievements always lead to doubts, perplexities and paradoxes. Reason finally undermines itself, and man is left intellectually hopeless and can only abandon reason and accept faith blindly. Bayle was delighted to attack any proffered theory "theoretically" and to show that intellectual activity always constituted "the high road to Pyrrhonism."

Besides being a sceptic, Bayle also wrote history. He was impressed by all the false and dubious historical information being disseminated, and by the fact that the same data could be used to argue for opposite religious views. Bayle was impressed by the prejudiced behavior of scholars and by the unreliability of participants in historical events as witnesses. The *Dictionary* was originally supposed to be an antidote to the misuses of historical data. The project was later reduced to just correcting and supplementing Louis Moréri's dictionary. The core of Bayle's work was the historical and sceptical digressive footnotes.

Patiently Bayle straightened out the historical record. With the scholarship of a Richard Simon, he attacked the Protestants for believing there was a Pope Joan and the Catholics for charging Calvin with sodomy. Patiently he set the historical record straight by going back to the sources and adjudging their veracity. Also, he kept posing problems to show that man was unsure about everything, developing a complete epistemological scepticism involving all data, including historical ones. At the same time, he insisted on trying to determine what happened when and where, and how this could be ascertained. In so doing, Bayle showed how we actually decide what data are reliable.

Critical history, when all is said and done, reveals just *la comédie humaine* – man, naked and alone. By employing practical standards, which may be theoretically indefensible, the unprejudiced examiner can set the record straight by separating the lies, rumors, and errors from the facts. Bayle's scholarship and wit, his insight and his tireless quest for data led him to uncover fully the human comedy: man seen in a non-Providential world – man seen apart from any role in a Divine Drama. Spinoza had set forth the metaphysical structure of such a world. Bayle showed it in living colors, in his bawdy articles on Old Testament figures, portrayed in terms of their lust, greed, chicanery, etc., apart from any Providential roles. The Biblical heroes are all too human, like European royalty – all part of the *comédie humaine*.

When the Providential framework was removed, Bayle put in its place a picture of man's development based on human moral failings. On the theoretical level, he kept developing a complete doubt however, about all kinds of knowledge, historical or otherwise. Bayle sought to show that all theoretical attempts at understanding anything always seem to fail. If the theoretical world is fundamentally unintelligible, what can one turn to? Bayle's answer was: to Revelation and to the *comédie humaine*. Unfortunately, there are, as the polemicists had shown, very great difficulties in finding and comprehending the Revelation. All that man seems capable of discovering, in spite of the sceptical difficulties, is himself and his foibles. History, Bayle said (before Gibbon), is the miseries and misfortunes of mankind. The basic causal factor influencing historical affairs is located between the navel and the knee. Man, in Bayle's views, makes no real moral or political progress. The historical study of man, from Biblical and classical times onward, shows what human nature is like. This picture constitutes the final sceptical argument. Any confidence there may have been in human accomplishments is now completely undermined. When the theoretical trappings have been discarded, what is left to examine is fallible and failing man. The examiners are also fallible and failing, and use methods that cannot be defended theoretically. A survey of man milling about in history eliminates whatever confidence we may have had in man in the past or in ourselves. All we can do, Bayle said over and over again, is abandon reason and turn to faith.

Bayle's fusion of the various sceptical themes produced a full-blown historical Pyrrhonism which many learned professors tried valiantly to refute. But his combination of epistemological scepticism extended to history, and a scepticism of the Judaeo-Christian historical framework was not easy to overcome. Voltaire called Bayle's achievement "The arsenal of the Enlightenment", part of which was the historical critique of man's rational endeavors, part the theoretical critique of his faculties and achievements, part the historical presentation of man outside Providential history.

David Hume was the culmination of this transition of sceptics against history to sceptics as historians. He took over Bayle's dialectic and used it to destroy rational philosophy. He then proposed replacing philosophy by the "Science of Man", whose laboratory study would be history. In a thoroughly secularized world, without Bayle's blind faith, man studied in secular history was offered as the principal way for evaluating and undermining man's pretensions and achievements. Since Hume saw above the navel, he introduced political and psychological factors for interpreting man's historical record and evaluating his purported achievements. Employing the practical tools and standards of historical research of Bayle, and Bayle's historical Pyrrhonism, Hume could, in a new age in which the religious concerns of the past were gone, develop a philosophical and historical picture of secular man as the constructive result of his basic scepticism. The world of history, for Bayle, was basically meaningless unless one had faith. Hume was one of the first to try to portray human history as meaningful and comprehensible *in its own secular terms*, in terms of a complex of human and natural factors. Edward Gibbon, building on his two mentors, Bayle and Hume, further developed this new history: philosophical history—the study of man historically in purely naturalistic terms. Whether Hume, Gibbon or others succeeded in this is, at least to me, still open to question. I am not fully convinced that we have escaped from Bayle's world of doubt into a better understanding of ourselves. Perhaps by re-examining how we came to see ourselves as the result of natural and secular historical development, we may better understand how we came to certain present views and dilemmas.

DISCUSSION

RÁNKI: Just what is historical scepticism? It seems to me Popkin uses the notion in different senses, first concerning the problem of the reliability of history and second, to refer to a sceptical view about a given position in history. I think the two are very different. One is a sceptical attitude toward history as a form of knowledge, and the other is a sceptical *attitude* about a given dogmatic interpretation of history.

It is my opinion that, while the second form is a type of scepticism, it couldn't be regarded as a general form of historical scepticism. I can't see it as a form of historical scepticism in the seventeenth and eighteenth centuries. I think it was rather the beginning of the foundation of modern critical history, based on methods which had their beginning in the early nineteenth century with the work of German and French historians. This problem was not especially clarified.

POPKIN: I certainly agree that scepticism as an epistomological attack upon certain kinds of data and scepticism as an attack on a certain system of belief of things are two different notions. I claim that in the seventeenth century, which generated the sort of development I was tracing, the same people were doing both of these things at the same time, criticizing documents, observations, interpretations, and so on, and also criticizing the Judaeo-Christian framework of history. In my opinion, if they had just done the former, one wouldn't have attained the result that ended up in Bayle and Hume. If they had simply criticized the religious aspect, I don't think we would have gotten modern historiography either. But we are probably just beginning to explore both of these types of scepticism, and possibly several others like scepticism about moral beliefs, which may have also played a role in changing the seventeenth-century outlook of the nature of man. But these two strands, I would argue, although quite distinct, were fused by men like Bayle, Richard Simon, and others into a criticism of documents, a criticism of Judaeo-Christian historical belief, and into a new picture of what man's development might have been.

LAKATOS: Popkin's paper continues his vast and fascinating work on the history of sceptical thought. His most important work shed so much new light on the history of modern philosophy that one feels that all the textbooks on this subject should be radically rewritten around the Popkinian scepticism-dogmatism axis.

However, I think the Popkinian vision should be taken with some Popperian salt, unless we want to end up as mystics. For Popkin is not only one of the most distinguished historians of scepticism ever, but also a most distinguished sceptic. But scepticism is an anti-intellectual movement which, with the help of the sharpest intellectual methods, discredits intellectual effort. Its main argument is that there is an infinite regress in pinning down the meaning, and another infinite regress in pinning down the truth, of any proposition. So we can never specify what we say, and therefore intellectual endeavors, including rational theology, are pointless. So we open up the door to mysticism and to irrational theology. This is how some of the greatest sceptics turned into mystics. Now I think that only if we temper the dogmatic-sceptic attitude with Popperian critical philosophy can we escape mysticism.

From the Popperian point of view scepticism and dogmatism are dialectical poles of *one and the same* doctrine: "justificationism". "Justificationism" (or, as we may call it, to annoy Popkin: "sceptical dogmatism") has two main characteristics. The first is that only *justified*, proven knowledge is knowledge. As Locke put it: "To know and to be certain is the same thing: what I know, that I am certain of; and what I am certain of, that I know. What reaches to knowledge, I think may be called certainty; and what comes short of certainty, I think cannot be called knowledge."[1] The second characteristic of justificationism is its conception of criticism as formulated by Descartes: "Reject anything about which your imagination can entertain the slightest doubt as if it were absolutely false."[2] The main form of justificational (I shall call it henceforth *classical*) criticism is proliferation of inconsistent theories; since a *possible* alternative to a proposition kills that proposition, *unless is can be proved* (i.e. a premiss found which is of unquestionable authority and which is such that the criticized proposition logically follows from it), proliferation of explanations destroys them all.

Classical criticism convicts a proposition on sheer suspicion—the slightest suspicion is enough for the epistemological inquisition to crucify the victim. It is up to the defendant to prove his innocence beyond *any* (not only reasonable) doubt. In this inquisition the executioner—the *"sceptic"*—and the victim—the *"dogmatist"*—are both bound by the spell of the chimera of perfect knowledge and cooperate in the ritual.

What if no proof can be found for the theory under the attack of sceptical heuristic? The answer of classical sceptico-dogmatist heuristic is this: You have then to abandon your proposition as if it were false and, if you can, withdraw to what looks like a weaker version of it that has less content but that is not doubtful any more. This rule of classical criticism indicates the most important mechanism of justificational thought: it inexorably leads to a step by step reduction of the field of rational inquiry. The most important pattern of such reduction is this: First cast doubt on any source of indubitable knowledge about the world except for sense-experience and arrive at (realistic) empiricism; then cast doubt on whether sense-experience was a proof of the external world and slip into solipsism; then cast doubt on the coherence of your ego and replace solipsism by the mist of "fleeting bundles of perception"—the infinite abyss of doubt leaves one with nothing spared.

How can one then save Knowledge from sceptico-dogmatist heuristic? Can one get from Nothingness to rich, established truth about the Cosmos? Classical criticism leads to an infinite reduction of knowledge. But perhaps there are some fixed points where one can stop this dizzy fall! If so, having established a foundation however modest, one may turn back and regain, slowly and cautiously, the whole lost paradise of scientific, religious, and moral knowledge.

However, the arguments of the sceptics in the early seventeenth century seemed to be overwhelming. But in the seventeenth and eighteenth century a miracle happened. The miracle was the victorious growth of scientific knowledge,

1. Locke, Second Letter to Stillingfleet, 1697, *Works*, Volume IV, p. 145.
2. Descartes, *Discours de la Méthode*, 1637, 4ᵉ partie.

and, in particular, the overwhelming success of Newtonian theory. Few doubted that Newton discovered the *Truth* about the universe. But how could one *prove* it? How could one fit science into the framework of justificational thought?

The solution of this problem was the ambitious dream of all classical philosophers from Descartes to Kant, Russell, Carnap. They may have differed about the location and the mechanism of the turning-points—where and how to stop sceptical criticism; but not about the end result. They all, as Popkin put it, dreamt about emerging "as so many Saint Georges, prepared to slay the sceptical dragon,"[3] by first descending through the torments of Cartesian doubt (or Carnapian methodological solipsism) to Nothingness, and then returning victoriously, climbing (cautiously stepping from established truth to established truth) up the epistemological solipsism (and beyond) to God's blueprint of the Universe. These efforts, together with occasional devastating sceptical counter-attacks on the turning-points or on the ladder, make up the history of classical epistemology.

But classical epistemology failed. It failed, because scientific heuristic cannot be reconstructed within justificationism. Scientific knowledge is speculation, it is conjectural—it is non-knowledge, according to classical epistemology. Since science has no claim to knowledge in the classical sense, it dismisses criticism from suspicion, but accepts criticism from refutation by "hard facts". If the scientist meets refutation, he immediately starts off to look for a theory that explains both the success of the previous theory and the refutation and more. He reacts to criticism by proceeding to a stronger and more refutable position instead of retreating to a weaker and less refutable one.

Classical criticism is part of justificational, content-reducing heuristic. Scientific criticism is part of critical, content-increasing heuristic. *Classical heuristic weds criticism to the destruction of knowledge. Scientific heuristic weds criticism to the growth of knowledge.*

This is how *classical sceptico-dogmatist heuristic spawned modern philosophy, scientific heuristic modern science.* One is tempted to turn the tables on Hume and pass the judgment on the twists and turns of classical epistemology: It is all sophistry and illusion; commit it to the flames!

Of course, the extension of scientific heuristic beyond science to other fields, like mathematics, prescientific speculation (what Popper calls "metaphysics"), ethics, etc., is a challenging and difficult task: one of the most important, I think, philosophy has ever had. Its success may also yield a new appraisal of the sceptico-dogmatist controversy: it would remove it from the central position it still occupies in the mind of those who have not recognized the basic unity of the opposites and the possibility of their dialectical "*Aufheben*" (cancellation?). The sceptico-dogmatist dialectic would then be replaced by new opposites: the rational and irrational, where "rationality" would stand for a generalized logic of scientific discovery.

And this brings me to my final critical comment, viz. on Popkin's appreciation of science. Popkin would like to persuade scientists to the instrumentalism of the "mitigated sceptics", to save it from destruction in the dogmatist-sceptical

3. R. H. Popkin, *The History of Scepticism from Erasmus to Descartes,* 1960, Chapter X.

warfare. But real science — *if Popper is right* — is alien to the kind of instrumentalism that the mitigated sceptics and Popkin advocate.

So I think the crux of Popperian criticism of the Popkinian conception of the history of ideas will hit it at its appraisal of science.

But all this will be a long intellectual battle; this informal note was only meant to be a first shot.[4]

POPKIN: Lakatos has raised some very complex issues. First, on the question of mysticism, the relationship between scepticism and mysticism needs much clarification. Often the mystic is a sceptic with regard to usual so-called "rational" knowledge claims, in view of his possession of a special type of knowledge and/or experience that clarifies problems in a way that ordinary "knowledge" and experience do not. Sometimes the sceptic becomes a mystic, as Pascal did, not because his doubts imply irrational or supra-rational belief, but because of a type of illuminative experience he has had over and above his sceptical disputations. It may be, as Montaigne and Charron said, that it is easier to receive the (or a) Divine Message after scepticism has had the effect of eliminating dogmatic beliefs from one's mind. Then, as Pascal said, one can hear God. But there is obviously no necessary connection between scepticism and mysticism, since many thoroughgoing sceptics like Gassendi, Bayle, and Hume definitely did not end up as mystics. My own mystic views, as far as I know, are not so much the product of my philosophical scepticism, as they are of my own personal religious experience, and I suspect that this is the case with other sceptic-mystics too.

The other point I will comment on is Lakatos's purported solution to scepticism. It seems to me that "scientific criticism" is much like, or maybe even the same as, what I have called "constructive" or "mitigated" scepticism, as represented by Francisco Sanches, Gassendi, Mersenne, Pascal, and Bishop Wilkins and Joseph Glanvill of the Royal Society. After accepting the sceptical critique of dogmatic knowledge claims, these thinkers then tried to develop a way to proceed to gain acceptable information about the world that might be open to sceptical criticism if considered as genuine knowledge of reality, but which serves as a guide to living if taken as hypothetical, or as about what seems to be the case. Their way led to the continuous enlargement of information and theories without claiming to yield any metaphysical knowledge of the real nature of things. I think this kind of scepticism has played an important role in the development of modern science. Unlike Lakatos, I do not see this as a step beyond the "sceptico-dogmatist" controversy, but as one result of scepticism.

Lakatos may be right that "real" science is something different from the science developed by the constructive or mitigated sceptics. But, up to this point, I can't see the difference epistemologically or metaphysically.

NAESS: It is unfortunate that there has been very little systematic study of Sextus Empiricus during the last fifty years. As a consequence, two fundamental kinds of positions, academic and Pyrrhonian scepticism, are not kept apart. The term 'Pyrrhonian' is now, in accordance with a bad terminological tradition,

4. Also, see my comment on Mercier's paper; and for a more elaborate note on the same topic see my paper on "Criticism and the methodology of scientific research programmes," in *Proc. Arist, Soc.*, 1968–69, pp. 149–186.

often used for "academic". At least four of the activities or claims which Popkin, or those he quotes, attribute to Pyrrhonian sceptics are typical of academic scepticism according to the greatest of all authorities in these matters, Sextus Empiricus.

1. The argument against the *possibility* of historical or other knowledge. This is not typical of the Pyrrhonian who suspends judgment about the possibility of knowledge, and who continues to search for truth. Both these attitudes are stressed by Sextus Empiricus as essential parts of Pyrrhonism. In an environment in which the *impossibility* of historical or other knowledge is taken for granted, the Pyrrhonian will by his counterarguments undermine confidence concerning the thesis of impossibility.

2. The view of Hervet that, since nothing can be known, the propositions of Calvin cannot be known to be true, is again academic scepticism, not Pyrrhonian. Two theses are affirmed, namely, the impossibility of knowledge and the validity of a rule of inference. The Pyrrhonian suspends judgment in these matters.

3. "The argument would go on forever." This is tantamount to postulating there would be an infinite series. But this is academic scepticism, not Pyrrhonian scepticism. The Pyrrhonist suspends judgment in these matters.

4. That man *is* fallible is academic, not Pyrrhonian, scepticism. The Pyrrhonist has found decisive arguments neither for nor against the thesis of the fallibility of man.

The distinguishing character of the Pyrrhonian scepticism, according to Sextus Empiricus, is that if a proposition *p* is asserted, the Pyrrhonian invites you to compare it with non-*p*, listing *pro* and *contra* arguments of lower and higher order. After careful comparison the Pyrrhonian suspends judgment in relation to the two theses: "The *pro*-arguments are decisively stronger than the *contra*-arguments" and "The *contra*-arguments are decisively stronger than the *pro*'s." He ends up with his famous utterance (phoné), "not more this than that". A phoné is an utterance that is not uttered as a proposition but as a kind of exclamation on a par with saying "sweet" when tasting honey. By saying "sweet", as Sextus avers, the Pyrrhonian is not asserting the proposition that honey is sweet or that I have a sensation of sweetness. He does not get involved in subjectivism or objectivism.

In this century some versions of *academic* scepticism have been formulated precisely within Anglo-American philosophy. A fairly precise and comprehensive discussion is going on, mostly with a negative conclusion: academic scepticism as a fundamental epistemological position is inconsistent. But the possibility of Pyrrhonian positions or postures is largely unknown. As seekers, the Pyrrhonians are more likely to be of interest to philosophers of science. They are researchers in a most radical sense.

If somebody says, "But Pyrrhonism is not a *position*," I would retort: "Have philosophies always been positions containing propositions asserted to be true or valid? Socrates, was he not a philosopher in the first dialogues?" One should not require of "isms" that they be packets of assertions or rules. In any case, Pyrrhonism is highly relevant to the critical investigation and assessment of positions which do include propositions and rules.

I expect that my systematical remarks can be fitted into Popkin's historical framework with ease.

POPKIN: Naess has some important points, and some of the confusion between the two types of scepticism in my paper was a result of trying to condense a wide range of material in a very small space. But an even greater confusion stems from the historical fabric with which I was dealing: very few people in the seventeenth century drew the careful distinction Naess drew. It is quite hard to disentangle the influence of Sextus Empiricus from the influence of Cicero in the academic sceptical tradition. Many researchers took their material from both, and, as the Jesuit polemicists, they were not holding either position. They were just looking for arguments that would do the most damage, so they tended to make the strongest claim, "all human beings are fallible", and the like. Thus, by studying the material itself it is hard to tell whether some of these scholars are strictly academic or Pyrrhonian. The term "historical Pyrrhonism" did develop in the seventeenth century and referred to all kinds of sceptics — whether or not they were in fact Pyrrhonists, or even going further: to the iniquitous dogmatism of the academics.

HASAN: I tend to agree with Popkin about the period in the history of scepticism on which he laid the most emphasis, but I don't think the fact that he ignored what happened before 1560 can be wholly excused. I do not personally know very much about what happened between the time of the Greeks and the rise of the Islamic Empire. However, it is certain that much of the European knowledge of Greek philosophy came through the Arab world and the Islamic civilization.

The term "Islamic civilization" does not necessarily mean that it springs entirely from a Muslim or Arab origin, but rather that it was the product of the interaction of Islam and Arab culture with the cultures of the peoples whom the Arabs dominated. This wide range of external influences was not restricted to new converts who came from Syrian, Coptic, Jewish, or Persian backgrounds, but included Judaeo-Christian subjects and their traditions and the vast literature that the Arabs translated from the Greek, Persian, Syriac, Indian languages, etc. All these influences were assimilated into the Islamic civilization, and in many instances a new synthesis was produced. This vast amount of knowledge found its way into Europe through Muslim Spain, Sicily, the Ottoman Empire, and other channels of communication. One may therefore say that between the ancient Greek sceptics and the Renaissance lies an important human heritage in which the Islamic civilization played an important role.

Now, if I may go on to something which is peculiarly Islamic: the development of a critical approach to historical data. Although pre-Islamic Arabia was not a literary society and one does not expect from it a developed historiography, one does find a primitive sense of history in the oral traditions of the Arabs in their genealogy and tribal battles or "days".

With the appearance of Islam a more scientific approach to history began to emerge. The Muslims began to commit to writing the basic literature of their religion, i.e. the Koran and the *Hadith* or the traditions of the Prophet. It was about three years after the death of Muhammad that they collected the Koran and cast it into the written form. The time taken for this endeavor is indeed much shorter than the period Popkin mentioned before the Bible received similar attention.

To establish the validity of the prophetic traditions from the masses of sayings attributed to Muhammad, the *Traditionists* adopted a critical attitude towards their data and to the people who compiled them. They tried to establish the authenticity of the *Hadith* by establishing the competence and integrity of a transmitter or transmitters through a chain of authorities. This method was soon adopted by professional historians in recording oral traditions pertaining to history. Since this method did not question the content of the historical material itself, the validity of such data was doubted by other learned men such as theologians. This attitude, however, did not hinder the expansion of historical writing among a people who became historically conscious of their military conquests and intellectual achievements in their vast empire. In this context and in the third century of the Muslim era (ninth century A.D.), world histories were written.

During the same century, probably as a result of Hellenistic influence, more searching questions were posed. Was the Koran *created*? Did it depict God's own words, and if so, in what language were they uttered? "The creation of the Koran" stirred a lively intellectual discussion or controversy that lasted for some time.

Such type of discussion does not seem to have influenced the writing of history markedly; however, by the time Ibn al-Athir (died 1234) had produced his world history *al-Kamil fi'l-ta'rikh,* there was a definite improvement in the quality of presentation.

Muslim historiography developed slowly until it reached its peak in the fourteenth century in the writing of Abd al-Rahman Ibn Khaldun (1332 – 1406). Ibn Khaldun, a North African savant who traveled widely in North Africa, Spain, Egypt, and Syria, is best known for his *Muqaddima Fi'l-ta'rikh* or introduction to his seven-volume comprehensive world history *Kitab al-'Ibar.* In his *Muqaddima* he cast doubt on the techniques used by his predecessors in writing history and adopted a very critical attitude not only towards the sources of his material but towards its content as well; he began to seek the "internal factors" that move a society. Thus history was no longer treated as a dull statement of facts but rather as a dynamic appraisal of factors causing them. Ibn Khaldun's new approach to history was hailed by historians, philosophers, and sociologists. He expounded the study of human society, its rise, its decline and downfall. And he adopted this dynamic attitude of explaining history long before some modern historians tried to put history into cycles. Although Ibn Khaldun had at least laid the foundation of a more impartial and scientific approach to history, his example was never adopted or developed by his successors in the Muslim world.

Now, I do not mean that Ibn Khaldun's approach to history is the most objective method one could follow, nor that it is comparable to the scientific approach which historians would adopt today in their studies. Ibn Khaldun did much towards achieving this goal, but much more had to be learnt from the Renaissance and the following period, as Popkin has shown.

HINTIKKA: I wonder how Popkin justifies his implied claim of a kinship between historical scepticism and the methodology of early modern science. I would like to say that these two traditions were, to a surprising extent, foreign to each other, not only in the sense that their representatives constitute two classes which do not overlap very much, but also in the deeper sense that the spiritual aims of the two movements were markedly different. One can perhaps illustrate this by reference to the intellectual optimism and belief in progress of somebody like Bacon on the one hand, and on the other hand by reference to the pessimism and agnosticism which often accompanied historical scepticism. I would go so far as to suggest that the two traditions were assimilated to each other only by fallacious hindsight when the methodology of early modern science was misinterpreted by men like David Hume.

POPKIN: It seems to me that, in the seventeenth century, a certain part of the sceptical movement plays a great role in the development of modern scientific methodology, and that some of the sceptics in this period are in the forefront of the scientific revolution.

The first I would put in this class are Francisco Sanches (who was Professor of Medicine in Toulouse), Pierre Gassendi, and Father Mersenne, all of whom used sceptical argumentation against any form of metaphysical dogmatism. They insisted that although one can't accomplish what the metaphysicians were trying to accomplish — or in a Pyrrhonian sense, it is not yet known whether one could accomplish this — there is at least something else one can do. And what they considered excessive scepticism to a reasonable scepticism, i.e. a doubt concerning all matters metaphysical and theological and, at the same time, under-probably in the eighteenth, not just in proposing Baconian empirical science, but a science with a hypothetical model, predictions from the model being checked by experimentation. Here this is not taken as a form of knowledge but rather what Gassendi called a *via media* between scepticism and dogmatism.

Something like this is involved in the early works of the Royal Society, of people like Bishop Wilkins, Joseph Glanvill, and some of the earlier theoreticians, when they claimed they were not doing what the metaphysical scientists like Descartes were doing, but that they were studying just the phenomena of nature, putting them in order as a predictive device, and also as a way of moderating what they considered excessive scepticism to a reasonable scepticism, i.e. a doubt concerning all matters metaphysical and theological, and, at the same time, undertaking a constructive study of nature in a hypothetical fashion. So scepticism provides, or at least did in the seventeenth century, a way of formulating a methodology of modern science and a way of making it into something other than a part of a metaphysical system.

Turning to Hasan's comments, I think the Islamic world is very important in understanding where the Renaissance world comes from, and where some of those problems I was dealing with come from, and it wasn't through any prejudice of any sort that I omitted all this. I just started somewhere and moved some place else on the historical drawing board. But it seems to me that if one traces back where some of these problems of doing critical history arise, he finds a good deal of it in previous developments, especially in Spain, and a lot of it was initiated, in fact, by scholars who were educated in the Spanish universities and continued the Islamic and Jewish traditions of medieval Spain. The sort of work which is involved in the Polyglot Bible, I think, is a continuation of what had happened much earlier; and a good deal of the inspiration even of somebody like Bayle stems from Medieval, Islamic, and Jewish texts.

I have argued elsewhere and often that we western historians of ideas suffer greatly from a lack of knowledge or awareness of the tremendous intellectual achievements that took place in the Islamic world. All of us who work on the development of European ideas from the Renaissance onward need to give much more attention to the Moslem, Judaic, and Near Eastern roots of many of the problems and theories that have shaped our intellectual world. With regard to the history of modern scepticism, I think much needs to be understood about developments, especially in Spain up to the end of the fifteenth century, to comprehend the period I have studied: that of the revival of scepticism in Western Europe from the sixteenth century onward. Figures like Montaigne and Sanches in France probably derive some of their sceptical ideas from previous Spanish sceptical currents. Many issues previously fought over by Moslem and Jewish thinkers became live issues for Renaissance and seventeenth-century philosophers. On the matter of critical and sceptical Bible scholarship in the seventeenth century, I do not know whether previous Moslem developments played much of a role, but medieval Jewish discussions definitely did.

KEYES: The purpose of my comment is neither to criticize the paper that we have heard, nor actually to raise a question, but rather to add an observation about the relation between dogmatism and scepticism.

This relation between faith and doubt must not be thought of merely as a *tension* intrinsic to theology, though there exists a theological tradition that deliberately tries to consider this tension characteristic of its method of inquiry. To cite a few instances: in Kierkegaard's *Philosophical Fragments* and in Jaspers' *Von der Wahrheit* we encounter, at least implicitly, the contrast between the given world and transcendence.

But for the theologian of the twentieth century who adheres to the Christian tradition, the relation between (humanistic) dogmatism and scepticism, or between faith and doubt, becomes a *positive* element. I have Tillich's *Systematic Theology* in mind. On the theistic level, this relation (or tension) is brought out by contrasting God, as the ground of all being, with the abyss of all being, or — on the anthropological level — by the trust that the theologian has in faith and not in scepticism.

Well, the intention of my remarks is simply to suggest that this question of the relationship of the two dialectic elements, dogmatism and scepticism, while it is not pertinent to theology alone, can nevertheless become a positive element, as in Tillich's interpretation.

POPKIN: In my opinion, one of the great transformations that occurs in the seventeenth century is the development of scepticism against faith. Here scepticism starts off in the Renaissance as critical orientation quite close to Kierkegaard and Tillich and is called "The Ways of Faith," "Preparation for Revelation," and so on, and it is only, I think, with people like La Peyrère and Spinoza that scepticism then becomes an attack on faith instead of a justification of it.

TENNESSEN: Let me comment upon the rather peculiar Kierkegaardian form of the doubt-dogma or doubt-faith relationship, which roughly could be stated something like this: Since it is clearly the case that we can't believe anything we know, we should not try to know what we believe. I mean to say, you don't believe you have your clothes on. You have in point of fact more belief, the more *unlikely* that which you believe is. Consequently, what rational theology does is destroy the possibility of faith. So the champion of belief is he who believes that — particularly if he is also an 'expert' in geology and archeology — the world was made in the year 4004 B.C., the 24th of October, at 9:00 a.m.

YOURGRAU: How can one ever establish the veracity of historical documents if one maintains scepticism *vis-à-vis* the scrutiny of documents? I am not a historian. I don't know how to go about establishing beyond reasonable doubt the genuineness of any historical document in the seventeenth or sixteenth or any century, for that matter.

POPKIN: Bayle is very concerned with what looks like the most obvious problems of data, and devotes the longest footnote in the whole dictionary (it runs in present-day print probably a hundred pages) to an enemy of his who declared in a sermon that God commands us to hate our enemies. Bayle straightway jumped on this remark and said it shows an un-Christian attitude. The enemy immediately replied, "I never said it." There were 1,200 witnesses. . . .

Bayle then examines the problem: How does one really ascertain whether the man said it or not? The man is a witness; he says he didn't say it. Twelve hundred people say he did say it. To what extent can you be sure at this point? Bayle now plays a fugue on the problem of coherence of witnesses, plausibility, and so on, and it ultimately amounts to what has been the standard of the Anglo-American law courts: reasonable doubt.

Similarly with the problem in Huxley's *Devils of Loudun*, which Bayle was the first one to solve — whether the nunnery was infested by demons. He goes through the court testimony and tries to show how much of this is superstition, what he would throw out of court, and it becomes a way of ascertaining what, in fact, people who deal with this sort of material will accept. Bayle studied what, in fact, people who deal with this sort of material will accept. Bayle examined techniques of harassment. He studied experts from the Renaissance onward, in addition to some of the medieval Islamic and Jewish scholars, as to what the functioning criteria are. He also constantly searched for counter-cases. You find a place where these criteria break down, and if these criteria break down too often, then you give them up and rely on others. So veracity comes to mean whatever the epistemological limitations allow for.

BRECK: It would seem to me that many of the self-avowed sceptical historians of the Enlightenment were "true believers" in the perfectibility of the human race, in some sort of physical progress. In this sense they are similar to H. G. Wells, who once said, "I have no theory of history. I see the facts go marching by on the horizon, and I run as fast as I can to keep up with them and point them out." Even a cursory reading of Wells does disclose that he had a firm commitment to a theory of human behavior and a dedication to a sort of progress. And it would seem that similar sentiments could be ascribed to these historians.

POPKIN: Bayle doesn't believe in the improvability of mankind. For him, any slice of the world's history is the same. You are just dealing with immoral man doing the same sort of things. Whereas, by the time of Hume and Gibbon, I would say that not only has there developed a progress doctrine, but also a revolutionary doctrine. History has become a way of criticizing existing institutions and reforming them, just by showing they have had a development that is rather sordid, and that therefore something should be done about it. As to how this doctrine of improvability was developed, well, it requires sweeping away the Judaeo-Christian notion that all is providential, and further the additional conviction that something can be done about it.

HINTIKKA: What I was concerned with in my remark was the relation of two historical traditions. Labels like "scepticism" are not important here, nor are the distinctions between the several types of scepticism. What I have been impressed with repeatedly is the great difference between the ideologic viewpoint of the really creative early modern scientists and that of the representatives of the sceptical tradition, and I don't think the new names mentioned explain this state of affairs.

POPKIN: During the summer of 1965, in France, I gave a paper entitled "Scepticism in the Sciences", in which I was making the claim that sceptics were responsible for the methodology of sciences, and somebody rose immediately and said, "Every famous scientist isn't on your list, and everyone on your list contributed zero." This is an overstatement, of course. I would argue that the constructive or mitigated sceptics helped to provide the methodology of modern science, but that the creative scientists of the sixteenth and seventeenth centuries, quite often, were people dominated by metaphysical dogmas.

Chapter XIV

CAUSATION IN HISTORY

Robert S. Cohen
Boston University

Herodotus starts his work by saying he has two purposes. The first is to preserve the memory of Greeks and barbarians in their struggles; and the second, "In particular, beyond everything else, to give the cause of their fighting one another." While I share this viewpoint, there are grave difficulties with such a causal understanding of history.

Causation is a difficult concept in philosophy and in the sciences. Common agreement has yet to be reached among logicians, physicists, historians, etc., even about what the difficulties are; much less have they been resolved.

History, too, is a difficult concept. I will assume, for the sake of discussion, that 'causation' might be seen as a conceptual bridge connecting our three disciplines. But it is a bridge which raises as many difficulties in its construction and crossing as it appears to settle. Its help is problematic. And 'history' is itself another bridge. I would argue that *causation in history* is a phrase that refers not to these two bridges but to one, that the two taken together may provide a way of surmounting the differences among these various disciplines. Moreover, the accepted role of philosophers in elucidating concepts may apply by showing us what 'causation' means and what 'history' means.

My personal route in coming to history may help, not as apologia, but as clarification. My interest in physics became an interest in foundations; and the word "foundations" was soon a synonym for "philosophical understanding" of physics. Such understanding was seen to require some logical reconstruction of physics, both as theory and as experimentation; and then it widened to the general logic of science. 'Reconstruction' was for me a most difficult word, because I did not know whether what emerged was in any genuine sense related to what went into the logical reconstruction machine. Yet it did seem that a logical sausage grinder would make something intelligible, even if arti-

ficial, out of the physics that was put into the forward end; and what
came out was at least interesting, if not always physics. Why was I
interested? How did it clarify, as was claimed?

Now, my difficulty in understanding why I was interested took a
second and even more problematic form, which concerned the part that
was not clearly, or not completely reconstructed. Briefly, I had difficulty
with the residue of unreconstructed notions, ideas which seemed to me
unclear even after logical analysis. I was provoked into genetic, or historical, studies of some physical ideas. Still an amateur, I was brought to the
history of ideas; but these ideas were not political or philosophical or
moral; they were called "scientific" ideas, although there were particularly
troublesome ones whose claim to that epithet varied with epoch and with
reconstructionist criteria.

It was easy to find historical examples such that the *causes* which
made men think certain ideas and find them useful were evidently not to
be seen as the *reasons* which could justify those ideas, within a logical
reconstruction. This then meant that the history of scientific ideas gives
examples of the well-known distinction between causes and reasons, or
between causes and rational justifications; but it left me unclear about
just what 'causes' are. This time I was trapped, for what were needed
were *historical* causes of the 'existence' of scientific ideas rather than
physical causes. 'Physical causation' was merely one of the ideas at
whose history I was looking. And in view of the residue, it seemed that
'historical causation' had to be understood in order to understand the
meaning as well as the history of 'physical causation'. 'Historical causation' is self-referential, since we might propose a causal theory of the
origin of the notion of 'causation in history'; 'physical causation' is not so.

Thus, a bog appeared before me. From what other scholars have written of the history of ideas, we see that there often are hidden meanings
involved in ideas. That of course is what I had been guessing, but the
metaphor, 'hidden', was itself puzzling. Could ideas have a subconscious
aspect in their origin or function? Was there, to put the matter loosely, a
historical content in those ideas and theories for which logical reconstruction and axiomatization had failed, or at least had been incomplete?
Is there still a historical residue, or a latent content, remaining in their
usage? The historical content was intriguing, because many ideas had
their genesis in sources which were not themselves scientific (at any rate
not in any way which I had recognized in my own training) but rather
outside the historical development of science. So the extra-scientific
causes of science became important.

Now, two concepts arise in looking at this: ideology and interests.

For it became clear that looking at the history of scientific ideas was only a special case of looking at the history of ideas in general, and that ideas in general can fruitfully be understood by using the concept of ideology. Science, although distinct, would nevertheless, I thought, be part of the general history of all ideas, which includes those innumerable others which are not scientific.

True ideas are historically caused. This is, as Mannheim and Merton have so neatly put it, the "Copernican revolution" in understanding ideology [11-13]. It is the theory that belief in true ideas is just as much causally and historically brought about as belief in prejudice, myth, or illusion. It is very easy to say that your friend's crazy ideas deserve an explanation, let us say one derived from his craziness. He may be drunk or drugged, or motivated in ways he doesn't know, but at any rate, he believes as he does for reasons which should more correctly be called causes. But your own ideas, not your friend's but your own correct ideas, surely they are not caused but are responses to reasons, to clear evidence and clear thinking? Not so. The new Copernican revolution is one which would treat all ideas and beliefs as ideology, all mental activities as caused. And with this sociological and historical theory of knowledge comes the disturbing possibility that scientific knowledge is acquired, tested, refuted, judged, accepted, rejected, or ignored—as an effect of causes which lie partly, or at times wholly, outside rational canons. Can we believe that the criteria of explanation, and of science itself, are established only relative to an historical situation? Is scepticism about knowledge, perhaps in its most ancient form, the proper response to such historical relativism? Is the new Copernicanism to be abortive?

I think not. Not epistemological scepticism but a new empirical problem seems to me to be the proper response. Can we establish, by historical inquiry, what is the most likely social ensemble of human need, desire, conflict, prejudice, and training to produce a search for, and attainment and recognition of, true knowledge? We must try to know the circumstances such that truth is in man's interest and such that he knows this to be the case. Otto Neurath, in what seemed to him an optimistic passage, once asserted with Marx that the proletariat, as those who benefit least from industrial society, would be "the bearer of science without metaphysics," of true knowledge without illusion or distortion. Suppose (as seems likely) that this is not characteristic of the proletariat but that a far more complex analysis will be needed to establish what are historically-situated *perspectives on truth*. Then the Copernican revolution would be fulfilled. But I suspect that the research in this matter of the historical sociology of science is largely ahead of us. What, then, can we say about the notion that all ideas, even scientific, are ideological; that they are caused, and serve particular and immediate interests?

The idea of ideology is muddled, I fear, when it seems to demand that *science* be construed as "true" *ideology*: that is, we would define *science* as what is believed (and practiced) when truth serves a given human interest.

In fact, when I looked at science this way, I found peculiar problems in the theory of ideology. Because, I suppose, science is a peculiar ideology. Of course, scientific ideology is peculiar because of the claim for objective bases of scientific knowledge. We do have some grounds for thinking that our ideas (hunches, hypotheses, conjectures) are correct or rationally probable; they have been tested and supported by experience, on one view, or subjected to refutation by experience, on another. Whether we are justifiers or anti-justifiers, we believe they do explain phenomena, to an extent which is itself understood. But even *the history of the idea of scientific explanation*, of what are criteria and standards of understanding, has to be understood. Otherwise we, seeking logical understanding of the cognitive products of our science, are isolated from our own historical roots. Why did our predecessors have different standards of explanation, and, indeed, why do we have ours? These are problems not in some metascience of science, not in the philosophy of science, *but in a science*. But which science? An ordinary science, I think. These problems form an aspect of the science of history, namely the science of the history of science.

As concepts develop, they undergo changes in connotation, practical relevance, metaphorical associations, and theoretical and cultural context. Often they mean more than they purport; to understand scientific conceptions profoundly and thoroughly may be beyond us. In physics, we speak with respect of those whose understanding is, as we say, "deep" or "intuitive". While such words usually describe (but fail to explain!) the creative labor of men of genius, they also suggest that there is still an incomplete understanding of conceptions of nature in our textbook expositions, our overtly clear research papers, and our logical form of philosophical reconstruction of scientific theory. The historical development of concepts is also a development of our understanding of concepts; not nature, but that social phenomenon, man's idea of nature, changes in history. What parts of the earlier meaning remain in the later? How do ideas, beginning perhaps as metaphor, both effect and also survive great changes in epistemological presuppositions, logical apparatus, technological vocabulary, and empirical technique, as well as the distortions of personal and cultural interest and prejudice? Above all, perhaps, what remains when criteria of explanation have changed? When does a concept of mystical or mythological origin become "de-mythologized"? And how may we offer an answer unless we try to understand the total temporal career? A history of ideas is necessary for an adequate logic of ideas.

History has the task of completing theories of scientific explanation, because each such theory (just as each explanatory idea) is historically situated, historically evolved, and historically referential. It seems to me that only historical analysis is likely to elicit the hidden content of ideas. It remains an open question to what extent ideas in our present advanced stage of self-conscious logical and empirical sophistication and of conceptual liberation carry such content.

Now, not only was I a physicist to start with, but I found out that, while the history of science (in this case the development of scientific ideas) is a branch of scientific investigation, it is not part of physics; there had to be some other science. I think history is just such an investigating and explanatory science. So I was driven to reflections about history, determined to try to understand the history of scientific ideas as a kind of history, to give an historical explanation to physicists.

But then if one offers a historical explanation of why scientific ideas have occurred, whether we say they occur in brains or minds, one is compelled to say that this is 'merely' describing the genesis of ideas, this is saying how come they have come about. Is this committing the genetic fallacy by confusing causes, sources, and rational analysis? Surely analysis is always permissible, but does it suffice? To put it differently, do rational and historical analyses coincide? Anyway, only historical analysis can tell whether contemporary rational explication is complete, or incomplete. And the weight of evidence that there *is* a relativism of epistemological perspective suggests great caution before we label genetic analysis as fallacious. Edel showed that the theory of ideas is an intersection of the analysis of content and context; and context is historical. The fact is that, in answer to the question *why* ideas occur, we give the sources. But the word 'sources' here is something of an epistemological equivocation. I, at least, am asking for causes.

Yet, I do not mean how can one classify causes, or effects, or cases of causal nexus, for I am sure one can, and I am reasonably sure it is unimportant to do so. Carr has a pertinent little story in the third of his Trevelyan lectures, *What is History?*[1]. He says when he was a child he was much impressed by finding out that "appearances notwithstanding, the whale is not a fish," and then he goes on to say that nowadays such questions move him less. I, too, and I suppose anybody with any sense, was once moved to find out about whales; it was an adolescent move, but still I *was* moved to find that there could be a fish which isn't a fish, and, moreover, that a dry scientific tactic like classification, has something "deep" to it. But classification, however interesting, is not what I meant by historical explanation. I wanted what many theorists will consider to be sufficient conditions; I didn't believe that causal explanations would require necessary conditions, but I did want more than sequential classifications. Now, looking at the history of science, it is evident that there are two kinds of conditions which bring about the existence of somebody thinking in a certain way, or experimenting (which, after all, is merely a kind of practical thinking) in a certain way; that is, there are internal and external causal conditions, those internal to the science being looked at and those external to it.

One might ask whether other sciences are external to the given science one is looking at, but this is probably an unprofitable question, as it calls for arbitrary decisions as answers. Boundaries are often imprecise, and it appears to me that this is especially the case when we seek to distinguish systematic interactions within what may loosely be called a "discipline," from interactions of the discipline with other human enterprises. The internal-external distinction of causes poses a hazy border problem which we need not pursue. I take "external" here to mean external to the scientific enterprise.

Now, there are great questions and there are small questions about the relationships between the development of scientific ideas and the external culture: social, economic, political, religious, aesthetic – indeed all the aspects of human culture are involved. As an example, do the commercial history of grinding lenses for spectacles and the scholarly history of scientific optics have something to do with the commercial demand as well as the aesthetic criteria for Dutch realist paintings? A very small question, to which I think the answer is "yes". Whether interesting or not, this is a legitimate problem in the history of optics. To use a well-worn historian's phrase, it is "no accident" that the lens-grinder Spinoza, the scientist Huygens, and the painter Vermeer were Dutchmen. A small problem.

There are larger problems too, some painfully obvious. Why is it that ancient Greek science declined when it was so good? And "so good" may mean merely interesting; also, it undoubtedly had imaginative theorists, profound mathematicians, sound experimenters, steam engines, and so forth. Why then did Greek science, on the edge of modern mathematics and the mathematical-experimental approach to nature, nevertheless decline? Was there anything internal which was self-destructive to science? Perhaps, but it seems to me that the problem fairly cries out for an external historical explanation.

So one can ask: Can we give an explanation of why science declined at that time? Why was there so long a period of little progress in European science? And yet again, why was there such a sudden brilliant bursting forth of many fields of science under Islam, particularly mathematics, logic, medicine, jurisprudence, optics, and a continued fine tradition in astronomy? But within four centuries, once more science declined. Why didn't Islam become seventeenth-century Europe? Or, why did sixteenth- and seventeenth-century Europe make the unique breakthrough to the foundation of modern scientific methods, both of investigation and of explanation?

These are genuine scientific questions which can be answered by causal conjectures. But my own concern with such questions of genesis and influence is, of course, not unique. Others have been dealing with them, at least for a century. Moreover, the contexts of such gross genetic analyses suggest that we must deal with trends, not with precise particular sequences. Historical causal theories tell conditional stories; they deal with potentialities of a situation in its changing state, but they cannot claim to specify fluctuations of chance and of spontaneity, nor always to include the historical potentialities of still other co-existential situations. As is the case in other sciences, historical theory treats models of reality, and reality, in its turn, realizes the models in varying degrees. Historical theories are such crude idealizations (what a piquant juncture of words!) that they must predict trends of events, and not events. The nineteenth-century social theorists, not least Marx and Comte, saw this clearly. Not inevitability but probability and alternative possibilities are the most severe achievements to be hoped for.

Thus we cannot expect a precise causal analysis of the decline of Greek science, stage by stage; nor can we rigorously explain what exactly produced Galileo and Newton, with their talents, opportunities, and achievements. But being satisfied for the moment with explanatory theories of trends, I would like (for socially practical reasons) to know what conditions did tend to generate the rise and fall of sciences. This question interests me, and in addition it is topical. Perhaps the answer would be applicable to underdeveloped cultures today; what happened then in southern Europe and England and Holland seems to be an important issue now. Since our Western science had sufficient social force to flourish and produce effects in such overtly different times as our ancient world, our Islamic period, and our late and post-Renaissance, it ought to be worthwhile to examine scientific efforts to comprehend or master nature, cross-culturally. Under what cultural conditions may science emerge from non-science (as in Greece)? May it do so by puzzling and nonscientific transmission from earlier science (Islam, perhaps)? Under what conditions by transmission but with a qualitatively revolutionary change (sixteenth-century Europe)?

Generally, physics does not deal with trends. How might we approach a causal explanation of trends within historical developments, and do so with some similarity to physics nonetheless? How can we preserve conceptual clarity and empirical control? Two fundamental proposals should be mentioned. Aristotle offers, with "efficient cause," the elementary and intuitively appealing metaphor of the craftsman, who has ability, who acts, and then produces. And, more generally, this revealing view of man as a creative worker who forms as well as transforms much of his world, coming down through Hegel and Marx, serves as a pictorial model for causal efficacy in nature and in history.

We have seen that one fundamental question of the history of science concerned external influences, external not in the sense of cross-disciplinary but in the sense of social and cultural background. And here Aristotle's concept of efficient cause may be useful.

One aspect of the craftsman's causing is that he is an external influence on raw material, and his control is typically tested by repetition, a simple and probably pre-philosophical idea; but this metaphor of external influence suggests, in a way, that causal explanations might be laws, or law-like generalizations.

How can *laws* govern *trends*? Most likely, we might say that law-like generalizations are idealized, and assertions about trends refer to the insufficiently accurate "embodiment" of the laws in actual historical circumstances. Trends, then, are objective potentialities which need not prevail; and if they do not prevail, the cause of that failure must itself be found. But such discourse about tendentious trends in actuality, alongside other discourses about idealized models (typical of such different social theorists as Marx and Hayek, and so carefully explicated by Popper), can puzzle a scientific methodologist. It is puzzling because there is seldom any occasion to discover repetition in history, and yet we can hope to find causal explanations; at least it is not counter-intuitive that we may. Causal hypotheses could be exemplified in quite genuine but unique ways, perhaps exemplified only once. Indeed, the occurrence of modern science is an illustration, however complex the causal texture may readily be admitted to be.

Granted causal hypotheses about how science tends to arise in history, how may we test these hypotheses? This would depend on our theory of tests. Suppose we begin with Mill's theory; then his methods might be applicable, since one is looking for causal factors, trying to isolate them, to identify them, or to compare the effects of their presence with that of their absence. And yet any straightforward use of Mill's procedures must be abandoned, again because of the puzzling characteristics of historical studies. These puzzles are, in simplest statement, of several sorts. Variables cannot be isolated; identifiable sequences of events observably do not repeat; accidents occur and are often decisive. Mill's elementary formalisms apparently do not apply.

One might hope to find more than the unique instance of a general hypothesis, all other examples being of *absent* properties, with perhaps only the single one wherein the hypothesis is positively exemplified by a presence. Without observed repetition, causal understanding in historical matters would be a combination of hypothetical or conditional statements with a difficult procedure of analysis, whereby certain variables are isolated from the multitude which might be looked at. We see that a historical hypothesis combines several independent conditional statements.

But this hope of stretching Mill's methods seems to collapse, since we cannot easily use these methods in estimating situations of *absent* properties, nor can Mill's procedures hold their force when the essentially irrelevant action of historical accident is so easily demonstrated. Either we fortify Mill by a supplementary account of accident, or we can no longer use causal explanation in history. So let us turn briefly to what I understand "accidents" to be, and to their possible interference with causation.

Most striking about our recognizing accidents, perhaps, is the undeniable observation that they provide genuine historical understanding. And further, but not obviously, such understanding is consistent with causal explanation. No, why?

To say why, we must state what accidents are, or to put the matter more generally: What is the role of chance in history? It should be evident that one can have a causal understanding of what chance is; at least it was evident to Lucretius that everything that occurred to the atoms as they fell through the void could be traced and comprehended, but the intersections of causal lines, he says in effect, could not themselves be predicted. Let me clarify my point. If one looks to the well-worn example of the man who walks down a tree-shaded street and is hit on the head by a falling apple, both the fall of the apple and the walk of the man can separately be causally explained, and the collision between them can also be explained; but the intersection of the two courses needed to explain the collision is *for each* an accident relative to its own previous variables. One can perhaps look at Lucretius' account of accidents and say that his account may indeed *not* be an integrated part of his philosophy, but it is an essential part of Laplace's.

So in some fundamental sense, the intersection of causal lines may be an accident; and even if such an intersection is not ontologically an accident (not even in the slightly modest sense of a large-scale unified theory of the world), at least relative to one's understanding of the man and of the apple, it is so. The intersection, not surprisingly, is not predictable in terms of psychology, which may describe the man's motivation to walk down that street, but not the apple's. The intersection is not predictable in terms of botany either, though botany tells us of the ripening of the fruit, which makes it detachable from the tree. Physics plays its own essential but limited role too. The drama may yet be predictable, but in some discipline which is an intersection of psychology and botany and physics; and there is no such science. Nor need there be. And so, I am scientifically content to call this an accident, not separately predictable; ontologically, my judgment is still the reverse.

Now that ends the hackneyed dispute about Cleopatra's nose. That the size of her nose is an accident, so far as the history of Rome and of Egypt and of Antony are concerned, is obvious: it is a matter of biology. That it is historically causative, no one denies. In fact, accidents are the joining of any two independent causal lines, whether belonging to different disciplines or within what we normally call one discipline, as the history of the Incas and that of Spain illustrate. My earlier examples are cross-disciplinary merely for clarity.

Thus, epistemologically, the theory of cross-causal links explains accidents satisfactorily. And yet we all feel that the matter is not fully disposed of. Indeed it still is disturbing methodologically that there are accidents; but they raise no epistemological problem. It is methodologically disturbing for an obvious reason: accidents are unique and purely relative to the causal lines that they join. They are, as such, simply particular, and they can lead neither to statistical nor to universal generalization. (Thus, in the alternative sense that accidents *are* statistically generalizable, we should not speak of them as *historical* accidents: our problem at the moment concerns the repeatability of historical events.) Historical generalizations, on the contrary, or historical explanations of history (not biological explanations of accidents, but historical explanations of history), insofar as they are attainable, show the relationship of the individual and the unique to something more general. History should show, that is to say, what is general in the unique — if it can be shown.

Well, this is nice but perhaps impossible. One of the jobs of science, we are often told, and I am even inclined to think it is the most important social task of science, is to show which of our ends can be realized and which cannot – to tell us which of our aspirations are vain. An important task of history then, would be to show that there *can be no purely historical understanding* of why a particular man has been hit on the head by a *particular* apple, or of the appearance of Cleopatra's nose on the *historical* scene. Only insofar as purely historical understanding is possible at all can there be causation in history; it is the role of the historian I am delineating, not the extent of his possible success. Accepting this limitation of his possible success, we can see why there is no logical conflict between cause and chance in history.

How then can we use causal explanations in the study of history? We need not worry in any particular manner about the multiplicity of causes. Such multiplicity or complexity is common to other sciences. But the often bewildering mixture of variables, causes, and accidents may lead us to despair of any practical use of causal analysis in history. Perhaps my positive remarks about trends will let me rest with general and conditional long-range propositions. But these are equally productive of despair, and they taste of speculative metaphysics. Moreover we *can* occasionally predict, plainly predict. As I read Farrington's impressionistic essay [6], I suppose he was right, for example, in describing Francis Bacon as a prophet of what, given certain conditions, might occur when science became applied. Another striking example: Well before the first World War, Hilferding [9] predicted the tendency to a Fascist ideology in capitalist countries; not merely Fascism – that was predicted by Lenin and by Jack London [10], too – but even details of its ideology. (I have just mentioned trends and predictions in the same breath. It is because trends are conditional that they can serve as foundations for predictions. But I am not certain; a preliminary difficulty must be faced.)

Historical prediction may be *self-frustrating* or *self-fulfilling*. This may interfere with the test of predictions. Knowing the prediction about their own lives, people may deliberately avoid what is predicted to occur. The more popular of the two is, perhaps, historical prediction that is self-fulfilling. Knowing the prediction, people may deliberately make it real and climb onto the bandwagon. But, after all, these are not insurmountable problems in the logic of historical prediction; one can explain both the "self-frustration effect" and the "bandwagon effect", and predict them too. Popper [16-19] has lumped these two effects under the title of "the Oedipus effect", in allusion to the myth wherein the oracle's predic-

tion to the parents of Oedipus had the "bandwagon effect". He argues
that trends are conditional and that prediction is neither certain nor
complete. In fairness to Popper, I must mention another conclusion that
he draws from this; I regret I cannot discuss it here, though perhaps I
ought to, since I both appreciate it and do not fully accept it. His point
is this: The future growth of science is essentially unpredictable; and it
affects the future of man most profoundly. For Popper, even conditional
trends would not be of much use.

Nevertheless, I think that we should take Popper's ardently argued
notion of methodological individualism very seriously, and I should
like to approach it this way. I have explained why I consider the idea of
conditional trends to be the principal operative notion in historical
studies. Now this brings me back to my initial view of the similarity
between physics and history. After all, secular astronomy seems to be
pretty much the study of conditional trends, assuming no outside interference. Yet there is a fundamental difference. And it is located in the
conception of the properties of a group of entities. With Popper I find
no scientific way to say that there are properties of a group which are
self-caused, or uncaused, and with him, I agree that, nevertheless, there
are group properties that are not shared by the individual members of
the group. Also, whether we are naturalistic or supernaturalistic, we
should further agree that groups may have emergent properties.

Yet, much as I agree with Popper about these characteristics, I
seem to differ greatly with him concerning their function in historical
explanation. Now once again it is necessary to state more clearly what I
mean by historical explanation. To begin with, while I cannot discuss
reductive explanation here (for that would take me far beyond a discussion of history), I must nevertheless mention that the *problem*, "Can we
find any reductive explanation of group properties?" is fairly well known;
yet there is no general awareness, much less agreement, about the *solution*
to it. My concern here, however, is first to try to clarify the problem with
historical relevance and then to offer a tentative solution. I think it is
true to say that practically all writers, Popper included, hinge the explanatory problem just posed on the ontological question decided above.
Yet the problem "Is reductive explanation possible?" is distinct from and
independent of the problem, "Are properties of groups objectively com-
posite properties of their members?" We can be ontologically anti-reduc-
tionists (with Popper) and yet methodologically reductionists (against
him). Moreover, though I believe it is likely that the emergence of group
properties may be reduced by individualistic explanation, once groups
are there, these properties may without contradiction be subject to laws

of their own (trend-laws) which are not individualistic in the least. And so, curiously, I find myself ontologically agreeing with Popper, yet methodologically disagreeing with him both from the right — regarding the emergence of group properties — and from the left — regarding the life of groups. The logic of emergent explanation, I confess, is still unclear, even when seen in the light of modern research on the logic of causal relations. But I believe that emergent explanation is a variety of causal explanation. Some people are understandably shy of using a mode of reasoning not yet captured by the formal logician. But we should give tasks to logicians rather than follow behind them. After all, even the *reductio ad absurdum* had not yet been formalized a century ago. And almost everyone uses intensionality, knowing that it still baffles formal logicians. All that may perhaps be said at present is that the logic of emergence, when it arrives, will incorporate and explicate the idea that new group properties result from the structural achievement of higher levels of complexity.

Thus I think the program of using trends and trend-analysis in historical explanation has not yet been exhausted; on the contrary, with the help of logicians, methodologists, metaphysicians, social theorists, and historians, it may yet reach its maturity. We want causal explanation in history, and we want to make it scientific. That is to say, as I have suggested, that we want both causal theories of the emergence of group-properties and conditional laws of their development; and we want all of them to be testable. I have discussed a few of the main objections to this kind of historical scholarship. What may emerge is that, though historical trends are marred by accidents, complexity, clashes amongst themselves, and the Oedipus effect — nevertheless, testing specific trend-theories by the use of relatively simple comparative methods is a *scientific* challenge which is manageable.

I may concede just one more difficulty and yet remain optimistic: just as there is, as yet, no logic of emergence, so there is no logic of comparison (*pace* Mill with his concomitant variational method of similarity and difference). So far as I know, no one has solved this problem in a general way — not even Hegel. To put the matter modestly: The emergence of properties which are existentially (or ontologically) unique to a given level of complexity is not a miracle, by which I mean not unexplainable, although often unexplained, and its logic is still in abeyance. Levels of complexity have explanatory connections; the properties, however, are not thereby reduced, *but anticipated.*

Suppose we fulfill the program outlined above. Suppose we have causal explanation proper in history. As I said at the beginning, I would be very happy in that case. But others would not. Their apprehensions demand some consideration.

Causal explanation is often considered to be utterly deterministic, and thus to deny human freedom. Is this true? There is a long and distinguished tradition in Western thought which claims that freedom is compatible with causation. I cannot and would not wish to add to the literature on this matter. But I shall merely state my impression that the question is still open. And (need I say?) I side with this tradition. I think I can even hope to use a thesis of one of the major opponents to this tradition in order to ameliorate apprehension. It was Popper who admitted that metaphysical determinism would be harmless so long as the determinist admits, if only on the common-sense level, the existence of choice, and of action through the exercise of choice. Certainly those in the tradition which finds no incompatibility between causality and choice *a fortiori* qualify under Popper's minimum requirement. Indeed I should emphasize (with Plekhanov) that the main practical function of a causal understanding of historical processes is to enable intelligent people to act effectively and reasonably and to be inspired by a sane hope of realizing their goals and of abandoning their illusions.

Let us claim, then, that genuine freedom is compatible with causality. It seems to me we must be even more ambitious: If we want significant causal explanation in history, one of the items on the agenda—in my judgment the crucial item—is to explain the emergence of freedom from necessity. It was Popper who said that history has no meaning, no purpose; and he concluded that we are able to put meaning into it. The question is a step deeper than Popper indicates: What has produced this ability of ours? The question, by Popper's own canons, is a quest for a causal explanation. Need I say more? But more may be said: By admitting the legitimacy of this quest, Popper's maxim, oddly enough, enters the mainstream of materialist thought; for a fundamental requirement of a materialist understanding of history precisely concerns the emergence of freedom in a previously unfree world.

History is itself a phenomenon in history; that is, knowledge of history is one of the liberating causes which can produce free acts. This is the determining, enlightening, and humane role of *feedback as a human phenomenon.*

Can one give a causal-historical theory of how men might be nurtured so as to exist in a conscious and active state wherein they are free to choose? This liberation is not only a question of history. It is what Rousseau meant when he said that men could be forced to be free, that they are born helpless, unfree, and dependent, that those who are forced can also be educated. The great task of education, of that tyrant, the teacher, is to produce students who are themselves free of the tyrant, and thenceforth self-determining. When that is done, present causes bring freedom into a deterministic and causal context.

The practical outcome of historical causal theories would be tasks, rather than passive knowledge. We need causation in history in order to attain reasoned mastery over nature and over the worst unfeeling social forces which those who oppose causal theories most fear. Hence, I suggest, causal theories of history need generate neither fear nor despair nor inactivity nor indifference — they may even generate new hopes.

Acknowledgment

I am grateful to my colleague and friend, Joseph Agassi, for his detailed and critical assistance with the final draft of this paper.

REFERENCES

1. E. H. Carr, *What is History?* New York, 1962.
2. R. S. Cohen, "Alternative Interpretations of the History of Science," in *The Validation of Scientific Theories* (P. Frank, ed.), Boston, 1956.
3. R. S. Cohen, "Dialectical Materialism and Carnap's Logical Empiricism," in *The Philosophy of Rudolf Carnap* (P. A. Schilpp, ed.), La Salle, Ill., 1963.
4. R. S. Cohen, "Is the Philosophy of Science Germane to the History of Science? The Work of Meyerson and Needham," in *Ithaca: Proc. Xth Int. Congr. Hist. Sci.,* Paris, 1964.
5. A. Edel, "Context and Content in the Theory of Ideas," in *Philosophy for the Future* (R. W. Sellars, V.J. McGill, and Marvin Farber, eds.), New York, 1949.
6. B. Farrington, *Francis Bacon: Prophet of Industrial Science,* New York, 1949.
7. B. Farrington, *The Philosophy of Francis Bacon,* Liverpool, 1964.
8. F. Hayek, *The Counter-Revolution of Science,* Glencoe, Illinois, 1952.
9. R. Hilferding, *Finanzkapital,* Berlin, 1909: the crucial passage from p. 427 is translated in Paul M. Sweezy, *The Present as History,* New York 1953, p. 65-66.
10. Jack London, *The Iron Heel.*
11. K. Mannheim, *Ideology and Utopia,* London, 1936, especially Chapter V.
12. K. Mannheim, "Towards the Sociology of Mind," in *Essays on the Sociology of Culture,* London, 1956.
13. R. K. Merton, "The Sociology of Knowledge," reprinted in his *Social Theory and Social Structure,* Glencoe, Illinois, 2nd ed., 1957.
14. A. Mieli, *La Science arabe et son rôle dans l'évolution scientifique mondiale,* Leiden, 1966.
15. O. Neurath, *Lebensgestaltung und Klassenkampf,* Berlin, 1928.
16. K. R. Popper, *The Poverty of Historicism,* London, 1957.
17. K. R. Popper, *The Open Society and Its Enemies,* Princeton, 1950, rev. ed.
18. K. R. Popper, "Prediction and Prophecy in the Social Sciences" and other essays in *Conjectures and Refutations,* London, 1963.
19. K. R. Popper, *Of Clouds and Clocks: An Approach to the Problem of Rationality and the Freedom of Man,* St. Louis, 1966.

DISCUSSION

NEF: The study of history has come to be concerned with all sides of human life and thought, and I believe it might be helpful if we would seek to better understand the connections between various sides of historical evolution and (to use a trite phrase) the good, the true, and the beautiful.

Such possibilities are offered in historical study by comparisons, and still more by the discovery of interrelations between historical evolution in many fields.

For example, there are the relations between beauty and the conquest of the material world. A number of men in the realm of beauty in our time have, by their art, done much to establish anew a sense of tradition: Stravinsky in music, Derain in painting, Eliot in literature, and more boisterously, Frank Lloyd Wright in architecture. All have shown in their work the value of what Stravinsky, in speaking for music, calls the framework for "the musical constructions which are of the greatest interest", which enabled musical ideas to be expressed in something approaching perfection. The only example, Stravinsky says, we have had is the framework which was hardly evolved before the middle of the seventeenth century and which hardly lasted beyond the middle of the nineteenth.

It does not follow that a new system is impossible. In recent times, Schoenberg and Stravinsky themselves have been working in their music toward a new framework. That was also the aim of Derain and, odd though it may seem to you, of Chagall in painting, as the latter showed in a contribution made to the symposium of the Center for Human Understanding three years ago, in *Bridges of Human Understanding*, published in 1964. Such was also the purpose of Eliot in poetry and in criticism. It has been the purpose of Wright and Mies van der Rohe in architecture. History might contribute to understanding if it could show relations among these recent efforts at perfection in the arts and other efforts elsewhere: in the building of institutions, in the building of philosophy, in the constructions of science, in efforts like the United Nations and UNESCO, to find a basis for living together without total destruction.

YOURGRAU: I think that Cohen wants to make two points. One of them is that you have to decide upon the sense, the meaning, of what actually is — by its very nature — without meaning (as illustrated by T. Lessing's *Geschichte als Sinngebung des Sinnlosen*).

The second point he made was that you cannot have a look at history without some sort of value assessment, and that this requires, among other things, a causal appraisal. Although I like his argument, I don't like the main thread, and for this reason: I can't imagine how history is possible without causal appraisal; in this respect it resembles biology and medicine. But in spite of our naive faith in causality (if this should really affect the future of our reasoning) physics today shows that we have rather more grounds for acausality than for causality. Actually, causal analysis is only one of the possible approaches to getting order out of a seeming chaos. In any case, we couldn't claim today that there is in history an *innate* causality in the sense in which people used to believe for 2000 years that causality is immanent in historical events.

RÁNKI: I agree with Cohen's appraoch. Natural sciences shouldn't be roughly copied and arbitrarily transposed into history writing, as has at times occurred.

But I think every science (and I look upon history as a science) has its particular characteristics, appropriate for its object.

Regarding causation in history, I miss in Cohen's approach any analysis of economics and technology. How shall we write in the twenty-first and twenty-second centuries about the history of the twentieth century without considering the so-called Keynesian Revolution? How can we write about the development of the twentieth century, for instance, without looking at technical development such as computers?

TENNESSEN: Cohen tries to escape certain traditional predicaments by using the word 'law-*like*' rather than just 'law'. But the problem is: *how* law-*like* do you want to make your generalizations? Cohen is surely right in saying they are law-*like*. But *how* law-like? Do they have to pass the counterfactual test or the test of subjunctive conditionals? Should they be as law-like as, say, Kepler's laws, for instance? It is not *just* a coincidence, I suppose, that our planet moves in the way it does; in point of fact: if anything *were* our planet, then it would be bound to move in this way. Where does one draw the line here? I think Cohen is begging the question by saying law-*like*. It is a wise thing to do, but it does not contribute to the solution of the traditional problems of "covering laws", etc.

BONDI: As a physicist, I want to add to what Cohen said so clearly and so well, by trying to drive a little deeper into the meaning of 'accident', which seems to me quite crucial to this whole issue.

When one mentions 'accident', then one must realize straightaway that the motor insurance industry lives entirely on the predictability of accidents. On the other hand, when we speak about accidents in history, we mean something else. And so I try to make myself a model for this purpose, a physicist's model of history. What I want to consider, then, is, on the one hand — though I don't want to enter into the theoretical problem of free will — the *practical* unpredictability of human behavior, and for this I use, as the physicist's image, the *practical* unpredictability of behavior of particles in certain situations. What I have in mind is much more the attitude of the kinetic theory of gases in the nineteenth century rather than indeterminacy, just as I have in mind the practical unpredictability of human behavior much more than free will.

On the other hand, we know that in certain circumstances the summation of many such unpredictabilities leads to definite prediction, e.g. in the motor insurance business or in the calculation of pressure in the kinetic theory of gases. The basic question then is: How much or how little of history is predictable in spite of the unpredictability of individuals?

What we are looking for as a vital ingredient of history is something which we can also construct in physics, namely, amplifiers of unpredictability; although the unpredictability is basically a portion of the microstructure of the individual human personality, social and political structure may contain amplifiers for it. We can, in physics, imagine a bomb exploded when the random noise fluctuations of a suitable amplifier exceed eight times the average, which may or may not happen in a given time. Although the "cause" lies in the unpredictable behaviour of the microstructure, the electrons — the result, i.e. the explosion, is quite clearly in the "macro-world".

Thus I like to view history as a gas into which are put a number of amplifiers of unpredictability. If we suppose that the number of such amplifiers is zero, then we have the Marxist interpretation of history. If we say that it is only the amplifiers that matter, then we have what you might call 'biographical history'. To try to specify where and how important are the amplifiers of unpredictability seems to me to be a major task in history.

Moreover, these amplifiers give us a chance to follow causation through. In the physical picture we know that in a completely deterministic system we can never trace any causal links because the whole pattern is so tight and unique. Causal links can only be traced in systems that have uncaused events, real 'news', whose effects we can follow through. And so if we can go on to identify the amplifiers of unpredictability in history, then we can follow the flow of consequences that come from such amplified random events. I think this may be the most fruitful way of looking at causation.

TÖRNEBOHM: I should like to have a clarification of the differences or similarities between causal hypotheses, on the one hand, and general law-like hypotheses and theories such as economic theory in economic history, on the other hand, as well as an explanation of their functions or the roles they play in the enterprises of historians.

HASAN: It seems to me that the cause of a thing lies in its nature, status, or meaning. And the combination of all these makes it a broad subject of study. The combination of these factors should be sufficient to explain the causes of the subject matter that we are studying.

The more one thinks about this, the more he finds that historians, unlike scientists, cannot always have all the facts or data in front of them. Before the nineteenth century, there was always a dearth of data. And thus what historians claim to be a scientific and objective approach is different from what natural scientists would describe.

If historians would ask themselves the basic questions as to what caused the French Revolution or what happened in the French Revolution (or for that matter any important event), they would probably get more or less the same answer. However, although many historians do ask themselves what happened on the day the French Revolution broke out, very few would bother with the details of the private life of Louis XVI. The fact that his son died at that date is of some significance to some and of no significance to others.

A historian's understanding of what happened and why it happened is made more difficult by the fact that human nature is very difficult to gauge and understand, and at the same time almost unpredictable. Therefore, to find a perfect explanation of what happened in the past and of how and why it took such a turn, is not easy.

However, despite the differences among historians in matters of detail and their explanation of such details, they occasionally do agree about the major causes of a particular event. They sometimes explain such few basic agreements by what they call 'laws of history'. For instance, it is almost a law of history that whoever rules Egypt will always endeavor to extend his right hand to Syria.

Now, if I may touch on another point, the difficulty in my mind is not only the dearth of historical data, the criteria which each historian applies in finding out about facts, nor the difficulty of understanding human nature, but also the difficulty caused by the implicit attitude which one takes when applying his own evaluation, his own moral judgment toward a particular set of facts. This, to my mind, is an important point which causes a lot of different interpretations among historians.

It seems to me that at present it is even more difficult to understand objectively and scientifically what is happening. The difficulty is not due to dearth of data but rather to the fact that the abundance of data and the complexity of factors leading to an event need greater effort for understanding.

RÁNKI: The approach to history as a science made by Bondi is rather interesting. But it is a misunderstanding to say that Marxism denies absolutely the role of the so-called amplifiers. The distinction between Marxism and other historical theories is not whether it does or does not deny these amplifiers, but how much it stresses them.

KEYES: If we take the statement, "Our task is to make the world rational," we could conceivably understand it in the light of Nietzsche's aphorism that the world can be justified only as an aesthetic creation. And further, if we interpret it theologically, we now have a paradigm for what I mean by the "Tillichian Paradox", that is to say, the dialectical relation of faith and doubt, which in my view has nothing to do with the juxtaposing of a mythological chronology for creation and a scientific account of the origin of the universe.

This paradox is rather the positing of reason by faith where there is none, a theological program which I say is the opposite of de Chardin's attempt to find these theological conclusions latent in the natural context.

BRECK: Historians live in a workaday world of events, persons, movements, and are inclined to shrug their shoulders when philosophers claim to see connections, causes, and patterns of some sort in history. Indeed, for many years it was a chief distinction of the historian that he disclaimed any "theory of history", leaving such ideas to people who did not need the sobering experience (they believed) of the naked fact and the solid description of events and institutions. Recently I perceive some shifts in thinking about the past on the part of formal historians, at least an admission that they have assumed or even hidden values, by which they select the "facts" of history and organize the past.

The distinction is often made between sociology and history, that the former deals with trends and general laws, whereas the latter is concerned only with the unique event. Barbara Tuchman, for instance, speaks glowingly of "history by the ounce", by which she means description of actions without any reference to a "theory" of historical causation. However, it is apparent that theories about man's past are necessary and much in need of development. Surely the professional historian has much to learn. If, for instance, I declare my disapproval of a particular doctrinaire theory of history, do I not then need to reveal my own biases and the hidden (though firm) sense of what history "really means"?

COHEN: I shall try to respond to various commentators and questions briefly, in turn.

To NEF: I am inclined to agree that my two notions of "bridge"-concept and of comparative-historical analysis need to be pressed and widened. Certainly 'causation' has no monopoly in connecting diverse fields of human concern, and indeed comparative analysis of alternative searches for perfection should be pursued. History can help. But I believe that causal analysis in understanding diverse phenomena, in their historical development and in their factual interactions, will have a key role.

To YOURGRAU: It is true that causality is rather neglected in the workaday vocabulary of physical scientists, except in imposing a commonsense temporal order of cause first and effect second upon some esoteric theorists – and even this 'commonsense' is thought to be a speculation which may be violated. And it is also true that intrinsic probabilities govern the most accepted understanding of fundamental microphysics. But I must say that this is irrelevant to my espousal of causal understanding in history. Is classical macrophysics causal? Is classical statistical physics causal in principle? And is not history a classical macro-subject? In any case, history should neither emulate the latest fashion in physics nor too uncritically take the theoretical entities of physics and their (hypothetical) properties as the ontological substructure of historical reality. Within the multistoried house of emergent levels, one could imagine a world in which causality would reign at one level, acausality or pure chance at another. Having uttered this defense, I must admit a certain regret that causality may find its home in historical study while becoming fundamentally homeless in the land of physics. Some decades hence, a historian may have to defend 'causality in physics'.

To RÁNKI: While I applaud methodological democracy among the natural and social sciences, I should want to add that comparative method of analysis also should be applied to the various methodologies of the specialized disciplines. To learn from each other is not to be subservient; and *comparative methodology* is, to my mind, the major critical task of epistemologists in this cultural age of science.

To TENNESSEN: Historical tendencies *are* 'law-like', and they are attended by a coterie of "other-things-being-equal", contrary-to-fact dreamy speculations, and the entire logico-epistemological morass of model-making. But this is no worse in principle with historical generalizations than with those in experimental psychology or in the natural sciences, once you admit that my discussion of accidents and of comparative testing is plausible. Indeed, to answer Tennessen's blunt question "*how* law-like?", I should like historical knowledge to be *very* law-like.

To BONDI: Amplification of fluctuations is not a close model for intersection of those independent causal chains which are on the same 'level', and it is the latter which seems crucial to me in understanding 'accidental' factors in history. Even in crude form, what we need generally is not *amplification* (since the two causal chains may be not at all in a micro-macro, or other, subordinate relation) but *coupling* or relevance-matching. The analogy of biography-society to molecule-gas generally does not illuminate the most typical chance phenomena in history (accidents), but it does provide a way to picture those accidents which arise from the intersection of genetic-biological or pure psychological causal sequences with those of history proper. (As a small comment, I should suggest that Bondi meant 'amplification effects' rather than 'amplifier'; otherwise he would have to interpret the latter as some kind of entity.)

To TÖRNEBOHM: To my mind the distinction between 'normal' causal hypotheses and law-like hypotheses is one between predicting events and predicting trends. Insofar as trend-laws and normal causal laws (as in natural science) are pure models whose application is only empirically in doubt, there is no difference. Thus when I say that a model predicts trends rather than events, I am referring to its use and interpretation, not its own structure or logic; and I thereby take account of looseness of empirical fit, and our inability to identify genuinely independent variables in the complexity of phenomenological properties, and also the essential limitation on use of laws which is due to quite contingent accidents.

Chapter XV

RELATIVITY THEORY AND HISTORICITY OF PHYSICAL SYSTEMS

Hans-Jürgen Treder
German Academy of Sciences

It is one of the most fundamental and elementary experiences in physics that all macroscopic bodies, especially the stars and apparently also the visible part of the cosmos, have a history. This means first that the state of any macroscopic body changes irreversibly in time, so that its state depends on its whole history. Moreover, it means that there exists one well-defined direction of time in the cosmos, from past into future. The irreversible changes of the cosmic bodies take place relative to this direction.

It is further a well-known fact that all the elementary processes in physics are reversible and that their laws are symmetric to a change in the sign of time. (It seems possible that a weak breakdown in this symmetry takes place in the K-meson decay, but that should be of minor importance to the issue under discussion.)

The best known attempt to reconcile irreversibility with reversibility is by Boltzmann and Gibbs, who developed a plausible theory of statistical thermodynamics. Therein one distinguishes between macroscopic and microscopic states. A macroscopic state corresponds to many microscopic states. These microscopic states are all equivalent, but the statistical weight of the macroscopic state is directly proportional to the number of corresponding microscopic states. The probability of a macroscopic state increases with the number of corresponding microstates. Irreversibility of macroprocesses implies transition from a macroscopic state with few corresponding microstates into one with more corresponding microstates.

But it is also known that this attempt represents only a partial answer to our problem. In particular, the connection of irreversibility with the unique direction of time leads to a vicious circle. The theory of ergodics has shown that a well-defined direction of time is already needed to conclude a time-directed evolution of physical systems in statistical physics. Such a sharply defined time direction is required for any explicit statement of the initial and boundary conditions.

Further, we know from astronomy that there is a general evolution of the universe (of the meta-galaxy), and it seems impossible on logical grounds to explain this evolution by invoking the main concepts of statistical physics. The universe is definitely not a closed system; it seems that new matter is continuously created, probably in the form of new cores of heavy galaxies.

Hence, one can contend that exactly the time-directed structure of the cosmos induces the time-directedness of all physical processes. If this claim is correct, then the direction of time is not the result of a statistical average, but of a dynamical law induced by the structure of cosmic space. This conception has been developed in recent years and we want to apply it in the first part of our consideration. We shall later see that such an answer too does not include the whole solution of the problem; cosmology will eventually demand radically new ideas.

I. In special relativity theory, all events E causally connected with the event E_0 at time $t = 0$ at the world-point P_0 ($\mathbf{r} = 0$) are given by the boundary and the inner region of the Minkowski light-cone:

$$x^2 + y^2 + z^2 - c^2 t^2 = 0. \tag{1}$$

For retarded fields (solutions of the field equations with the argument $r - at$ and $0 < a \leqslant c$) the half-cone with $t < 0$ contains all events E_p which had causal influence on E_0. The half-cone $t > 0$ contains the events influenced by E_0. For advanced fields (solutions of the field equation with the argument $r + at$ and $0 < a \leqslant c$), both half-cones change their meaning. The half-cone $t < 0$ here is the locus of the events E_A influenced by E_0, and the half-cone $t > 0$ contains the events influencing E_0.

If there exist symmetrically both retarded and advanced fields then there is no difference in the physical meaning of both half-cones — past and future do not differ and are not defined.

In reality, only retarded fields exist. The definitions of past-cone and future-cone are unique. The half-cone $t < 0$ contains the events influencing E_0 and the other contains the events influenced by E_0. Along the world-line $\mathbf{r} = 0$ of the observer in P_0, we assume the events $E_{-2}, E_{-1}, E_0, E_1, E_2$ at times $t = -2, -1, 0, 1, 2$. The passive half-cone of E_0 then contains the passive half-cone of E_{-1}, which contains the passive half-cone of E_{-2}, and so on. On the other hand, the passive half-cone of E_0

is contained in that of E_1, and so forth. The size of the passive-experienced events increases with eigentime t, as the size of the events which can be influenced decreases. The sequence of the relations of the so-called "containing" are perceived as a directed flow of events, defining the direction of time from passive-past to active-future. In regard to the observer, it is essential for the existence of a subjective direction of time that the half-cones of the Minkowski light-cone differ from each other, and this difference is implied by the non-existence of advanced solutions of field equations in reality.

One can see also that a world containing only advanced effects is not distinguished from the one described above. The size of the passive-registered events increases with decreasing time. The direction of time, defined by the flow of events, is opposite to that of the first case. It is given now by $t = -t$. Relative to this subjective time, the advanced fields in t are now retarded, and vice versa:

$$r - at = r + a\bar{t}, \qquad r + at = r - a\bar{t}. \tag{2}$$

In a plethora of research papers it has been shown that the advanced solutions of field equations are in reality eliminated by the expansion of the universe owing to its non-stationariness.

If the Einstein equations are postulated, a stationary, homogeneous, and isotropic cosmos is not stable. Even an infinitesimal perturbation leads to expansion or contraction. Further, for a stationary universe to be consistent, it would have to be in radiative equilibrium. Each region of space emits and absorbs the same quantity and quality of radiation. In such a universe no direction of time can be defined. But, indeed, the Einstein theory of gravitation does not permit a stable stationary universe — nor therefore a radiative equilibrium.

If in the universe there exists a system of coordinates in which relative velocities and accelerations are zero in the mean — a hypothesis signifying that the universe does not rotate globally — then one can define a global cosmic time by construction of a synchronous system of reference, corresponding to Einstein's definition of synchronism and simultaneity. This system is the rest-system of cosmic matter. It is mathematically realized by a Gaussian coordinate system:

$$g_{00} = 1, \qquad g_{i0} = 0, \qquad (i = 1, 2, 3) \tag{3a}$$

in which the world-lines

$$x^i = \text{const.} \tag{3b}$$

are the geodesic world-lines of cosmic matter. The geodesic distance

$$x^0 = c(t_1 - t_2) \tag{3c}$$

between two space-like hypersurfaces $t_1 = $ const. and $t_2 = $ const., which describe the state of the universe at t_1 and t_2, defines the difference in time between two events E_1 and E_2.

If matter in the universe is homogeneous and isotropic, then the synchronous system of coordinates is given by the metric of Robertson-Walker;

$$ds^2 = c^2 \, dt^2 - R^2(t) \, d\sigma^2, \tag{4}$$

where $d\sigma^2$ is the metric of a 3-dimensional space of constant curvature. This form of the metric is induced only from geometrical and kinematical assumptions, not from field equations. According to astronomical observation, the universe is indeed nearly homogeneously and isotropically filled with matter, especially if one substitutes in the place of systems of stars more continuous matter. The very points of this continuum define an Einstein synchronism, provided the world-lines of the particles are selected as lines of coordinates of the world-time t. Such a universe contracts or expands according to Einstein's gravitational theory.

In a universe expanding sufficiently rapidly, where

$$R \sim t^n, \qquad (n > 1)$$

a "world-horizon" appears in consequence of this expansion. In an expanding universe, a world-horizon means that a given world-point P_0 of 3-dimensional space can be attained by retarded actions only when they originate from a source located within a certain 3-dimensional neighborhood of P_0. This neighborhood is limited by a sphere, the world-horizon of P_0. Events occurring outside the sphere cannot bring about any retarded action of P_0. Such a world-horizon results from a weakening of all retarded actions which increases with the distance covered. The weakening is induced by a Doppler effect lowering all frequencies as well as by a simple weakening of intensity with distance. The result of both effects is given for radiative sources not too far away from P_0 by the equation

$$I \sim \frac{\text{source strength}}{4\pi r^2 R^2(t) \left(1 + \dfrac{\Delta\lambda}{\lambda}\right)^2}. \tag{5}$$

The advanced actions, as mentioned before, can be interpreted as retarded actions with respect to the subjective time t, oppositely directed to $\bar{t}$. Now, with respect to the time $\bar{t}$, a universe contracts and expands with respect to the time t. Therefore, the exactly reversed effects occur for all advanced actions as for the retarded ones. The Doppler shift of the frequencies here generates an increase of the frequencies; the radiative density grows permanent with progressing time. No world-horizon exists for advanced actions in an expanding universe — the advanced field leads to divergences in such a universe in contrast to the retarded actions. The hypothesis of an expanding universe with advanced actions leads to logical inconsistency. Indeed, Hogarth showed that in an expanding universe no advanced fields can conceivably occur.

Because of the convergence of the retarded, and the vanishing of the advanced, actions in an expanding universe with world-horizon, no radiation equilibrium is possible; the famous Olbers paradox no longer exists. The non-existence of advanced actions means for an observer at rest in the universe (who is at time $t = 0$ at the point P_0) that the half-cone with $t < 0$ represents the passive past which cannot be influenced by the event E_0 any longer; the other half-cone $t > 0$ represents the future, actively influentiable by the event. Therefore, in an expanding universe a subjective direction of time exists for each observer. This direction is determined by the expansion of the universe — the radius of curvature $\sim R(t)$ of the space V_3 increases with the subjective time of each observer at rest in the universe.

The equal-directedness of the subjective time with the time defined by the increasing world radius R verifies a *Gedankenexperiment* which is to change the expansion of the cosmos into contraction. For a contracting universe, the world-radius decreases with time. For such a universe — in full analogy to the above considerations — a weakening of intensity and a Doppler effect into lower frequencies follows for advanced fields. Further, a world-horizon exists for each observer too, but for advanced fields only provided the universe contracts sufficiently quickly with time t. The retarded fields would lead to divergences and are therefore to be excluded. In fact, in a contracting universe all retarded actions are suppressed.

Because of that, for an observer at rest, the half-cone $t > 0$ at point P_0 and time $t = 0$ includes all events E_p which have influenced the event E_0; it is therefore the locus of the passively experienced past. Contrariwise, the half-cone $t < 0$ includes all events E_A that can be influenced by E_0; consequently it is the active future of the event. The observer accordingly experiences a flow of events leading from positive to negative values of t. As noted above, the subjective time is given by $t = -t$. But with respect to this reflected time t, the world radius R grows, and all fields in the universe are retarded with the argument $r - at$.

An observer recognizes, therefore, a cosmic red shift for the fields retarded in his subjective time t. Consequently, generalization yields: the subjectively-oriented time of each observer is given by the expansion of the cosmic space V_3, that is, by the increase of the world radius R. This thesis implies no information about the state of motion of the universe, but is a simple consequence of the definition of the subjective-time direction. For each observer there always exists, by definition, a universe expanding in his subjective time. A contracting universe is logically excluded.

II. It is thus a matter of fact that the direction of time is determined by the kinematics of the cosmos, namely, by its historical properties following from general relativity.

Several physicists have proposed the question whether in connection with the historicity of the cosmos there are also to be expected historic properties of elementary processes, which could be, in their turn, an essential feature of the history of the bodies in the cosmos. Such an influence (of the evolution of the universe in time) on the elementary processes would mean that the eigenvalues of a microscopic system which characterize its state, especially also some universal constants, are time-dependent and, more exactly, also history-dependent. First, this fact can imply that some dimensionless constants vary with the cosmic time, but second, it may suggest that eigenvalues and universal constants depend on the history of the elementary particle in question.

The latter would signify, I suppose, that at the cosmic time $t = 0$ a number of elementary physical systems (for instance, elementary particles of the same kind) are at a point P_1 of the cosmic space and that they are then undistinguishable (especially rest masses, lifetimes, and all charges agree). We assume further that the world-lines of the particles differ and that they meet again at time $t = a$ and point P_2 of space. Then historicity of universal constants would imply that the particles now can be distinguished because the characteristic values of rest mass and charges differ.

In mathematics we describe that as follows: the relativistic eigenvalues are scalars. If we consider the force-free transport (parallel transport) of a vector along a time-like curve $x^i = x^i(\tau)$, then the dependence of eigenvalues on the history means the following:

From $A^i_{;k} \, dx^k/d\tau = 0$ (parallel transport of A^i) does not follow $(A^iA_i)_{;k}$ $dx^k/d\tau = 0$ (constancy of the scalar), but at least

$$(A^iA_i)_{;k} \frac{dx^k}{d\tau} = A^iA^k \phi_{ikl} \frac{dx^i}{d\tau} \tag{6}$$

if we choose the simplest statement. But this statement is equivalent to the cancelling of the Lemma of Ricci. Instead of this lemma,

$$g_{ik;l} = 0,$$

which expresses the metricity of the affine connection, we have to put into our cosmological questions a general statement:

$$g_{ik;l} = \phi_{ikl}, \quad \text{with} \quad \phi_{ikl} = \phi_{kil}. \tag{7}$$

Therefore, one can introduce in cosmology the concept of historicity by the assumption that the geometry of space-time is not metric with respect to the affine connection. This is the manner of reasoning already followed from a discussion between Einstein and Weyl in 1918 about Weyl's conformal field theory. It is easy to see that the concept of the aging of some universal constants can be interpreted as a special case of our general hypothesis.

Here one has only to notice that (corresponding to Weyl's postulate) the world lines of cosmic matter can be brought into the form

$$\frac{dx^r}{d\tau} \delta^r_0 \quad \begin{array}{l} \text{with} \\ \text{and} \end{array} \quad \begin{array}{l} g_{i0} = 0 \ (i \neq 0) \\ g_{00} = 1. \end{array}$$

Then the aging is described by a non-metric affine connection which, in a synchronous system of reference, has the form

$$g_{ik;l} = \phi_{ik0}\delta^0_l \ , \tag{8a}$$

with

$$\phi_{ikl} = \lambda g_{ik}\delta^0_l \ , \tag{8b}$$

This concept is a generalization of the known hypothesis of Dirac, Milne, Jordan, Hoyle, and others. At least for cosmological speculations, it seems to be an interesting possibility of thought to introduce geometric structures of the cosmos with non-metric affine connections. Evidently the historic field ϕ_{ikl} (7) has to be very weak, because the astrophysical facts prove that the universal constants in the neighborhood of our galaxy, at least in a good approximation, have the same values as on earth. It is also possible to think that the non-metric part of the linear connection

$$\Gamma^i - \{^i_{kl}\} = \tfrac{1}{2}\, \phi^i_{kl'} \tag{9}$$

which describes the historicity, is a functional of the other physical fields. This would imply that for the different types of fields the same metric is clearly defined but that we have different affine connections. In this way one could represent the change of the ratios of the different coupling constants in a geometrical manner.[1] Moreover, on this occasion the historic field can also be a macroscopic average about singular elementary processes.

The following idea could be of interest. Let us assume the historic field to be of the form (8b) in a synchronous system of reference; then the values of the universal constants, such as rest mass and charges, depend on the time, since the particles existed in the cosmos. The matter of newly emerging galaxies could differ from the matter of older galaxies exactly by these constants.

DISCUSSION

MERCIER: Treder is dealing with the question as to why there is a direction of time. To answer it, he confines himself to the consideration of physical processes. He notes that, in spite of the connection between directedness of time and irreversibility, one is led to a vicious circle by taking this connection literally, because of the necessity to state boundary conditions. Then, starting from Einstein's equations, he explains that if divergences are to be discarded, only retarded fields can propagate in the universe, and that by definition of the subjective time of any observer, the universe must be expanding.

The existence of a subjective time for all observers in those expanding universes for which Treder has given certain general conditions seems to be an important statement. This (subjective) time is given by the increase of the radius of three-dimensional space.

Therefore the cosmos appears to be a historic cosmos. Treder calls it "historic" and this term may be subject to some questions. Treder puts the question — which very closely relates to the paper by Gamow —: Are the microphysical constants also historical? I think that Treder was able to give general conditions under which such constants will be dependent upon cosmic time, viz., that space-time must be non-metric with respect to the affine connection, and specified that according to observations made, the fields eventually responsible for a time-dependence of such constants must be, at most, very weak.

---

[1] The concept of different linear connections for different types of fields is already introduced in relativistic physics by the spinor connection. The spinors are all scalars in space-time; they are not to be distinguished by algebraic means from scalars, but their affine connection differs from that of Bose-fields, corresponding to the formulation of Einstein and Mayer that spinors are scalars with a different law of linear connection.

YOURGRAU: I do not quite see the necessary connection between Boltzmann's and Gibbs' treatment of the Second Law, statistical physics, Boltzmann cosmology, and relativity theory. If one intends to provide a plausible model, I don't think it is necessary to enter into that very tricky connection between the microcosmos of statistical thermophysics and the macrocosmos of general relativity—we might easily get into tremendous difficulties of a mathematical and a physical nature.

In addition, I think that Treder tries to show that Einstein's universe is in a way no longer contigent upon the model of the universe of the astronomer or astrophysicist, since it purports to furnish some sort of super-theory, which could embrace many possible types of universes, some in a trivial manner, others in an informative and physically significant way.

Now, what happens, for instance, if certain phenomena, accounted for by experimental physicists or observational astronomers, can be interpreted in a non-relativistic manner? Bondi, Synge, *et al.,* have tried to show that red shift does not require a relativistic explanation.

And another question: What happens if—as some physicists predicts—we should one day discard field theory altogether? After all, we know that there are several valuable, essential, and positive insights in Einstein's general relativity theory. And thus I ask myself: What will happen in the future if we may be able to establish by reason of cogent arguments that our formalisms do not require the field concept at all? Will then the formalism you presented today still be valid? What will we do with ϕ and our symmetries, once this idea has been laid to rest?

Treder says further that he will also allow in his model for a contracting universe, although he considers this only as a possible, very special case. I am sure Treder is well aware of the fact that most physicists today accept some sort of expansion-of-the-universe theory, and therefore, the increase in the radius. And yet Treder seems to be absolutely convinced that the Einstein model fits beautifully into an expanding universe. But Einstein had to resort to other models, e.g., to that of de Sitter. One wonders to which model Treder is actually referring.

May I perhaps suggest a possible answer? Einstein venerated Archimedes. As a student, I couldn't quite understand why. (Einstein did not know Greek science very well, be it pure or applied.) However, now I do. When Archimedes was asked how many grains were in the universe, he replied that it depended entirely upon which of the two (plausible) universes one had in mind. If one had one kind of universe in mind, he would arrive at a certain figure; if one preferred the other type (model), the numerical value would be different.

TREDER: I agree that the Boltzmann cosmology is one of the most sophisticated ideas in physics. In Boltzmann's time, his cosmology was the only possible way of understanding the cosmic time-order. However, Boltzmann's conception in this particular realm is rather strange. Since we are acquainted with general relativity, we have other models for the cosmic time-order and these models are not so strange as Boltzmann's original idea.

The interpretation of the cosmic red shift as a Doppler effect is not only based on concrete astronomical data; it is the only possible explanation of the Olbers paradox! In fact, this paradox provides *the* fundamental support for our belief in an expanding universe.

In the hypothesis proposed in my paper, the expansion of the universe is a consequence of the direction of subjective time. In other words, according to the direction of subjective time, the universe expands *per definitionem*.

COHEN: If the time-order is so closely connected with, or even defined by, the rate of change of R, is it possible to consider different time orders within such an Einsteinian universe, what some people (Reichenbach prominent among them) call "branching time", or would that be forbidden? And connected with this question, I am not sure that using the rate of expansion is really a suitable way of defining the order of time. It seems to me there must be hidden here in some logical underground a previous notion of time-order in terms of which the rate of change of R is expressible; or does Treder mean to say that the time variable is actually defined by the rate of change of R? Put in other words, can this part of Treder's paper be formulated in terms of a causal theory of time? I have some doubts.

TREDER: In the space-time which I discussed in the first part of my paper, there exists one congruence of space-like hypersurfaces. The surfaces are universal rest-systems of the quasi-continuous and irrotational cosmic matter. If we give the surface-congruence the equation $x^0 =$ const. and if we construct on these surfaces a Gaussian coordinate system, we have a universal Einsteinian synchronism with a cosmic time $x^0 - t$.

In this time, the time-difference Δt between two events E_2 and E_1 is the geodesic distance between the hypersurfaces $x^0 -$ const. which are containing these events. This is the time-difference $\Delta t = x^0(E_2) - x^0(E_1)$. In the Robertson-Walker metric this also means: $t = R(E_2) - R(E_1)$.

My theory of time is a causal theory in the sense of Reichenbach and my definition of the cosmic time-order is a special case of Einstein's definition of synchronism and is possible for a universe with a quasi-continuous distribution of matter without rotation.

BONDI: I am glad to see that Treder takes the Olbers paradox so seriously. I have been preaching this paradox for nearly twenty years, and it is very nice to find some fellow-feeling.

I am a little less happy about the idea of constants changing with time, on the grounds that the perfect definition of the universal time in the Robertson-Walker system with universes assumed to be completely uniform does not very easily carry over into the actual non-uniform universe. In particular, there will be non-integrability. If I start out from here and draw a conceptual circuit so that I move from the starting point to a simultaneous point (and so on), then it follows that when I retrace the circuit to the starting point it will not be simultaneous with the initial point of departure. In other words, if every bit of the universe carries its clock with it since the year dot, and if the constants of physics depend on this, then we should have here and now a mixture of such constants; my feeling is that, contingent on the kind of variation you assume, this mixture may be more than our experiment can tolerate. I think this should be very closely investigated.

There are only two ways out from this difficulty that I can see: one is repulsive and the other is complicated. It is repulsive to assume that there is a universal time irrespective of the motion of matter, because this makes space and time absolute again, which I think no theoretician would like very much.

The only other possibility is to introduce a universal world scalar defined by a field equation like Hoyle's C-field or the field that Dicke works with, which introduces a new complication for every limited purposes indeed; therefore, I don't like it either.

TREDER: I agree, of course, with Bondi as to the paramount role of the Olbers paradox.

To the question of the constants changing with time: It is the main point of my discussion of this problem that "history" of a physical elementary quantity means that the value of this quantity *hic et nunc* is dependent on its space-time path. Historicity of universal physical constants means indeed non-integrability of the transport of scalars formed from vectors, etc. This non-integrability is the consequence of a non-metrical affine connection.

YOURGRAU: Perhaps it would be helpful to restate the nature of the Olbers paradox. More than a century ago, the German astronomer Heinrich Olbers discovered the following puzzle: Suppose the universe is infinite and crowded uniformly with stars. The cogent *sequitur*? The heavens at night should be ablaze with light. The reason is simple: distant stars will of course be faint; still, they will be so numerous that the total illumination due to remote stars in a shell of thickness dr would equal that by nearby stars in a shell with the same dr:

$$dI \propto 4\pi r^2 \rho dr\left(\frac{1}{r^2}\right) = 4\pi\rho dr.$$

And yet, the stunning fact is that the nights are almost dark. How can one account for this paradoxical state of affairs? Well, one might say that, with increasing distance from the earth, density distribution of the stars decreases. But this explanation never found many advocates, because it accords to our earth the hardly justifiable fate of being situated in the densest corner of the universe! This inference, moreover, has never been verified. Thus, cosmologists propounded the theory that galaxies are ever *receding*. And this theory has been empirically corroborated by solid observational data.

GAMOW: I am going to make some general remarks concerning two different ways in which cosmological problems are treated. One approach is that of "postulatory cosmology", in which one asks what the properties of matter and radiation *should* be in order to obtain philosophically desired cosmological models. Another approach may be called "factual cosmology," in which we accept the physically established laws governing matter and radiation and look for cosmological models which are derived on the basis of these laws and are consistent with astronomical observations. As a physicist, I prefer the second approach, since today we know the laws of matter and radiation well enough to use them in broader regions than those covered by direct experiment. When an engineer wants to design an automobile, a jet plane, or a spaceship, he starts with the well-known physical and chemical properties of the materials he uses and looks for the arrangement of these materials which would satisfy his purposes. Similarly, a physicist looks for the universe which will satisfy known laws, be these obtained by direct experiment or theoretical deductions.

For example, the time-dependent solution of basic equations of general relativity, which satisfies the requirements of homogeneity and isotropy, is

$$\frac{dl}{dt} = \sqrt{\frac{8\pi G}{3}\rho l^2 - c^2\alpha^2}\,,$$

where l is the distance between any two selected points, and $\alpha = l/R = $ const., R being the curvature radius of the universe.

How does the density ρ change with time? Well, ρ is composed of matter-density ρ_{mat}, which (due to the conservation of matter) is inversely proportional to l^3, and ρ_{rad}, which is inversely proportional to l^4. Thus one concludes that during the early stages of expansion the radiation predominated over matter. This gives us the temperature dependence on time and permits us to calculate various thermonuclear reactions on the basis of well-established laws of nuclear physics. Similarly, using the laws of gravitational fluid dynamics, we can predict the masses and sizes of the condensations which became galaxies.

It is interesting to notice that the temperature obtained in such calculations for the early stages of the expanding universe, being extrapolated to the present time, gives for the residual temperature today the value $7°K$[1] in rather good agreement with recent measurements.

I want to state very strongly that, in my opinion, the problems of cosmology should be solved on the basis of known physical laws and not vice versa.

TREDER: The Newtonian theory is an approximation for weak gravitational fields in a quasi-flat space. In the cosmogony of the universe, this condition is not fulfilled for all times and domains.

The possibility of a Newtonian interpretation of the cosmological facts is not clear *a priori*. But we have to prove the conditions for a Newtonian interpretation from the results of Einstein's exact gravitation theory.

---

[1] G. Gamow, *Det Konglige Danske Videnskabernes Selskab,* **B27**, No 10, 1953.

Chapter XVI

GENERAL RELATIVITY AS AN OPEN THEORY

Hermann Bondi
University of London

The question of the openness, or comprehensiveness, of a theory is very much related, though not identical with, certain questions that became apparent very early in the development of modern physics. It is, I believe, only an apocryphal story that when Newton first proposed his theory of gravitation and the attraction of the earth on the moon, there were objectors who said that it was absurd, when people knew next to nothing about what went on *inside* the earth, to pretend to be able to infer its attraction over a far distance. They felt that a knowledge of the constitution of the earth was a necessary prerequisite to the discovery of the laws of satellite motion. Nowadays we might tend to laugh at this, but, of course, there was a grain of justice in what they were saying.

If we look at the very long-term effects, such as the secular acceleration of the moon, then they are indeed closely determined by the imperfection of the elasticity of the earth. But the major phenomenon — the general motion of the moon around the earth — is one of those many cases in physics where we can say *something* without knowing *everything*. I think that this is an extremely important point in science throughout, that you must so ask your questions that you can get *some* answers in spite of a great deal of ignorance. I sometimes call this the principle of maximum shoddiness, that you should always so design your equipment (as is, of course, well known to experimentalists) that the errors and the imperfections which are bound to be there matter least. And it is one of the major tasks of science, in my view, so to formulate its questions that a great deal can be said (not everything, but a great deal), in spite of a very substantial lack of knowledge.

In my great admiration for Newton there is perhaps nothing that impresses me more than the neat way he cut the problem of the solar system in two. After all, at that time there was no obvious reason to think that it was easier to answer questions as to why the planets moved in ellipses than about why Jupiter has so many times the mass of the earth. There was a single complex of questions about the structure of the solar system and the way it behaved, and this complex of questions was indeed very neatly cut in two by Newton. One set of questions, namely, "given the masses, positions, and velocities of planets, how are they going to move?", was for all practical purposes solved by him. The other half of the problem, as to why do the planets have those masses, positions, and so on, is essentially the problem of the origin of the solar system, and 300 years later we aren't very much wiser. Thus to pick out what you can say in a state of extreme ignorance strikes me as a vital feature in the whole pursuit of science.

We have, I think, a lot to learn from this and should keep it in mind when we consider general relativity. General relativity is a theory that in a certain sense is comprehensive. It is comprehensive in the sense that it allows for laws like the conservation of momentum. Now, we know of course that conservation of momentum is a super-principle. It applies to a system irrespective of what goes on inside that system. And so you can again say something without saying everything.

I would like to give a rather prosaic example. This was a problem we faced many years ago when we had a baby in its carriage asleep in the long hall corridor of our house. Some time or other the baby would wake up and would start kicking, and through various complicated effects this tended to move the well-braked baby carriage until it hit the wall and then scraped the paint off. If you try to see how you can prevent this happening, then there are two ways of doing it. One, you can try to get a very high-grade baby psychologist who tells you when the baby will wake up and how it will kick and what position and initial velocity you should give to the carriage to avoid its bumping the wall. We rejected this method. The other method is to take the brake off, because when you take the brake off, no horizontal force is exerted on the baby carriage. Therefore the center of mass of the system (baby plus carriage) must stay put irrespective of what goes on inside. If the baby is strapped in, it cannot move far relative to its carriage, and so the carriage cannot move much relative to the center of mass of the system which, as we have seen, is fixed in space. Thus the carriage cannot bump the wall. Here we have usefully applied the principle of conservation of momentum in spite of our total ignorance of just when the baby will wake up and how it will kick. And such a method works.

This seems to me exactly the way science should operate. However, sometimes scientists are more ambitious, and I think we have had a great deal of such ambition in regard to general relativity. General relativity is the modern theory of gravitation. We know gravitation-producing sources from observations, and these sources are always large bodies. Whether or not we can apply this knowledge on a small scale, we do not know. Sometimes our theories work well beyond the range of experiments on which they are based, and sometimes they don't. Our experimental knowledge of relativity, certainly as far as the production of gravitational fields is concerned, has as its limit quite a substantial size of body. At that size of body the atomic structure of matter is not relevant and, indeed, the theory, *qua* theory, is a continuum theory. I don't think that you can, without a great deal of caution, go much further than to say that the way mass, momentum, and energy appear in relativity, viz. through the energy-momentum tensor, is in the form of continuous functions and not as associated with particles.

The problem of particles has, of course, been raised frequently in general relativity from two quite different points of view. One is to expect relativity to hold absolutely and for all sizes because it is a very beautiful theory. Thus you may expect the theory to say a lot since it is mathematically so terribly complicated. Some scientists therefore apply it down to particle sizes, trying to see what the results may be.

The second point of view stresses that the main application ot gravitational theory lies in the astronomical field where the separation between bodies is always very much larger than the bodies themselves, and therefore, again, we can regard the sources of the gravitational field as point masses.

Now, I do not think of either of these views as particularly convincing. When we use relativity rather than Newtonian theory on the large scale, then we do it because we want to be very precise. Only very accurate measurements reveal the difference between the two. When we want to go to this degree of precision, then matters like the imperfection of elasticity of the earth (which leads to the secular acceleration of the moon) are of relevance; the earth acts in this manner only because it is an extended body and not a point.

Another objection stems from the fact that when the point-source concept is used, what is really done is that the ordinary description of energy and momentum in a continuous body is replaced by a singularity, and this again does not seem to me to be terribly sound because we are still dependent on the structure of this singularity.

To show how important this structure is, remember that there is nothing in the theory to prevent the existence of bodies with negative mass in the sense of producing a negative, i.e. a repulsive gravitational field. But, of course, the theory is based on Galileo's principle that all bodies fall equally fast, and "all" in the language of the theoretical physicists means *all.* So if we have an ordinary mass and a negative mass close to it, then the negative repels *everything*, including the ordinary. The ordinary attracts *everything*, including the negative. Thus the two will shoot off, under suitable circumstances, at the same acceleration. If we bring the two together to make a dipole, we have a self-accelerated dipole. Thus the motion of a body depends on its structure even if it is only a point singularity, and therefore we need a complicated, if you like, singular equation of state for the body.

Granted that bodies are so difficult in the theory, there are two ways one can go about trying to improve this. One is to abolish bodies. This is the kind of method that Vigier discussed. Relativity is a very complicated nonlinear theory. We can apply the empty-space conditions everywhere in such a way that we simulate matter through just having gravitational fields although there is no matter. The gravitational field, as it were, holds itself together through its own attraction, forming the geons of Wheeler or the kind of particles which Vigier mentioned. Then, by applying the empty-space equation everywhere, we can mimic the behavior of real bodies.

I feel that we are restricting ourselves very greatly by doing this kind of thing. It may be successful, I do not deny. But my own temperament is just not that ambitious. It is clear that if we rely on these self-concentrated bundles of empty space to make up our matter, then it will in certain ways be limited, and if those who try this technique are as lucky as they are ambitious, then these limitations will be just the actual limitations of the matter we know. I must confess I think the chances are against it, but if others wish to spend their time on this idea, it is their time, not mine. We must always follow our own hunches, and that is the beauty of science. If we all thought the same way, science wouldn't advance at all. It is only because there is variety in our hunches that we get anywhere.

The other method is to restrict the empty-space equations to the part of space that is actually empty. In other words, we say that there are sources of the gravitational field. They can be anything; matter can, indeed, have many different properties and shapes and sizes, but we only say that it doesn't extend beyond certain regions. In this sense, then, you can use the theory as an *open* theory. Here I use "open" with the same meaning as when I state that Newton's second law is an open theory: when you say the rate of change of momentum equals the applied force, then in no way do you specify the applied force. You are quite free to put in what you like: elasticity, gravitation, electric forces, etc. In relativity we can do very much the same by putting in an energy-momentum tensor and not restricting it at all. Alternatively, one could restrict it in a suitably vague way corresponding to our vague knowledge of what matter is really like. For example, we never find pressures without density, and the density of matter tends to be positive, etc. But still we would regard general relativity theory as applicable.

I myself have been engaged for the last few years in this kind of attack, where in fact we have considered what can happen to the gravitational field outside a body of finite size; never mind what that body does. In other words, we have here a finite bounded region. Outside it space is empty. The nature of what is inside the region is not examined. We want to discover what gravitational fields can exist outside as a consequence of having this finite but otherwise unrestricted source inside the region.

It has sometimes been argued in the past that this isn't a very good method because if there is a finite system there, and we don't want the after-effects of its being put there to be appreciable, then it must have been put there very long ago. If it was put there very long ago, nothing can happen to it now. This criticism is incorrect, as will be clear to those acquainted with time bombs. A time bomb can be put in position very long before it goes off, and changes its shape and structure. I might say that a time bomb is of great interest to me because it is one of the very few practical examples of a material whose equation of state contains the time explicitly. I think this is quite an important concept which plays very much the same role in gravitational theory as an active, as opposed to a passive, system in electrical network theory.

Now, I would like to say a few words about our successes and our failures in this direction. We have assumed that we live in an expanding universe, and that therefore only retarded effects exist. Thus one finds that if the body changes its shape or structure, then its gravitational field must change accordingly. The changes convey information about the changes of shape and do indeed travel with the same velocity as light. Moreover, this transmission of information implies a transmission of energy and therefore a loss of mass of the source. All this came out rather neatly and nicely from the equations and showed us, so we thought, that in general relativity we have really a sensible theory of gravitation, where changes in the gravitational field are propagated with the fundamental velocity, and where transmission of information requires transmission of energy, which on general physical principles is just what one would expect.

Unfortunately, Newman and Penrose recently made an absolutely disastrous discovery. In my own field it seems to me as disastrous as Rabi's discovery of the quadrupole moment of the deuteron which, of course, was one of the great discoveries in physics, since it showed the world to be a much more complicated place than had been thought.

Now, what Newman and Penrose found was that, irrespective of emission or non-emission of waves, a certain quantity is conserved. This is a strictly non-Newtonian quantity in the sense that it is not linear in the field variables. In the static case, when the body is at rest and when we know what the Newtonian field should be and the relativistic field is practically identical with it, we can link this conserved quantity with certain properties of the body, basically with its quadrupole moment. Now, let us imagine that we have a body which is originally static and has a nonzero quadrupole moment. Later the body begins to change its shape in an oscillatory manner, thus varying its quadrupole moment. It therefore emits waves and so loses mass. Eventually it settles down into a beautiful sphere which, of course, has zero quadrupole moment. Then we would expect the external field eventually to be that corresponding to a sphere. But since we have the Newman-Penrose conservation law, we know that this conserved quantity is different for the fields due to a static sphere and due to the original static body of nonzero quadrupole moment — and that can't happen, because the quantity is conserved. Therefore, the body cannot settle down completely into a sphere at the end, or at least the external field will never be identical with that of a sphere.

In some sense this is not utterly surprising. We are dealing here with a problem in the propagation of waves, and it is only in some particularly simple cases of wave propagation (e.g. sound in three dimensions or light in three dimensions) that all of an emitted pulse travels with the velocity of the disturbance. If I clap my hands, then a wave goes out in the shape of a sphere, and all the perturbation is always confined to the surface of this sphere. Inside the sphere there is no perturbation at all, because the whole effect travels with the velocity of sound.

With a very much more complicated equation, as we have in general relativity, it would have been too much to expect such simple behavior, as happens in very many other wave problems. Although the main disturbance travels out on the surface of the sphere, there is something lagging behind (a so-called tail to the disturbance). But in all cases in which we know of such tails, of such remnant effects, they gradually decline to zero. According to this Newman-Penrose conservation law, however, in our case these tails cannot decline to zero. The field can *never* become that of a sphere.

Now this strikes me as extremely serious. It means that the gravitational field of a body depends not only on its present state (or, rather, its state back-dated with the velocity of light), e.g. if I observe the sun from here, then, of course, what matters is what the sun was like eight minutes ago when the light that comes here now was emitted — it also depends, according to Newman and Penrose, on its earlier history right to the year dot, because their conservation law is strict. This seems to me a very unhappy situation. When Newton established the law of gravitation and applied it to the solar system, identifying the sun as the source of the gravitational field, then he had the enormous advantage that the sun was a body that was perceptible by means other than its gravitational field. The Newtonian idea of a source is like a brick that one actually strikes when one gets to it. Moreover, the fact that these sources have certain conservation properties is very useful. Thus Newton told us that the gravitational field is entirely and totally due to bodies that can be perceived by other means and are permanent.

Now we learn that according to general relativity there is at least a little bit of the gravitational field that does not only depend on the brick that I see there now, but on what went on long before, and I find this most disagreeable. It is true there is nothing against effects traveling more slowly than light: there is a rule only against certain effects traveling *faster* than light. But it would be a very much more convenient universe if *everything* traveled with the speed of light, and this dependence on history is something which I think we can now definitely identify in the general theory and which makes it a markedly less attractive theory. Frankly, I don't think we can do any better than this theory; we have to live with it at least for the time being. It is interesting to see the results when we regard it as an *open* theory; it shows itself to be a little nastier than might be expected.

DISCUSSION

VIGIER: First let me comment on the question of the laws of motion. As Bondi has shown, this is related to the question of the nature of matter. Now, I think that in favor of Einstein's point of view (meaning the attempt to describe everything as fields) there are two types of arguments, an aesthetic argument and a "plumber's" argument, to borrow from Yourgrau's vocabulary. I will start with the aesthetic argument. It is always disagreeable to utilize two uncorrelated models within the same level of physical reality. What really (and finally) exploded classical mechanics was Maxwell's discoveries of the field properties of electromagnetic interactions.

Well, it is always nice when one can unify various aspects of matter at a higher level. That is what Einstein did. It is not satisfactory to give first the field equations, and then, on the other side, to postulate the laws of motion of test particles within these fields. The description of matter as singularities or bunched solutions of fields is very tempting because you unify in a single set of laws two completely different points of view. We had been accustomed, owing to quantum mechanics, to think of bodies like bulls in an arena. If you can conceive of the whole situation as a field with bunched aspects, then you have unified in a single series two types of laws which had to be assumed independently. The fewer assumptions one makes, the better it is. That was Einstein's main argument, I believe.

The second point I want to make is the following: isn't the Hamiltonian description of nature all that we really need? The fact that there is a tail does not destroy the point of view that, given sufficient initial knowledge, one can calculate the behavior of phenomena, even if this should prove inconvenient. There is no reason at all to believe that nature has been built for the convenience of physicists!

BONDI: I think that on the last point Vigier and I are in agreement. I mean we have managed to live with deuterons having a quadrupole moment, so no doubt we can live with history in gravitation. But that does not prevent me from regretting both.

I also want to make it quite clear that I accept entirely the fact that the equations of general relativity imply equations of motion, and, like Vigier, I think this is a marvelous result. But I do not hold it is a particular advantage, as a question of method, to coagulate matter into points in order to demonstrate this. You can do it with extended matter, with jelly. It is difficult to assume equations of state. It seems that he also assumes equations of state for the particles, but his equations are singular ones, and I prefer the alternative; still, this is not really an important difference.

TREDER: First, the derivation of the equations of motions from the Einstein equation in general relativity means that the world-line of a particle (singularity) depends upon the structure of this particle. Only for "monopole"-particles is this world-line a geodesic line.

Now, concerning the point about Maxwell and the field. Maxwell undoubtedly was guided by the concept of the field, and through this arrived at his equations, which are what we test. But we know at least since the work of Wheeler and Feynman that Maxwell's equations can also be derived in a different way! And yet, the myth of the superior and unique importance of the field concept tends to persist.

Second, I agree with Bondi and Vigier in that it makes sense to prove the particle-like properties of geometrical structures (like geons, etc.). Yet I do not believe *a priori* that all matter-structures are geometrical structures; but I maintain that all geometrical structures have a physical meaning.

In the case of Vigier's particle model, I cannot understand the physical meaning of the strong gravitational field existing on the exterior side of the particle surface. The exterior metric must be the Reisner-Nordström metric, and the Schwarzschild constant is the particle mass. This field is strong on the surface only if the geometrical radius is of the same order as the Einstein gravitational radius.

VIGIER: The problem is quite simple. One field is very weak, the other very strong. Why should they mix? Well, for a very good reason: the gravitational equations are not linear and this thing must fit. I mean, it is all very nice and fine for the quantum physicist to say, "We are going to throw away gravitational interactions." You can always do that for a certain type of problem, but in reality, if you want to work correctly, you have to take them into account. Even when gravitation is very weak, in a way it is always important, and for a very simple reason:

in Einstein's point of view the equations are nonlinear; if one fits the $g_{\mu\nu}$ and the first derivatives, however weak the field is, it influences the behavior of any high concentration. That is all. It is a very well-known property of nonlinear equations since the time of Poincaré.

YOURGRAU: Let me make an imaginary experiment. Say I have a fictitious physical constant which I call a "Bondiday", symbolized by B. Now, I can express some physical constants, fundamental physical constants, such as the fine-structure constant α, in terms of other physical constants. I will write down:

$$\alpha = \frac{hc}{2\pi e^2} \quad \text{and} \quad B = \frac{R^3 \kappa e^2 h}{4\pi c^3}.$$

Then, suppose I have been able to measure the second (absurd, night-marish) constant empirically. (Great "plumbers" like Birge, DuMond, and E. R. Cohen have redetermined the numerical values of some fundamental constants; recently Dicke measured anew the gravitational constant G).

Well, I go to the experimentalists and learn that one or two of those constants on the right-hand side of my above two fundamental, physical constants have been redetermined with a finer degree of precision. Consequently, if one maintains that some universal constants can be expressed in terms of some other (experimentally) established constants, then I fail to understand how one can on logical grounds uphold the notion of fundamental constants in physics. All you can say, mathematically speaking, is that you have parameters. (This follows from Dirac's cosmological principle argument.) Well, we can't have it both ways: if we claim that the universe is a closed, static system, then the notion of a universal physical constant can be retained; if, however, one flirts with the idea of an expanding universe, then one can merely say, "All right, new empirical knowledge compels me to redetermine my constants from time to time."

This is a point which has been ignored in the literature. Why? Because we do not want to give up the notion of fundamental constants of physics. Did not Planck liken them to the fundamental bricks of the universe? Did he not compare them with *the* key-holes through which we can observe how the structure of the universe *really* looks? In my opinion, this attitude has to be dropped: There are no "elemental bricks", in an absolute sense, as physicists assumed when they worked, e.g. with the Newtonian G or Sommerfeld's α.

The fine structure constant is not a dimensionless constant but a hidden-dimension constant—it can be shown that it is a 'piggyback constant', so to speak, which only uses numerical values because of certain (theoretical and empirical) prior commitments. It is not, in the strict sense, the pure numerical constant $\alpha = 1/137$ that we read in textbooks.

I raise this issue because it concerns all who are engaged in questions pertaining to modifications of relativity and quantum theory or of any other physical theory that includes so-called universal, fundamental constants.

BONDI: Well, if our 'constant' is not nondimensional then we can always regraduate our measuring instruments to *make* it a constant. It is truly nondimensional constants that we can't monkey with. 10^{40} is a *true* constant. In a recent paper by Salpeter he bases the whole of stellar structure purely on the nondimensional gravitational coupling constant 10^{40}.

MERCIER: Bondi, at the end of his paper, marvels that it is remarkable how Newton created a theory of gravitation about the planetary system based on the extremely weak gravitational force; we can *perceive* the source of the gravitational field not by gravitational means, but only thanks to light. And thus Newton's law of gravitation tells us why we observe the inescapable "dependence on history".

I think this is a very important point. There are, or at least there seem to be, several types of interaction in the world; for a couple of centuries physicists only knew about two of them, gravitation and electromagnetic interaction.

Now, our watches are built according to various principles: there is in this house a beautiful clock which is a gravitational (pendulum) clock, whereas my wrist watch is not a gravitational one but an electric clock, because it has a spring with a certain structure in it; there is no reason whatsoever *a priori* why these two clocks should be regulated together. If one is declared regular, then one has to consider the possibility that the other is irregular. A radioactive clock or any other clock that might be difficult to make, even by some good mechanic, could be irregular too.

In my opinion, there arises from the variety of possible clocks a problem which is not completely cleared up in physics. Either it is possible to construct one huge, unified theory in which all the interactions are only aspects of one single physical phenomenon—and in that case there is only one time to be measured with any clock because all these clocks will be the same clocks—or it is not possible to do so. I think this question has some bearing on the most fundamental principles of physics and even on history, since history is a temporal phenomenon which finally is measured according to some clock.

BONDI: Of course, I agree entirely with Mercier. I think we owe these considerations basically to Milne, whose contributions tend to be very much underestimated nowadays. He is somewhat out of favor, but he has introduced some of the most important notions in this topic, such as the use of radar as the basic distance-measuring method. He initiated the idea that different types of clocks might measure different times. Because all this was hidden in a cosmology that nobody now likes, the value of these contributions tends to be underrated.

I would, of course, entirely concur that we have, then, two separate questions. Let us concentrate upon the purely observational one: are the ratios of these different forces—which are the same, more or less, as the ratio of the rates of these different clocks—changing in time? Do we have any evidence of such a change in time? Gamow gave us some evidence that a particular ratio between gravitational and electrical forces has been changed by less than a certain given amount in a few hundred million years. I think this is the correct interpretation of his remarks. We can try and amass more and more evidence about these relative rates.

The second point is that you can look simultaneously at theoretical "pictures" that contain variations of one sort or another. To assume no variation is to make as big an assumption as to assume any particular variation.

COHEN: It is interesting to see that Milne's ideas might, indeed, have some philosophical bearing when we consider different kinds of time measurement, particularly since the two orders, or rates, of time are related logarithmically.

Theologians, for example, might be interested to know that a finite origin in time on one rate would be infinitely far back on the other. They would, I suppose, prefer the creation hypothesis and therefore take one rate of time, namely that which starts with zero; and this was exactly the rate preferred by Milne (in his *Modern Cosmology and the Christian Idea of God*). Maybe he fell into disfavor because he preferred that time rate.

In addition, Bondi and Mercier raise the question of how this might bear upon the time of history outside of physics. Perhaps one should say outside of cosmology. I suppose it would now be clear that the time rate of human history would have to be the time scale which pertains to atoms and molecules with electrical vibrations, and presumably not the other one. I therefore do not see why any serious problem would arise for matter organized in atoms, molecules, and people. And hence, I don't see that there is much of an issue here for historians.

QUINE: Yourgrau argued that in his view a "Bondiday" must be regarded as a parameter. And yet it sounded to me as though what was changing was our knowledge or opinion regarding the physical situation rather than the physical situation itself changing through time.

YOURGRAU: Quine offers an interpretation which I did not have in mind. My argument is a mathematical one and physically quite naive.

We determine a physical constant in a laboratory, say, by measuring a constant quantity in specified units. We know furthermore that one can express certain fundamental constants in terms of other universal constants.

Now, if we determine new numerical values for any of the constants appearing on the right-hand side of the equations I presented earlier for α and B, then we are compelled to make corrections throughout, or we have to give up the possibility of expressing the α and B in terms of other constants, because simple mathematical rules force us to make readjustments. But these readjustments require new numerical values for several constants in order to retain the *same* ratio. In other words, once we obtain a refined numerical value, we have to adjust at least a few (if not all) numerical values of the constants in question. How else could we possibly recover the numerical value of the quantity on the left-hand side? All this is basic arithmetic. The problem was recognized by Birge, DuMond, E. R. Cohen and by myself, many, many years ago.

BONDI: The point is that we have very many cross-links in physics. We don't always use the same unit in every place. We use kilowatt hours to pay for our electricity and we use gallons to pay for the energy in our cars, and so on. If we seriously contemplate things changing, then we have to agree to use particular well-defined units. We have got to go to the filling station and ask the chap to put so many kilowatt hours into our cars. And if we do that, then no problem arises: the changes will automatically be carried through. We have, as it were, a very small number of independent constants; all the other are dependent constants, and, of course, if the independents change, the dependents will change.

YOURGRAU: But we are dealing with *measured* independent constants. And the dependent constants too are measured. I still contend: new numerical values in any one of these constants (independent or dependent) demand new (measured) numerical values in the other constants involved.

MERCIER: To comment on Cohen's question about the relationship between problems in choosing clocks and history: if you go to the North Pole where there is no day-unit (only a year-unit), and you have forgotten your ordinary (electric) watch, and you want to know what it is or how everything is going on within periods much shorter than the year, you are left only with your own body, with the beat of your heart, or the fact that you get hungry at certain different times. *That* is your own watch, and this watch is not regular with regard to the watch you have forgotten at home; there may be all sorts of watches in between, for example, the atomic watch and this very bad watch which your own body represents. One can even imagine a human society developing activities in cycles of various natures. All these cycles are watches. In this way, one gets a sequence of kinds of watches in the course of which history at the corresponding levels may be considered.

It would be desirable to have one single watch with the help of which any of these histories may be recorded. However, I think we encounter a certain problem here. Does there exist one universal watch with the help of which any history may be recorded?

BONDI: As far as the functionings of our bodies are concerned, i.e. the biochemical processes, we have surely every reason to believe that these are controlled by the ordinary electromagnetic time, just the same as that recorded by a wrist watch. But when you ask about the food supply we receive and how much our aging depends on the radiation of the sun, then of course there will be a mixture of electromagnetic and gravitational watches; and, indeed, if we think of Yourgrau's nuclear watches, if suddenly all the nuclear watches were speeded up at an enormous rate, the residual radioactivity in us would no doubt do us an awful lot of damage very suddenly. So we are balanced, as it were, between these three timekeepers; or, if you like to have weak interactions and strong interactions separately, four timekeepers. I think in some measure we depend on them all, but if you put a man into a closed, healthy environment and give him food, then essentially he would rely completely on the electromagnetic watch.

COHEN: I suppose history would go by calendars, which are a little difficult to observe since they might be gravitational watches.

YOURGRAU: Perhaps it is the wrong idea to pursue the notion of a universally all-embracing time in the sense indicated by Mercier. That is exactly the chimera we are facing: the urge to satisfy our aesthetic, formalistic preference for elegant theories rather than to struggle incessantly with partial, piecemeal knowledge reflecting physical reality.

Chapter XVII

WHAT IS RUSSELL'S THEORY OF DESCRIPTIONS?[1]

David Kaplan
University of California, Los Angeles

Russell expounded his theory of descriptions in a number of places, but perhaps the best known source is his 1905 article, "On Denoting" [1]. I think it may still be fruitful to discuss the doctrine of that article since some readers may disagree as to its main point.

Theories of descriptions concern the analysis of sentences containing definite descriptions. For example, "The present queen of England is shapely" or "The least prime number is even." Let us refer to the description itself as proper or improper according as there is or is not a unique object described. Thus the descriptions "The present queen of England" and "The least prime number" are both proper since they uniquely describe Elizabeth Windsor and two, respectively. But the descriptions "The present king of France" and "The author of *Principia Mathematica*" are improper since the first describes nothing and the second describes Russell and Whitehead equally well. The difficulties involved in the analysis of sentences containing descriptions (I will say "description" as short for "definite description") are most apparent in connection with improper descriptions. This should not be surprising, since improper descriptions are rarely used knowingly, and thus usage does not provide a clear guide. On even the most elementary question of analysis, the truth value of such sentences, we find disagreements. Consider the sentence: "The present king of France is bald." According to Russell's theory, it is false. According to the "chosen-object" theory of Frege[2], elaborated by Carnap[3], in which all improper descriptions are treated as if they uniquely described some previously chosen

[1]This paper has benefited from a number of sources, including a seminar at Princeton, a colloquium at Cornell, discussions with Montgomery Furth, and National Science Foundation Grant GP-4594.

object, the sentence is true (taking Yul Brynner as chosen object). According to the "truth-value-gap" theory of Frege,[2] elaborated by Strawson[4], in which improper descriptions are treated as having meaning but describing nothing, and sentences containing such descriptions are treated as themselves meaningful but having no truth value, the sentence is neither true nor false. I am aware of no theory according to which the sentence is both true and false, though no doubt such a theory has been or will be proposed.

This much is well known about Russell's theory: He takes the propriety of the description to be a part of the *content* of certain sentences containing descriptions and thus counts them false if they contain *improper* descriptions. More specifically, he claims that a paradigm sentence of the form "The such-and-such is so-and-so" is equivalent to "One and only one thing is a such-and-such, and that one is so-and-so." In symbols:

$$\text{``}F\imath xGx\text{''} \text{ is equivalent to ``}(\exists x)((y)(Gy \equiv y=x) \mathbin{\&} Fx).\text{''} \qquad (1)$$

This analysis does not provide a unique understanding of sentences of the form "The such-and-such is not this-and-that," which may be treated as equivalent to "One and only one thing is a such-and-such and that one is not this-and-that," thus assimilating "not this-and-that" to the "so-and-so" of the paradigm; but the given form may also be treated as equivalent to "It is not the case that one and only one thing is such-and-such and that one is this-and-that," thus applying the paradigm to "the such-and-such is this-and-that" and understanding the given form as its negation. This ambiguity is regarded by Russell as a simple scope problem on a par with the party invitation which reads, "Bring your wife or come stag and have a good time." And he introduces the terminology of *primary* and *secondary* (and by natural extension *tertiary*, *quaternary*, and so forth) occurrences to indicate the intended scope of the description in sentences of the given form. (Later, in *Principia Mathematica* [5], he introduces the more technically satisfactory device of scope indicators.)

These two features then, (i) that the sentence "The present king of France is bald" is taken to be equivalent to "One and only one thing is a present king of France and that one is bald," and (ii) that this leads to scope problems in the case of "The present king of France is not bald" which comes out true if the description is given secondary occurrence but false if the description is given primary occurrence, I take

[2]"Über Sinn und Bedeutung" [2]. See especially the discussion of sentences containing "Odysseus."

to be well known. And, possibly because they provide a convenient means of comparison with other theories of descriptions, they have sometimes been thought to constitute the distinctive feature in Russell's doctrine. But this, I believe, is not so. The peculiar and interesting feature is his position on denoting. For he claims that, in contrast to such proper names as "Elizabeth Windsor" which denote a certain person, a description like "The present queen of England" doesn't *really* denote anything. If the description is proper you may want to speak *as if* it denoted the unique thing described, and that won't get you in any difficulties (if you avoid oblique contexts), but it would be misleading. For it would lead you to believe that the two sentences:

$$\text{"Elizabeth Windsor is shapely,"} \qquad (2)$$

$$\text{"The present queen of England is shapely"} \qquad (3)$$

have the same logical form, namely subject-predicate form, with "Elizabeth Windsor" in one case and "the present queen of England" in the other case as subjects, both denoting the same individual, and "is shapely" as predicate, denoting, say, some class of individuals. Thus in both cases the sentence is true if, and only if, the given individual is a member of the given class. But nothing could be further from the Russellian truth. Sentences (2) and (3) do not have the same logical form at all, for if (3) were of subject-predicate form, so would

$$\text{"The present king of France is shapely"} \qquad (4)$$

be, by parity of form (as Russell would say). But according to Russell the truth conditions for (4) have nothing to do with any given individual being shapely.

And now, I believe, we are at the heart of the matter. Russell's article "On Denoting" is not about a theory of descriptions comparable to Frege-Carnap or Frege-Strawson. Russell's article is about logical form, and is in the tradition of those philosophers who have warned us of the dangers of confusing the grammatical form of a sentence in ordinary language with its logical form. Such philosophers have often sought to construct a *logically perfect language* in which grammatical and logical form would always coincide.

I am dissatisfied with Russell's theory about the logical form of sentences containing descriptions, and I will try to indicate the features I find unsatisfactory by comparison with an alogous theory which I call "Russell's theory of indefinite descriptions."[3] This theory is con-

---

[3]Russell's theory of indefinite descriptions is most clearly adumbrated in [15].

cerned with the analysis of sentences containing *in*definite descriptions.
For example:

> "A senator from New York is supporting Rockefeller."   (5)

Now (5) certainly has subject-predicate grammatical form in English,
but if you feel that its logical form is the same as

> "Jacob Javits is supporting Rockefeller,"   (6)

you can quickly disabuse yourself by comparing:

> "A senator from New York is supporting Rockefeller, and   (7)
> a senator from New York is not supporting Rockefeller,"

with

> "Jacob Javits is supporting Rockefeller, and Jacob Javits   (8)
> is not supporting Rockefeller."

Sentence (8) is a contradiction, but (7) is true. In fact, isn't it obvious
that indefinite descriptions do not even purport to denote a unique object
as names do? Accordingly, Russell's theory of indefinite descriptions
asserts that the logical form of a paradigm sentence like "A such-and-such
is so-and-so" is represented by the equivalent sentence "Something
is both a such-and-such and so-and-so." In symbols:

> "$F\alpha x G x$" is equivalent to "$(\exists x)\,(Gx \,\&\, Fx)$".   (9)

Note that this analysis does not provide a unique understanding of certain compound sentences containing indefinite descriptions, for example:

> "A girl danced with every boy."   (10)
> In symbols: $(y)\,(By \supset D(\alpha x G x, y))$.

Depending on whether the indefinite description is taken as having primary or secondary occurrence, the sentence will be equivalent to either

> "Some girl is such that she danced with all the boys,"   (11)
> in symbols: $(\exists x)\,(Gx \,\&\, (y)\,(y)\,(By \supset D(x,y)))$

or

> "Each boy is such that some girl or other danced with him", (12)
> in symbols: $(y)\,(By \supset (\exists x)\,(Gx \,\&\, D(x,y)))$

There are further parallels between Russell's theory of definite
descriptions and Russell's theory of indefinite descriptions, but I think

the point is made. The point is that Russell regarded definite descriptions exactly as we would regard indefinite descriptions. Grammatically, at least from the point of view of what is now called surface grammar, indefinite descriptions are terms and they function like proper names; but sentences which contain indefinite descriptions and appear to have subject-predicate form should be treated as idioms and expanded as in the paradigm.[4]

Russell's theory of indefinite descriptions seems to me both correct and natural. In fact, the analysis fits indefinite descriptions so perfectly that any disanalogies between definite and indefinite descriptions throw suspicion on Russell's theory of definite descriptions. Further, I believe that the leading ideas of Russell's theory of definite descriptions are more clearly seen in connection with Russell's theory of indefinite descriptions, and thus questions about the former might be more easily answered in connection with the latter.

With respect to disanalogies between definite and indefinite descriptions, I will just mention two ways in which Russell (and everyone else, I suppose) provides differential treatment. First, Russell invents a symbolic notation for definite descriptions and introduces them into the language of *Principia Mathematica*. So far as I know, neither Russell nor anyone else has ever given serious consideration to the introduction of indefinite descriptions into any formalism. Not that it could not be done. The foregoing brief remarks on Russell's theory of indefinite descriptions make it clear exactly how to do it. It is even clear how scope indicators could be introduced. It is just that nobody would think it worth doing. Why? Because such a notation, rather than providing a useful and succinct means of expression for investigating logical relations, would tend to obscure the logical form of the sentence and obfuscate the issues in question. This, of course, is exactly what definite descriptions of English are said (by Russell), to do but still he introduces them into *Principia Mathematica*.

The second respect in which Russell treats definite and indefinite descriptions differently is in connection with his offhand remark that one *might* speak of proper definite descriptions *as if* they denoted the unique individual having the property in question, although strictly speaking this would be incorrect. We certainly do want to treat proper descriptions in this way, and every other theory of definite descriptions with which I am familiar does so. Are there any cases (let alone the central cases) in which

---

[4]Compare "It is snowing," which also appears to have subject-predicate form.

it is natural to treat an indefinite description like a proper name, that is, as denoting an individual? Russell never suggests so.[5]

I wish now to raise the question of how to regard the fundamental equivalences (1) and (9) of the two theories. Russell called them *contextual definitions*. But what is a contextual definition? Ordinarily we think of definitions as being either stipulative or explicative (in an older terminology, nominal or real). That is, either a new expression is introduced and assigned the meaning of a phrase whose meaning is antecedently known, or else an old expression is given a more precise, or in some other way slightly adjusted, meaning in terms of some antecedently understood phrase. In the case in question, the new expression to be introduced is the definite or indefinite description operator. But what meaning is given to it, or even to the full description? None! For the central thesis of Russell's theory is that this phrase *has* no meaning in isolation.

Another notion closely related to that of definition which is, I believe, somewhat more appropriate to Russell's contextual definitions, is that of *abbreviation*. In an abbreviation a new expression is introduced to *stand for* an old phrase. But the phrase so abbreviated is not required to be meaningful; it may be any combination of signs. Abbreviation is purely a matter of syntax. Whereas only well-formed expressions can serve as definiens, any expression can be abbreviated. To understand the meaning of an expression in terms of the meanings of its components, we must first expand it into unabbreviated form (except, of course, insofar as the abbreviations are also definitions).

I prefer still another understanding of (1) and (9). We may treat them as rules for translating ordinary, logically imperfect language into a logically perfect symbolism. According to this conception the symbolic descriptions "$\alpha x F x$" and "$\imath x F x$" would be understood as not occurring in the perfect language at all, not even as abbreviations. They would appear only at a transitional stage in the translation process as an aid in representing the surface grammar of ordinary language. This seems to me a natural understanding of (9), since, as remarked above, no one would seriously consider introducing indefinite descriptions into a logically perfect language.[6] If Russell really took his doctrine seriously and were willing to completely abjure the 'misleading' surface grammar of definite descrip-

---

[5]At the beginning of "On Denoting," Russell suggests as an initial understanding of indefinite descriptions that " 'a man' denotes. . .an ambiguous man." But he quickly rejects this idea.

[6]Hilbert's ϵ-operator does not produce an indefinite description in the sense herein discussed.

tions, he should be willing to accept the present understanding of (1). His use of definite descriptions in *Principia Mathematica* indicates to me a lingering ambivalence.

Although I will not now attempt to make the notion of a logically perfect language absolutely precise, I would like to clarify it somewhat. The intuitive idea is that the logical form of a sentence should mirror its grammatical form. The grammar of a language is assumed to be given in terms of certain grammatical categories such as term, formula, two-place predicate, etc. Each atomic expression is assigned to some such category, and *formation rules* are given which tell us how we can form compounds of a given grammatical category from components of certain grammatical categories. The grammatical form of an expression is then determined by the formation rules. An expression is grammatically correct if it can be 'constructed' from grammatically simple components in accordance with the formation rules. Such a construction assigns a grammatical structure, or form, to the expression. To parse a sentence is to exhibit its grammatical form. Just as grammatical properties and relations, such as being a noun clause or being the subject of a given sentence, depend on the grammatical form of the expression in question, so logical properties and relations, such as being valid or being a logical consequence of a given sentence, depend on the logical form of the expressions in question. Logical form is determined by the *evaluation rules* of the language. These rules tell us how to 'construct' the semantical value of an expression in terms of the values of its logically simple components. (We here take the semantical value to be what Carnap calls "the extension," that is: a truth value for sentences, an individual for names, a class of individuals for one-place predicates, and so on.) Such a construction of the truth value of a sentence exhibits the logical structure, or form, of the sentence in a way analogous to that in which parsing a sentence exhibits its grammatical form. As shown by Tarski[6], the notions of validity and logical consequence can be given in terms of such constructions.[7] In ordinary language, replacements which do not change the apparent grammatical form of sentences, for example, replacing a proper name with "someone," may well introduce or obliterate relations of logical consequence between the affected sentences, thus indicating a change in logical form. This point, that sentences with the same apparent grammatical form can have different logical forms, is illustrated by (5)-(8). In a logically perfect language the logical form of an expression must always mirror the grammatical form. Therefore, for logical perfection we

---

[7]What I call evaluation rules are clauses in Tarski's definition of "satisfaction."

require that the logically simple expressions coincide with the grammatically simple (but well-formed) expressions, and that to every formation rule there corresponds a unique evaluation rule, such that any compound formed by applying the formation rule to given components is evaluated by applying the corresponding evaluation rule to the values of the components. This has the desired result that the semantical evaluation of an expression exactly recapitulates its grammatical construction.[8]

Given the grammar of a language, one semantical treatment may make it logically perfect and another not. Take for example a sentential language which contains: (1) the atomic expressions "P_1", "P_2", "P_3", etc., all belonging to the grammatical category *sentence*, and (2) three formation rules which allow us to form the compound sentences $\ulcorner(\Phi \supset \Psi)\urcorner$, $\ulcorner\sim\Phi\urcorner$ $\ulcorner(\Phi \equiv \Psi)\urcorner$ from any component sentences Φ and Ψ. Now consider two different methods for assigning values to the sentences. Method I consists of (1) assigning a truth value to each atomic sentence in accordance with some given interpretation of the atomic sentences, and (2) for each formation rule using a corresponding truth function to evaluate the truth value of the compound in terms of the truth values of its immediate components. Method I is the standard semantical analysis of such a language. Method II agrees with Method I for the atomic sentences, but provides two stages for the analysis of compounds, (a) 'eliminate' all biconditionals from compound sentences by replacing sub-sentences $\ulcorner(\Phi \equiv \Psi)\urcorner$ by $\ulcorner\sim((\Phi\supset\Psi)\supset\sim(\Psi\supset\Phi))\urcorner$, (b) evaluate the result as in Method I.[9] I call the language which incorporates Method I logically perfect, but the detour from grammatical form involved in Method II renders the language incorporating that method logically imperfect. This obtains in spite of the fact that the two methods assign the same values to all sentences.

Let us assume that the grammar of Russell's language distinguishes *term* and *formula* and contains among its formation rules: All variables are terms; all individual constants are terms; if Φ is a formula and v is a variable, $\imath v\Phi$ is a term; if τ is any term (a variable, individual constant, or definite description), $\ulcorner\tau$ is bald$\urcorner$ is a formula; plus the usual formation rules for identity, quantifiers, sentential connectives, etc.[10] Following Russell's informal remarks, we will understand his semantics as involving first the elimination of all descriptions by means of (1). In accordance with

---

[8]This relationship between formation rules and evaluation rules is developed in somewhat greater detail in Chapter 1 of my dissertation [7].

[9]It is important for the point of the example that all three formation rules are understood as primitive, and thus that biconditionals are *not* thought of as defined expressions.

[10]At this point we revert to the theory of definite descriptions and leave indefinite descriptions aside as an instructive amusement.

the conventions for dropping scope indicators in *Principia Mathematica*, we take the scope always to be the smallest possible. Such a semantical analysis will of course make (1) true, and in a trivial way. But it makes the language logically imperfect.[11]

This brings us to another question. Does acceptance of the equivalence (1) commit us to Russell's analysis of the logical form of sentences containing definite descriptions? The answer is "No." At least two different (but closely related) analyses can make the language logically perfect. We can follow the method of Frege-Strawson in claiming that improper descriptions simply don't denote, but in the Frege-Strawson evaluation rules for atomic formulas:

(i) $\ulcorner\tau$ is bald$\urcorner$ is true if and only if τ denotes something which is bald;

(ii) $\ulcorner\tau$ is bald$\urcorner$ is false if and only if τ denotes something which is not bald

(where τ may be any term: a variable, individual constant, or definite description); retain (i) and replace (ii) with:

(ii') $\ulcorner\tau$ is bald$\urcorner$ is false if and only if it is not true.

Alternatively, we can follow the method of Frege-Carnap in stipulating that some previously chosen entity will be taken as the common denotatum of all improper descriptions, but put this entity *outside* the domain of discourse (possibly by just letting the chosen entity be the domain itself). According to this method we would retain the evaluation rule (ii) and replace (i) with:

(i') $\ulcorner\tau$ is bald$\urcorner$ is true if and only if τ denotes something which is bald *and is within the domain of discourse.*

Note that in the unmodified Frege-Carnap theory the italicized phrase is otiose.

---

[11]It should be noted here that I assume the grammar to be given in the way described above, essentially what is now called an *immediate constituent phrase structure grammar.* It is not implausible to regard Russell's contextual definitions as providing his language with a *transformational phrase structure grammar.* Since Russell's implied evaluation rule for sentences containing definite descriptions, like that of our Method II, might plausibly be called a *transformational evaluation rule,* we might consider this mirroring of syntactical structure a kind of perfection. But in-so-far as there is a natural tendency to expect a language to exhibit the simpler immediate constituent form, a language whose grammar is described by transformational means might, on that account, be considered imperfect.

On both the modified Frege-Strawson analysis and the modified Frege-Carnap analysis the equivalence (1) is preserved, as is the accuracy of Russell's translation of " $\imath Fx$ exists" by "$(\exists y)(x)(Fx \equiv x = y)$" which in turn is equivalent to "$(\exists y)(y = \imath Fx)$". And since the evaluation rules require us to look only at the evaluation of the immediate constituent term and not at whether it is a variable, individual constant, or definite description, the language is rendered logically perfect.

It may be objected that all we have done is to find two ways of coding information about the syntactical character of a term (whether or not it is a definite description) into the evaluation rules. Thus if a term has no denotation or none within the domain of discourse, we know that it must be a definite description. But such an objection would be both inaccurate and short-sighted. Inaccurate, because it ignores the fact that the treatment of improper descriptions no longer distorts the treatment of proper descriptions, which now receive the natural semantical analysis. And shortsighted, because it ignores the possibility of using the semantical ideas of the modified Frege-Strawson method or the modified Frege-Carnap method to construct theories which depart still further from that of Russell but which allow a complete assimilation of definite descriptions and individual constants, and even provide for true atomic sentences about non-existing individuals. Indeed, such theories have already been constructed by a number of authors, among them Hintikka, and Lambert, and Scott [8].

Russell and Frege were both interested in removing the logical imperfections of ordinary language, but their methods were quiet different. (For what follows it is necessary to assume the 'translation-rule' interpretation of contextual definitions, according to which definite descriptions do not occur at all in the perfect language.) Where grammar called for entities whose nature was obscure, Frege attempted constructions, as with numbers, or a theory about the purported entities, as with propositions. Thus he sought to preserve the integrity of ordinary language by ontological ingenuity. Russell's response, at least in the case of definite descriptions, was by grammatical reconstrual and replacement. The two methods are easily contrasted by their analyses of "the number of planets is two" (I simplify for economy). Following Frege's method, we accept the apparent grammatical form (that of an identity sentence) and translate into something like "the similarity class of the set of planets = the set of all couples." Following Russell's method (though he might not follow it in this case), we dispense altogether with singular terms purporting to refer to numbers and translate into something like "there is a planet, and there is another, and there are no others." Either method may lead to logical perfection, but Frege's way seems to me more fruitful in the long run. I see Frege's method at work in Tarski's reduction of possible worlds to models,[12] Carnap's reduction of propositions to classes of models,[13] Wiener's reduction of ordered couples to classes [9] (and hence of relations to classes), and of course Frege's own treatment of numbers [10]. I would classify as applications of Russell's method: Stevenson's emotive analysis of "x is good" [11], Austin's reconstrual of the singular term "the-meaning-of-(the-word)-'rat' "[12], Quine's treatment of virtual classes [13], and, classically, Russell's own treatment of definite descriptions. The scope and limitation of Russell's method, in one of its applications, has received careful and extended discussion in Quine's *Set Theory and its Logic* [13].

Russell seems not to have viewed his linguistic replacements as being in sharp contrast with Frege's ontological constructions, for in his essay "Logical Atomism," he writes:

> One very important heuristic maxim which Dr. Whitehead and I found, by experience, to be applicable in mathematical logic, and have since applied in various other fields, is a form of Ockham's razor ... The Principle may be stated in the form: Wherever possible, substitute constructions out of known entities for inferences to unknown entities.
>
> The uses of this principle are very various, but are not intelligible in detail to those who do not know mathematical logic. . . .
>
> A very important example of the principle is Frege's definition of the cardinal number of a given set of terms as the class of all sets that are "similar" to the given set. . . . Thus a cardinal number is the class of all those classes which are similar to a given class. This definition leaves unchanged the truth-values of all propositions in which cardinal numbers occur, and avoids the inference to a set of entities called "cardinal numbers," which were never needed except for the purpose of making arithmetic intelligible, and are now no longer needed for that purpose. . . .
>
> Another important example concerns what I call "definite descriptions," i.e. such phrases as "the even prime," "the present King of England," "the present King of France." There has always been a difficulty in interpreting such propositions as "the present King of France does not exist." The difficulty arose through supposing that "the present King of France" is the subject of this proposition. . . . The fact is that, when the

---

[12]Implicit in "Über den Begriff der logischen Folgerung" [6].
[13]Reference [3], §40.

words "the so-and-so" occur in a proposition, there is no corresponding single constituent of the proposition, and when the
proposition is fully analyzed, the words "the so-and-so" have
disappeared." [14]

In so far as we find reconstruction of our grammatical intuitions and
reconstruction of our ontological intuitions equally congenial, this con-
flation of ontological construction with grammatical reconstrual is harmless. But in so far as our grammatical preconceptions continue to dominate our ideas, logical perfection achieved in Russell's way will remain
unsatisfactory.

REFERENCES

1. B. Russell, "On Denoting," *Mind,* Vol. 14 (1905); reprinted in B. Russell, *Logic and Knowledge,* London, 1958, and also in *Readings in Philosophical Analysis* (H. Feigl and W. Sellars, eds.), New York, 1949.
2. G. Frege, "Über Sinn und Bedeutung," *Zeitschrift für Philosophie und philoso-phische Kritik,* Vol. 100 (1892); "On Sense and Reference" in G. Frege, *Philoso-phical Writings* (translated by Geach and Black), Oxford, 1952, and also as "On Sense and Nominatum" in *Readings in Philosophical Analysis.* See especially foot-note 9.
3. R. Carnap, *Meaning and Necessity,* Chicago, 1946, especially §8.
4. P. F. Strawson, *Introduction to Logical Theory,* London, 1952.
5. B. Russell and A. N. Whitehead, *Principia Mathematica,* Cambridge, 1910.
6. A. Tarski, "Der Wahreheitsbegriff in den formalisierten Sprachen," *Studia Philoso-phica,* Vol. 1, 1936; and "Über den Begriff der logischen Folgerung," *Actes du Congrès International de Philosophie Scientifique,* Vol. 7, 1936. Translated as "The Concept of Truth in Formalized Languages," and "On the Concept of Logi-cal Consequence," in A. Tarski, *Logic, Semantics, Metamathematics,* Oxford, 1956.
7. D. Kaplan, *Foundations of Intensional Logic* (Dissertation), University Microfilms, Ann Arbor, 1964.
8. J. Hintikka, "Studies in the Logic of Existence and Necessity: I. Existence," *The Monist,* Vol. 50 (1966); D. Lambert: "Notes on E!III: A Theory of Descriptions, *Philosophical Studies,* Vol. 13, 1962; K. Scott: "Existence and Description in For-mal Logic," in *Bertrand Russell: Philosopher of the Century* (R. Schoenman, ed.), London, 1967.
9. P. P. Wiener, "Simplification of the Logic of Relations," *Proceedings of the Cam-bridge Philosophical Society,* Vol. 17, 1912-1914.
10. G. Frege, *Grundlagen der Arithmetik,* Breslau, 1884; translated as *The Founda-tions of Arithmetic,* Oxford, 1950.
11. C. L. Stevenson, "The Emotive Meaning of Ethical Terms," *Mind,* Vol. 46 (1937); reprinted in *Logical Positivism* (A. J. Ayer, ed.), Glencoe, Illinois, 1959.
13. W. V. O. Quine, *Set Theory and its Logic,* Cambridge, Mass., 1963.
14. B. Russell, "Logical Atomism" in *Contemporary British Philosophy* (J. H. Muir-head, ed.), London, 1924; reprinted in *Logical Positivism* (A. J. Ayer, ed.), Glencoe, Illinois, 1959.
15. B. Russell, *Introduction to Mathematical Philosophy,* London, 1919, Chapter XVI.

DISCUSSION

QUINE: (To Kaplan:) Concerning your notion of the perfect language, and the way in which the immediate context is determined by the immediate constituent: Is this determination effective?

KAPLAN: Consideration of the computability of the evaluation rules doesn't enter into my notion of logical perfection.

QUINE: It is not part of the definition?

KAPLAN: It is not part of the definition, although it is clear that in certain interesting cases you might want to impose that further condition, just as you might want to impose effective computability for the functions involved in building up the grammar, although in a very general case we do not.

YOURGRAU: I should like to raise a question in regard to the so-called logically perfect language. It is of course an artifact, because the language we talk is by definition not perfect. We therefore strive to develop an artifact.

Now, I learned from Carnap that if you have an idealized, formalized language, provided you have correspondence or transformation rules, you can express every situation, event, concept, etc. in an arbitrary language, colloquial or otherwise.

I would like to know whether you share the enthusiasm for, and the faith in, such a language which Carnap advocated so strongly. I understand this as an ideal goal, one that we would like to attain as logicians, but I am not convinced it can ever be be realized.

Twice in your paper you used the term "contextual definition," which is one of Church's favorite expressions. Do you mean to say that you can do away with the explicit definitions which the mathematicians desire and without which we have no physics? Are you satisfied with the contextual definition as long as we play the game according to the rules which you have just set out here? In other words, do you say, "I don't want to give any explicit, 'definite' definition?"

My last question is: Would you maintain that the defect in Russell's theory of descriptions, which had (with some justification) acquired the status of a scripture, is that it does not allow application to indefinite descriptions, although they are entirely legitimate and play such a large role in ordinary discourse? I mean not only in colloquial language but in specialized languages too, for instance in science. Would you therefore like to enlarge the range of Russell's theory of descriptions either by adding or patching up or even by eliminating whole sections of that "sacred text" of the logician?

KAPLAN: My comparison of Russell's theory of definite and indefinite descriptions was intended to be a *reductio ad absurdum*. It is quite clear that it would serve no useful purpose to introduce into the language an expression which grammatically functioned as a term but which had the meaning of "a girl," "a man," and so on, a term which was an indefinite description. I would not favor that.

On contextual definition versus explicit definition, I am all in favor of explicit definitions. Part of the point of my paper, part of the question that I was asking in the title, was an attempt to understand what a contextual definition really is. I suggested that contextual definitions don't satisfy one of the standard criteria for explicit definitions, that is, assigning meaning to the expression that is being introduced. Further, I suggested some other ways that we might think about these things. I think the phrase "contextual definition" is an extremely unfortunate choice for the kind of thing that is being done there. I prefer to think of the equivalences that are given in (1) and (9) in either the way that I suggested, as intermediate steps in translation, to help you to get the English under more control before you put it into the symbolic language, or else not to think of the description operators as being defined at all but to simply let this be an equivalence, an axiom. The equivalence would simply be an axiom within the language, and one would then try to find a way of doing the semantics.

Can everything be expressed in a logically perfect language? I don't see why not. This question suggests that the search for logically perfect languages — I have tried to make it clear that the notion of logically perfect languages is not a really terribly metaphysical notion — that the search for a logically perfect language is somehow an abandoning of the full expressive resources of the languages of everyday life. I don't think it is anything like that at all. The people who were after logically perfect languages were simply after languages whose syntax more clearly reflected what they were asserting. When we create a mathematical language, as can be seen quite clearly from some of my bilingual formulations here such as (11) and (12), there is an enormous increase in efficiency that is gained by going to symbolic languages. It is not that we can say something that was impossible to say before, but that we can say it more succinctly and more clearly, and the point about the logically perfect language isn't to make it shorter in any way, it is that it should make it clearer by having the grammar of the sentence somehow reflect more closely the meaning of the sentence.

You asked generally about Russell's theory of descriptions. I think a useful distinction that might be made is this one. Sometimes I distinguish what I call Russell's *theory* of descriptions from Russell's *theorems* of descriptions. Russell's theorems of descriptions involve a certain body of theorems formulated in a certain language. Russell's theory of descriptions involves a certain semantical analysis of that language. I think we can preserve Russell's theorems of descriptions without maintaining Russell's theory of descriptions.

TENNESSEN: I am always puzzled when we speak as though it was taken for granted that, for instance, Kaplan's formula (8) could be interpreted in the direction of P and $\sim P$.

It seems to me to be, at the face of it, a very unreasonable and implausible interpretation. It is clearly what I call an "analytical-form" sentence. Analytical-form sentences are never interpreted to be analytically either true or false. "A bachelor who isn't married" may still be interpreted as analytically true, but "a bachelor is a bachelor" can never plausibly be interpreted in that direction. It is clearly misleading (to say the least) to simply state: "(8) is a contradiction." What Kaplan probably intends to express is something like: "(8) is here meant to be interpreted as though it were an example of a (logical) contradiction." In other words, what he is really saying is: In spite of the obvious implausibility, let (8) be interpreted in such a direction that, if symbolized, it could be conveyed by $(P \cdot \sim P)$. Or is to be taken for granted that this is *the* interpretation of it?

KAPLAN: I agree that when we see somebody seriously asserting a sentence which is either trivially true or trivially false, the only plausible thing is to assume that he doesn't understand it in such a way that it *is* trivially true or trivially false. And so, looking at (8) (which happens to be a strikingly good example of just this kind of thing), I can see that the interpretation that Tennessen wanted to put on it is not this interpretation. So I agree on that.

However, I think the point I wanted to make with (7) and (8) could equally well have been made not by using the conjunction in the case of (8), which is in fact trivially false, but by describing a dialogue. I am trying to show that (5) and (6) are not parallel in their structure. And so instead of (7), I might have said: Imagine John saying, "A senator from New York is supporting Rockefeller," and Fred getting up and saying, "A senator from New York is not supporting Rockefeller," and, of course, they may be in agreement. On the other hand, imagine John getting up and saying, "Jacob Javits is supporting Rockefeller," and Fred getting up and saying, "Jacob Javits is not supporting Rockefeller." There the contradictory kind of interpretation is more plausible. I just wanted to point out that difference.

QUINE: Transformation and definition are related in certain ways, as Kaplan pointed out. If we bring in this transformation we no longer have what he calls the logically perfect system. However, we do have something that will fit into the pattern of transformational grammar. Also, if we could regard the transformation as a definition, we would have this relationship: The system would not be, as it stands, a logically perfect system, but it would be a logically perfect system plus a definition. Or, to put it the other way around, the definition would be instruction as to how to translate this non-logically perfect, this logically imperfect system into another which is logically perfect.

Then the question arises, whether in all transformational systems, the transformations could be regarded similarly as definitions, and therefore all transformational systems could be regarded as simply translatable into logically perfect systems. If the answer to that were "yes," I think people like Noam Chomsky would feel rather frustrated.

Then I am worried about contextual definition. We have to distinguish between contextual definition and implicit definition. Now, contextual definitions are explicit definitions as opposed to implicit ones, and, in fact, just about every definition that we ordinarily think of as an explicit definition is a contextual definition. We define "P or Q," for instance, as "not (not P and not Q)." I am not equating "or" to any expression. That would be an example of a definition which was not contextual. Or define "5" as "4 plus 1." This is not contextual either.

Here I am not giving anything that the "or" is equivalent to. I am explaining a context, and between this and Russell's treatment of descriptions I see no difference except of degree. Contextual definitions generally are explicit in the sense that they tell us exactly how we can translate a given notation into another notation that is free of the sign in question, whereas implicit definitions, so miscalled, aren't definitions at all; they are axioms, theorems, assertions of the kind that don't enable you, in the general case, to eliminate the signs or to translate into another language that is free of them.

KAPLAN: Concerning the question about the definition of "or," Quine claims that "$(P$ or $Q) \equiv$ not (not P and not Q)" is also a contextual definition. That is, we have on the left of the equivalence, or whatever it is that forms the definition, not just the sign being defined but also some other expressions, and he said that this is a contextual definition.

I don't believe that the difference between this and (1) is simply a matter of degree. I contend that there is a radical difference. In the case of "or," the definition could be thought of as really producing an explicit definition. In my view the crucial thing is something like this: As long as we introduce just the new sign plus variables for the argument expressions, it can be transformed into an explicit definition of just the new sign, provided that the logic which is available in our language is strong enough. If we have the lambda operator, for example, you can take the variables over on the other side, and really just define the connective. I don't think that anything like that can be done for Russell's iota without changing its grammar from a term-maker to a formula-maker, because the variables on the left in (1) include "F," which is not an argument expression of iota. Of course, the whole problem of defining variable-binding operators is a little unclear anyway. But I think that in any adequate treatment of the problem, there will be an important difference between, say, (1) and the definition of the existential quantifier, and I think that the difference is connected with how the grammar changes when the 'contextual' definition is transformed into an explicit definition.

Now, Quine raises a very interesting point about changing the transformational operations in the semantics into the kind used in a logically perfect language. Probably it can be done for things like the iota and things like the biconditional, but I don't think it could be done in general. The method that I suggested for getting a logically perfect language, which would handle Russell's equivalence, seems to me not to be completely obvious (although a number of other people have thought of it, including, I believe, Hintikka and Scott).

I agree very strongly with Quine's point about implicit definition. One of the points I want to make is that the word "definition" gets thrown around carelessly. Maybe I could make the point in a different way. We should carefully separate different kinds of definitions because they have different logical properties which are of some importance. The so-called implicit definitions aren't definitions in any sense. We have a number of primitive constants, write down axioms on them, and people say, "Well, you have implicitly defined them." You have, if the system has only one possible interpretation. If it is categorical, then there is a certain sense in which one might say that they have been implicitly defined. But in general one hasn't defined them in any sense — they have just been left open; one has merely said a few things about them. If somebody asks me to define the description, "The present queen of England," I don't define that expression implicitly or in any other way by saying, "She is more than 32 years old."

HINTIKKA: I would like to register a protest, not so much against the theory of descriptions in any of its forms, as against some of the uses that philosophers have made of it. I think that the original form of Russell's so-called theory is a very nice illustration of the philosophical principles that Kaplan quoted, and therefore, in a way, a very nice example of what can be done. But I think philosophers, especially philosophers of language, have pushed it into roles where it doesn't always help and sometimes confuses the issues.

What I mean is roughly this: In analyzing, not the kind of context Russell was in the first place interested in, but what are known as modal contexts, there is a great temptation to think that we can get rid of some of the problems by saying, "Well, let's just construe the free singular terms we have to deal with as definite descriptions. Definite descriptions we can get rid of in terms of quantifiers, so that problems concerning singular terms other than bound variables get solved in this way."

What we get in this way often doesn't help us to understand the underlying problems at all. Take, for instance, the problem of seeing how singular terms behave in the context of modal operators. It may be that by some superhuman ingenuity one could perhaps handle this by replacing other free singular terms by definite descriptions. However, as a matter of fact, I don't think we can see through the situation without bringing in, at least for heuristic purposes, free singular terms and without forgetting the theory of definite descriptions for the time being. On the formal level some of the difficulties with definite descriptions arise by way of scope conventions in modal contexts. But Russell's theory itself does not explain how these scope conventions — which really are not conventional at all — are to be chosen, or why they are chosen differently on different occasions. There are even good reasons for assuming that no scope conventions can do full justice to ordinary usage. Russell's theory of definite descriptions has been called, justly, a paradigm of philosophy. I am afraid that some of the uses that have been made of this "theory" are nevertheless paradigms of bad theory and bad philosophy.

LEJEWSKI: At one stage in his paper Kaplan mentions an interesting sort of tension within the Russellian theory of descriptions. He reminds us that Russell has a symbolic notation for definite descriptions, a notation which he is anxious to introduce into the language of *Principia Mathematica*. In fact, he introduces it by means of contextual definitions. But contextual definitions, and definitions in general, are treated in *PM* as typographical conveniences, which are outside the system altogether. Thus it appears that the descriptional notation is made use of in *PM de facto* although *de jure* it is not to be there at all.

One of the fundamental presuppositions of the language of *PM* is that at its lowest categorial level it has proper names. Improper names are not allowed as meaningful substituends for the lowest level variables. Improper names have to be analyzed away, and the same applies to improper descriptions. There does not seem to be any theoretical necessity for analyzing away proper descriptions, but, as Kaplan points out, Russell preferred to treat them on a par with improper descriptions. If we agree to do without improper names and improper descriptions, can we still legitimately introduce proper descriptions into a language such as that of *PM*? The answer seems to be "yes," but instead of "abbreviational" definitions we have to make use of definitions which, as regards their status within the system, are more like axioms.

A system comparable to that of *PM* can be constructed in terms of a language which presupposes that proper and improper names as well as proper and improper descriptions are all acceptable as meaningful substituends for the lowest level variables. The system is based on the following single axiom, $[a\ b]: a = b \cdot \equiv \cdot [\exists c] \cdot c = a \cdot c = b$, and it has the following rule of definition:

On the assumption that a thesis T is, at the given stage, the last thesis of the system, we can add to the system a new thesis of the form

$$D.[a. \ldots\,] ::a = \alpha(v_1, \ldots, v_n) \cdot \equiv \cdot \therefore \phi(a, v_1, \ldots, v_n) \cdot \therefore$$
$$[b]: \phi(b, v_1, \ldots, v_n) \cdot \supset \cdot a = b \,,$$

provided the following conditions are fulfilled: (1) α is a new constant functor; (2) a, b, v_1, . . . , v_n are distinct variables; in $a = \alpha(v_1, \ldots, v_n)$ each of a, v_1, . . . , v_n occurs only once; (3) $\phi(a, v_1, \ldots, v_n)$ has no free variables other than a, v_1, . . . , v_n; (4) $\phi(b, v_1, \ldots, v_n)$ is equiform with $\phi(a, v_1, \ldots, v_n)$ except that it has occurrences of b instead of occurrences of a; (5) $\phi(a, v_1, \ldots, v_n)$ is, with respect to T, a meaningful expression of the system.

In formulating this rule I have omitted, for the sake of simplicity, a few minor refinements.

Now, it seems to be obvious that the Russellian $\imath x F x$ is a function of F, which in turn is its argument. If we use $\imath$ as the functor, then the function can be symbolised as $\imath(F)$, and the functor $\imath$, which forms a definite description in concatenation with F as the argument, can now be defined as follows:

$$D1. \quad [a\,F] :: a = \imath(F) \cdot \equiv \therefore F(a) \therefore [b] : F(b) \cdot \supset \cdot a = b \cdot$$

Our definition is not a convention for abbreviating formulae. It introduces, into the system, definite descriptions *de jure*. It is to be noted that the problem of scope or the problem of primary or secondary occurrences of descriptions does not arise here.

Can our rule of definition be adapted to the language of *PM*? It seems to me that it can. We can, for instance, adopt the procedure whereby we are entitled to add to the system of *PM*, as a new thesis, the left-hand side of an equivalence satisfying schema D, provided the right-hand side of the equivalence has been proved to be a thesis of the system. This is only an outline of the procedure, whose details remain to be worked out.

Schema D shows how closely the problem of definite descriptions (and that of descriptive functions) is connected with the problem of definitions construed as means of introducing new vocabulary into a system.

KAPLAN: It seems to me worthwhile to investigate the consequences of admitting "definitions" of the form suggested into the system of *PM* and also into other kinds of systems.

I want to try to emphasize a certain aspect of my paper, though. Especially of late there has been a great interest in what is called semantics, that is, the connections between the syntax of a language and the interpretation of a language, and my paper was really on that topic. In particular, the notion of a logically perfect language seems to me to be a semantical notion concerning just this kind of relationship. The question of what is a definition and how different definitions can be introduced: contextual, implicit, abbreviation, and so on—these are all very interesting problems at the level of syntax. But I am mainly interested in the relation between writing out these different syntactical ways of doing it, the formulas, and the semantical interpretation that is put upon them.

QUINE: Concerning Hintikka's comment, I would just say this regarding the eliminability of definite descriptions and other singular terms (apart from variables themselves), which I insist on frequently: I am all for bringing them back in for heuristic purposes where they help. In so far as you can define something away, you also excuse its restoration. To define is to eliminate; it also is to exculpate.

KAPLAN: I think I am slightly on Hintikka's side on this matter and I believe that some of my remarks have something to do with this situation. Especially the idea of trying to develop the semantics so as to treat the language as being logically perfect — I want to leave out the whole business about transformational grammars for the moment. The idea is connected with a certain kind of integrity, taking the grammar of the language, the syntax, to be as it appears to be. Now, if we follow Russell's method, especially as developed by Quine — Quine has given us, in fact, the natural extension of it to take care of individual constants and has made the theory much more pleasing to the logicians — then the sentences that appear to have subject-predicate form, turn out on their analysis (whether it is by contextual definition or whatever, looking just at the semantics) not to have that logical form at all. Thus, so long as the original sentence remains in the language, it is not logically perfect. It is not that one cannot take a language without definite descriptions and get very interesting results for it, or that one cannot show that this kind of Russell-Quine elimination can be consistently performed. It is just the feeling that when that kind of transformation and elimination *is* performed, it is not absolutely clear that we really are talking about the sentences understood in the same way as they were in the first place. I suppose this applies, at least in part, to Russell's claim that the equivalence that he writes down, the so-called contextual definition, really does reproduce the sense or the meaning of the sentence that he is analyzing.

COHEN: I don't understand, as a non-logician, how this very important problem about equivalence is settled among logicians or among philosophers of the sciences who may offer logical reconstructions. When you disagree about whether what is to be defined and the proposed definition have in fact the same meaning, how does one settle this matter in logic? What is the methodology of settling a disagreement on whether the definition really does carry out what is claimed?

KAPLAN: I don't know that there is some way of settling these questions. Perhaps the thing to do is to develop the system, as Quine has developed it, and also to attempt to develop the system in which the language comes out to be logically perfect, or in some other way. Of course, you have to learn some logic so that you can understand what these developments are. You then look at the different developments, and if you want to, you can go ahead and be pleased more by one than by the other, or just be interested in both of them and choose the one that seems most appropriate for the problems that you have at hand.

It is always very interesting to discover that you may develop either of these systems in what seems to you to be an intuitively different way; and possibly at the end you can prove some interesting kind of equivalence between the two, or possibly not. But I don't think the question as to which of these really gives the correct sense of the word in the ordinary language is an interesting question until we have brought in things like the notion of a logically perfect language and said, "Well, this makes it logically perfect, this makes it contextual, this makes it so-and-so." Then we have some rigorous statements we can make, e.g. this definition has this property, it doesn't have that or another.

Chapter XVIII

NON-QUANTAL FOUNDATIONS OF QUANTUM MECHANICS

Alfred Landé
The Ohio State University

When dealing with the foundations of quantum mechanics one has to distinguish among the formal calculation rules, such as the Born commutation theorem for coordinates and momenta, as against their application to atoms and solid bodies, electromagnetic and other fields, and problems such as whether matter consists of particles or waves. I am going to deal here only with the *formal rules* of quantum mechanics. These rules are problematic enough in themselves. Most experts seem to be satisfied to know the technique of calculation, "the tricks of the trade," which work so well; they tell their students to "understand that there is nothing to be understood in quantum theory." The adept, however, and the philosopher of science may not be ready to accept this pragmatic attitude. He may ask for general and fundamental reasons *why* Nature restricts mechanical action by certain quantal selection rules, *why* physical data are obtainable from a matrix calculus in which ab differs from ba, *why* probabilities interfere in a wavelike fashion, *why* a mechanical momentum p is to be replaced by an operator $(h/2i\pi)\partial/\partial q$ acting on a complex function ψ, and so forth. In short, the student will ask for an *explanation* of the various oddities of quantum physics and mathematics. These oddities are sometimes traced to a fundamental antimony — not further explained — namely to the contrast between the wave "picture" and the particle "picture." Yet this "duality" is at least as problematic as its consequences. And elevating duality to the rank of a supreme principle or axiom appears, at least to this author, no better than accepting a *horror vacui* as the ultimate cause of liquids rising in evacuated tubes, or proclaiming an *élan vital* as the source of the multivarious manifestations of life. What is missing is an explanation, i.e. a deduction of the quantum theory from a *non-quantal background* which is acceptable without much further scrutiny as natural and almost self-evident, in the sense of Einstein's admonition in *Mein Weltbild:* "The supreme task of the theoretical physicist is searching for those general and elementary laws from which one may derive the world picture by pure reason—although those elementary laws themselves cannot be obtained logically but only by intuition based on a broad view of experience."

In the present case, the "broad view of experience," in my opinion, is the recognition that microphysical states and events are controlled by statistical or probabilistic, rather than by deterministic, law. I intend to show that the quantum formalism is a straightforward consequence of the general postulates of symmetry and invariance, familiar from classical mechanics, if these regulatory postulates are now applied to construct a general schema of *statistically connected data.* That is, my aim is to derive the general formalism of quantum theory from a statistical yet *non-quantal* basis. This may contribute to *demystify* a theory which for too long has had the reputation of being a set of strange rules imposed on microphysics by an inscrutable whim of Nature.

1. PROBABILITY MATRICES

Let us begin with simple considerations concerning a probabilistic schema of events. A certain observable quantity named A (it may be the energy, or the angular momentum, or the positional coordinate) pertaining to *a given physical system* may be capable of various values, $A_1, A_2, \ldots, A_m$. Another observable quantity B may be capable, in the same physical system, of various values $B_1, B_2, \ldots, B_n$. Whether these "eigenvalues" are discrete, or form a continuum, is irrelevant so far. However, let us assume for the sake of mathematical convenience that the "multiplicities" m and n are *finite*. But there may be an infinite number of observables $A, B, C, D, \ldots$.

Suppose now that an instrument built for determining A-values— let us denote it as an A-meter—has found the system in the "state" A_k. When the system is now subjected to a B-meter test, the resulting B-value cannot be predicted with certainty. Instead, following the previous state A_k, states $B_1, B_2, \ldots, B_n$ will occur with certain statistical frequencies or probabilities

$$P(A_k \rightarrow B_1), P(A_k \rightarrow B_2), \ldots, P(A_k \rightarrow B_n),$$

the sum of which is *unity*.

One now may compile a *matrix* of all the probabilities connecting various original A-states with various B-states revealed by a B-meter:

$$\begin{pmatrix} P(A_1 \to B_1) & P(A_1 \to B_2) \dots \\ P(A_2 \to B_1) & P(A_2 \to B_2) \dots \\ \vdots & \vdots & \vdots \end{pmatrix} = (P_{A\to B}),$$

with m rows and n columns. Each *row* sums to unity.

All this is self-evident as soon as one has accepted a statistically controlled schema of events as a "broad view of experience." Far from trivial, however, is the following *postulate of two-way symmetry of the probabilities:*

$$P(A_k \to B_j) = P(B_j \to A_k). \tag{1}$$

Equation (1) may be regarded as plausible, as being the statistical counterpart of the time reversibility of every classical deterministic process. From (1) follows that the columns of the matrix $(P_{A \to B})$ are identical with the rows of the matrix $(P_{B \to A})$, which rows sum to unity. Therefore, not only the rows but also the columns of $(P_{A \to B})$ sum to unity. And since the sum of all P's of the matrix is the sum m of all rows and also the sum n of all columns, m must equal n, so that the matrix $(P_{A \to B})$ must be *quadratic,* with the observable A having the same multiplicity m of values as the observable B and as any other observable C and D and so forth *in the same mechanical system.* Omitting the arrow signs from now on, the sum rules for the rows and columns read

$$\sum_j P(A_k, B_j) = 1 \qquad \text{for every} \quad k = 1, 2, \dots, m ;$$

$$\sum_k P(A_k, B_j) = 1 \qquad \text{for every} \quad j = 1, 2, \dots, m . \tag{2}$$

Notice the special case describing *reproducibility of a test result:*

$$P(A_k, A_{k'}) = \delta_{kk'} \qquad \text{or} \qquad (P_{AA}) = (1), \tag{3}$$

rendering (P_{AA}) a *unit matrix* with ones in the diagonal and zeros outside the diagonal.

From now on we denote as "observables" and deal only with such quantities A and B, etc., which fit into the schema above. Thus, position (x, y, z) is not an "observable"; but position (x, y, z) at time t_A is an observable A, and (x, y, z) at time t_B is another observable B. This mutual adjustment of theory and experiment exemplifies Einstein's words: "The structure of the system is the work of reason. But the empirical contents and their mutual relations must find their representation in the conclusions of the theory."

The next step in theory construction will be trying to find, by pure mathematical reasoning, a possible, or rather the only possible, *general* relation between the probability matrices.

2. INTERFERENCE OF PROBABILITIES

Why do the probabilities interfere in a wavelike fashion? They have to obey the law of interference as the only way of satisfying the postulate that there be a *general law of mutual interdependence* between the *P*-matrices. To be more specific: Suppose the two matrices (P_{AB}) and (P_{BC}) are given. We then *postulate that the matrix (P_{AC}) is either uniquely determined, or at least restricted multivalently, by a general theorem of the form*

$$(P_{AB}) \; \mathbf{v} \; (P_{BC}) \Rrightarrow (P_{AC}), \tag{4}$$

where **v** indicates a certain generating procedure still to be determined, and "=" stands for unique and "→" for multivalent determination. In both cases, the "intermediate" observable B is eliminated by the general procedure **v** so that B may be replaced by *any* other observable C or D, etc. Generality means that *any* two matrices P produce a third one as in (4), so that the interdependence theorem has *group character*.

Since the case → is too vague to start with, let us consider the case of unique determination first, applying it to matrices of quantities named ψ which later might either be identified with, or be closely related to the quantities P. The desired relation

$$(\psi_{AB}) \; \mathbf{v} \; (\psi_{BC}) = (\psi_{AC}) \tag{5}$$

satisfies the conditions of group law; first that of *association*, just because it contracts AB and BC to AC.

The second condition for group quality is that the quadratic matrices ψ_{AB} ought to include an *identity* member (e) so that

$$(e) \; \mathbf{v} \; (\psi_{AB}) = (\psi_{AB}) \; \mathbf{v} \; (e) = (\psi_{AB}) \tag{6}$$

If $(e) = (1)$ then **v** symbolizes multiplication $\times$. Therefore, since the P-matrices include the identity member $(P_{AA}) = (1)$, only the case

$$(\psi_{AA}) = (1) = (e)$$

can serve as a model for the ψ-matrices. The procedure $\mathbf{v}$ therefore must *symbolize multiplication*:

$$(\psi_{AB}) \times (\psi_{BC}) = (\psi_{AC}). \tag{7}$$

The only general and self-consistent symbolic multiplication of matrices, however, is the usual one of rows times columns, written out as

$$\sum_j \psi_{A_k B_j} \psi_{B_j C_n} = \psi_{A_k C_n}. \tag{7'}$$

The third condition for group quality is that each member (ψ_{AB}) has an *inverse* $(\psi_{AB})^{-1}$ so that their product yields (1). This condition is satisfied, (ψ_{BA}) being the inverse of (ψ_{AB}) because (7) together with (6) yields

$$(\psi_{AB}) \times (\psi_{BA}) = (1) \tag{8}$$

or, written out,

$$\sum_j \psi_{A_k B_j} \psi_{B_j A_{k'}} = \delta_{kk'}. \tag{8'}$$

In conclusion: The product theorem (7) with the special case (8) represents *the only possible way* of connecting quadratic ψ-matrices containing members $(\psi_{AA}) = (1)$ by a general and *univalent* interdependence theorem of group character. (Refer also to Postscript.)

The summation rule (8') does not determine the relation between individuals ψ_{kj} and ψ_{jk} (using an abbreviated notation from here on), although both ought to be *related* to the common positive quantity $P_{kj} = P_{jk}$. The *relation* between the ψ's and P's cannot be identity since the P's are all positive, whereas there must be positive as well as negative ψ's or perhaps complex ψ's, in order to yield zero in the product sum (8') for $k \neq k'$. And since the ψ-theorem (7) is the only univalent one (= rather than →), the interdependence between the all positive P-matrices *cannot be univalent*. Nevertheless, the P's must be closely connected to the ψ's, as seen from the following tabulation comparing the ψ-qualities established above with the desired P-qualities:

Established ψ-theorem:	group	$(\psi_{AA}) = (1)$	$\sum_j \psi_{kj} \psi_{jk} = 1$	
Desired P-theorem:	group	$(P_{AA}) = (1)$	$\sum_j P_{kj} = 1$	

There is *but one* possibility of connecting the ψ's with the P's, namely putting

$$P_{kj} = \psi_{kj}\,\psi_{jk} = P_{jk}, \tag{9}$$

hence either $\psi_{jk} = \psi_{kj}$ or, more generally, when admitting that the ψ's might be complex (with * indicating the complex conjugate):

$$\psi_{kj} = \psi_{jk}^{*} \qquad \text{(Hermitian quality).} \tag{9'}$$

In conclusion: The interdependence problem between the P-matrices has but one possible solution: defining P_{kj} as in (9) and connecting the ψ's through (7'), (8'), and (9'). If one admits complex ψ's, this connection is known as *unitary transformation*. It contains the special case of *orthogonal transformation* when all ψ's are real.

Probabilities are needed for calculating *mean values*. Thus, the mean value of the quantity B over its eigenvalues B_j, when starting from the original state A_K, is defined as

$$B_{\text{mean}} = \sum_j P(A_k, B_j)\, B_j. \tag{10}$$

Because of (9) one may write the same quantity as

$$B_{kk} = \sum_j \psi_{kj}\, B_j\, \psi_{jk,}, \tag{10'}$$

and rename it "the mean value of B starting from A_k and returning again to A_k," or simply "*transition value* of B from A_k to A_k." There is nothing new in this expression for a mean value save the mathematical and verbal notation. However, within the schema of orthogonal or unitary transformation there is but one possible generalization of (10'), namely

$$B_{kn} = \sum_j \psi_{kj}\, B_j\, \psi_{jn}, \tag{11}$$

defining the "*transition value* of B from A_k to C_n." According to ordinary ideas, one would replace the quantities ψ in (11) by quantities P. But this would not lead to (10) as a special case. Only (11) obeys the postulate of a *general* connection theorem between probabilities and then between mean values. Another special case of (11) is

$$B_{jj'} = B_j\,\delta_{jj'}\,. \tag{12}$$

Notice also that, if the ψ's are real, then $B_{kn} = B_{nk}$. In the more general case of complex ψ's, the B's, like the ψ's, are "Hermitian," that is, $B_{kn} = B_{nk}^*$. A frequent problem is that of finding the eigenvalues B_j when the transition values B_{kj} are known (similar to transformation to principal axes of an ellipsoid in geometry).

Altogether, the "interference law of probabilities" *via* the multiplication theorem (7) for "probability amplitudes" ψ, usually regarded as an inscrutable quantum feature, could have been anticipated by pure reasoning without quantum ingredients: It is a necessary consequence of the non-quantal requirement that the probabilities P connecting pairs of observables, A with B, B with C, and A with C directly, are interdependent by way of a general postulate of order rather than chaos, "order" meaning *group quality* of (4). The only possible order then is that of *unitary transformation,* which contains *orthogonal transformation* as a special case, the latter by means of real (positive and negative) ψ's, the former with complex ψ's. Why the quantities ψ are to be *complex*, in other words, why the imaginary unit $\sqrt{-1} = i$ enters the theory and leads to wavelike ψ-functions, will be shown next.

3. WAVE FUNCTION AND SELECTION RULES

So far the discussion has been only about "observables" $A, B, C, \ldots$ in general. Turning now to dynamics, we have to do with the special quantities $q = $ position, $p = $ momentum, $E = $ energy, and $t = $ time, forming pairs of "conjugates." In classical mechanics, "conjugacy" is defined by way of the Hamiltonian equations of motion. But this definition of p and q as conjugates suffers from having to assume some chosen function $H(p,q)$ as representing the energy E conjugate to the time t. In quantum mechanics, conjugacy of p and q is defined in terms of a relation between p and q themselves in three equivalent ways: by Born's commutation rule,

$$pq - qp = h/2i\pi, \tag{13a}$$

by Schrödinger's operator rule,

$$p \rightarrow (h/2i\pi)\partial/\partial q, \tag{13b}$$

or by the wave function,

$$\psi(p,q) = exp(2i\pi pq/h); \tag{13c}$$

as is well known, assuming one of the three leads to the other two. But all three are *quantum rules* and certainly of a most baffling character. After twenty-six years of trial and error, they were finally established in 1926 by the genius of Born, Heisenberg, and Schrödinger. Today they are accepted because "they work." In contrast, I contend that they are necessary consequences of adding to the previous assumptions (the two-way symmetry of the *P*'s and the postulate of *P*-matrix group order) the mechanical postulate of *Galilean invariance* in the following version, again of a probabilistic character:

The *mean value* of *any* function $F(q)$ calculated for transitions from state p to p', that is [replacing the summation (11) by integration],

$$F_{pp'} = \int \psi(p,q)F(q)\,\psi(q,p')\,dq, \tag{14}$$

shall depend on the *difference* $p - p'$ only. And, similarly, the transition value $G_{pp'}$ of *any* function $G(p)$ shall depend on $q - q'$ only. The same for the pair E and t of conjugate variables. The proof runs as follows.

If the "transition values" or "matrix elements" $F_{pp'}$ are to depend on $p - p'$ for *any* function $F(q)$, one may take for F the special case of a delta function, $D(q) = \delta(q - q')$ where q' is any fixed q-value. In this case, (14) reduces to the simple product

$$D_{pp'} = \psi(p,q')\,\psi(q',p'). \tag{15}$$

The right-hand side cannot depend on $p - p'$ for any *real* function $\psi(p,q) = \psi(q,p)$. It requires the complex case (9') with the second factor in (15) reading $\psi(q',p') = \psi^*(p',q')$ and with ψ necessarily having the *complex exponential* form

$$\psi(p,q) = a(q)\,\exp[ip\alpha(q)], \tag{16}$$

with *real* $a(q)$. Similarly, dependence on $(q - q')$ of the transition value $G_{qq'}$ of *any* function $G(P)$, including $\delta(p - p')$, requires

$$\psi(p,q) = b(p)\,\exp\,[iq\beta(p)], \tag{16'}$$

with *real* $\beta(p)$; hence
$$a(q) = b(p) = const., \quad a(q) = cq, \quad \beta(p) = cp,$$

finally yielding $\psi(p,q) = const.$ exp. *(ipqc)* or, writing $c = 2\pi/h$,

$$\psi(p,q) = const.\ \exp\,(2i\pi pq/h), \tag{17}$$

where h is an "action constant" of the dimension pq. In a corresponding fashion one obtains, requiring *Galilean invariance* for E and t,

$$\psi(E,t) = const. \ \exp\ (2ipEt/h). \tag{18}$$

(17) and (18) are wavelike functions of wavelength $\lambda = h/p$ and of frequency $v = E/h$, representing "probability amplitude functions" rather than "real" waves.

Immediate consequences of the *wave functions* (17) and (18) are the *commutation rule* (13a) and the *operator rule* (13b).

The *selection rules* for the exchange of E, p, and p_ϕ between bodies periodic in time, space, and angular space (with conservation of the total energy, total linear momentum, and total angular momentum) arise in the following manner:

A body periodic in time with period τ, e.g., a harmonic oscillator of frequency $v = 1/\tau$ or an atom displaying many (optically observed) periods τ_n, has observables A, etc., which depend on t in the periodic form (Fourier analysis)

$$A(t) = \sum_n a_n \ \exp\ (2i\pi t/\tau_n).$$

The transition value from E to E',

$$A_{EE'} = \int \psi(E,t)\ A(t)\ \psi(t,E')\ dt.$$

becomes, with the aid of (18)

$$A_{EE'} = \sum_n a_n \int \exp\ \{2i\pi\ [(Et/h) + (t/\tau_n) - (E't/h]\}\ dt.$$

It has vanishing values except when the sum of the exponents in one of the integrals is *zero*, that is, when

$$(E' - E)/h = 1/\tau_n \qquad \text{or} \qquad \Delta E = h/\tau_n, \tag{19}$$

which is Bohr's *selection rule* for "energy jumps" in connection with the frequencies $v_n = 1/\tau_n$ of atomic systems. Which is primary and which is secondary, the observed energy levels (Franck-Hertz experiment) or the observed frequencies (seen as spectral lines), is a matter of interpretation rather than of the formalism.

In a similar way, bodies periodic in a certain space direction q have observables $B(q)$ of the form

$$B(q) = \sum_n b_n \exp (2ipq/l_n),$$

hence have non-vanishing transition values $B_{pp'}$ only for

$$(p - p')/h = 1/l_n \qquad \text{or} \qquad \Delta p = h/l_n. \tag{20}$$

This is the little known selection rule for the momentum change Δp of space-periodic bodies with periodic (Fourier) components of length l_n. As Duane showed in 1923 (at that time for X-ray photons), it explains the selective angles of diffraction of particles through crystals, and then also through screens with two slits, as due to the *quantized momentum activity of the diffractor via* its periodic space components which stretch from one end to the other, covering both slits *simultaneously*. The Duane quantum rule (20) is apt to replace the old doctrine of "dual picture manifestation" (which disregards the question of how an electron can "know" when it has to switch from the particle picture to the wave picture manifestation) by the unitary quantum *mechanics* as a realistic physical theory. (For details see Ref. 1.)

However, I am dealing here only with the quantum *formalism*, not with problems of interpretation. I hope to have shown that the formal methods of quantum mechanics with all its pecularities is a necessary consequence of a few *non-quantal postulates*, known from classical deterministic mechanics, when these postulates are now imposed on a probabilistic structure of events. There is no mention of any special experiments of a quantal character; such experiments would have to refer to the special constitution of the tested objects, about which *nothing is presupposed in this discussion*. On the other hand, it is always easy to say "I could have told you so beforehand" and to ignore the historical path of discovery based on inductive reasoning.

POSTSCRIPT

The fact that the probabilities interfere by way of a superposition of probability amplitudes ψ can be derived from the postulates (1) that there is to be one general formal interdependence law connecting all probability matrices, and (2) that this general law should have the "ordinary" law $P(A \to C) = \sum_n P(A \to B_n) P(B_n \to C)$ as an average result.

This approach is more physical than the one in the paper above and at the same time mathematically conclusive.

REFERENCE

1. A. Landé, *New Foundations of Quantum Mechanics*, Cambridge University Press, 1965.

DISCUSSION

BONDI: I am delighted by Landé's paper, although I must also admit that it doesn't go quite as far as I hope we will get one day. It always seems to me very desirable to show that if one thinks very deeply through some simple experiment one can find, by reasoning, results that were historically first derived from much more complicated experiments. My particular love in this direction is the Olbers paradox (which was discussed elsewhere), where one can deduce from the observation that it is dark at night a great deal that historically was first discovered with big telescopes and spectroscopes. I always hope that one day we can deduce quantum theory from the mere fact that there are solid bodies. As a first step, it might be easier to deduce quantum theory from the existence of sharp spectral lines.

Landé's paper goes a good way in this direction, but I do hope that this work can be carried further and that eventually we will see quantum theory as a straightforward deduction from elementary observations.

MERCIER: At the beginning Landé said he wanted to explain how it is that the root $\sqrt{-1}$ comes into the theory. Now, there are other chapters of physics in which the complex numbers also enter. In his very interesting introduction to the third part of volume I of *Encyclopédie Française,* which was published at the beginning of the Second World War, Hadamard calls attention to the fact that we could not make airplanes if we did not have *i*, because the theorem of Kutta-Youkowski is demonstrated with the help of the theory of functions, which is the theory of functions of a complex variable. Therefore, there is at least one other chapter of physics, namely, hydro- and aerodynamics, in which these complex numbers play their role.

LANDÉ: The imaginary number $\sqrt{-1}$ enters in the following way (as to details, I refer to my paper). When you ask:

1. Which function $F(x)$ is such that $F(x_1) + F(x_2) = F(x_1 x_2)$? Answer: the logarithmic function, $\log x$.

2. Which function $G(x)$ is such that $G(x_1) \, G(x_2) = G(x_1 + x_2)$? Answer: the exponential function, $G(x) = \exp(x)$.

3. Which function $H(x)$ is such that $H(x_1) \, H(x_2) = H(x_1 - x_2)$? Answer: There is no such function. One must go into the complex domain to obtain $H(x_1 - x_2)$ as the result of a product. Indeed there is a complex function $H(x)$ such that $H(x_1) H^*(x_2) = H(x_1 - x_2)$ where * indicates the complex conjugate. It is the function $H(x) = \exp(ix)$. It is just this difference requirement of mechanics, for $x_1 - x_2$ and $p_1 - p_2$, and for $E_1 - E_2$ and $t_1 - t_2$, which makes for *Galilean invariance* and in the probabilisitic schema leads to periodic, complex "wave functions" which do *not* describe actual waves but probabilities for particle events.

To answer Bondi: My aim was precisely to derive the formalism of quantum mechanics without bumping my head physically into a wall and drawing inductive consequences from the effect, but by showing a blind and deaf-mute man how, by pure reasoning, without experience, he could arrive at the basic calculation rules of quantum mechanics, guided along by the postulates of two-way probability symmetry and Galilean invariance. One could almost say that the twenty-six years of experience and inductive effort which preceded Born, Heisenberg, and Schrödinger could have been spared if people had sat down for ten minutes of deductive thought on the consequences of the assumption of a probability controlled world.

POPPER: Landé quoted Einstein who mentioned intuition. Of course, Einstein certainly did not mean, when he spoke of intuition here, that this intuition is in any sense infallible. Certainly, he meant that intuition can very often be wrong. He himself said again and again how often he was wrong.

Now, what has Landé done here? He has tried to show that quantum theory can be derived not so much from an experiment, as Bondi suggested (perhaps this too may be possible one day) but can be derived from (a) the assumption that we have certain magnitudes which can be measured and whose measurements then are interconnected by probability functions, and from (b) the assumption that the probability functions are *reversible*, have two-way symmetry. That was the main point.

Now, I want to go only one stage back from this point of the explanation. What I have to say is this: The *problems* which quantum theory sets out to solve are probabilistic problems, and it is for this reason that we have to bring probability theory and probabilistic functions into quantum theory. The original problem of Planck was a probabilistic problem. The problem of Bohr turned out to be a probabilistic problem. Especially in the later development of the Bohr theory, when he introduced the correspondence principle, it became very clear that his was the probabilistic problem of explaining the *intensity* of spectral lines, which includes, of course, spectral lines altogether; but the intensity was to be measured by the corresponding probability. If spectral lines are somehow or other explained on the ground of experiments in which photons hit a photographic plate, or something like that, then the intensity of spectral lines depends on the number of photons which hit the particular plate. The intensity thus becomes a frequency number and, therefore, a probabilistic number.

So I would say one could go one little step back beyond where Landé started. The quantum theory was in practically all stages a theory which tried to answer frequency or probability questions. No doubt from this we can derive, as the next stage, that what we have to find is a probabilistic theory with probability functions. Adding reversibility, which we take over from classical physics, one can go, as can be seen in Landé's paper, incredibly far. Landé did not, of course, reach in this one paper all the stages which he can actually reach with his theory; and I also think he doesn't really say that he can actually reach, from this starting point alone, the whole theory. Somewhere or other the magnitude of h has to be obtained experimentally.

YOURGRAU: Most physicists know that Landé is today the most powerful and certainly the most devoted and devout prophet of the monistic and realistic interpretation of quantum mechanics, in the sense that he advocates the particle structure of the universe rather than the wave structure. He doesn't accept the old myth that on Sunday we have waves, on Monday particles, and so on.

Landé says, and I support this viewpoint, that quantum mechanics requires acausality, in contrast to people like Bohm, who try to introduce causality through a trick *via* the back door by simply suggesting that, to put it in the extreme, $\Delta x \Delta p_x < h$.

Landé's derivation, which I think is very impressive, requires the two-way symmetry postulate for the probabilities. If that symmetry postulate is omitted, the whole edifice crumbles.

I am somewhat confused by the mention of the order in the micro-universe, whereas the second law of thermodynamics postulates disorder in the micro-universe.

Finally, certain of Landé's ideas remind me somewhat of Gibbs, of the microcanonical ensemble, and so forth. There would seem to be some sort of conceptual link.

LANDÉ: The probability schema is more general than having applications to thermodynamics alone. As far as order and disorder are concerned, I assumed that the probability matrices are interdependent rather than chaotic. Of course, this may be regarded as a metaphysical requirement, and it does not have a counterpart in classical theory. But it leads to the interference theorem as a consequence, so that this general statistical order is the main point.

TREDER: I think that Landé arrived at the momentous conclusion that quantum mechanics is the necessary consequence of two presuppositions: the probabilistic interpretation of all laws of nature and the postulate of the validity of all symmetry principles in classical kinematics. In addition, I should like to concur that the Gibbs paradox possesses a significance for quantum physics similar to that of the Olbers paradox for cosmology.

LANDÉ: Parts of quantum mechanics can be derived from trying to overcome the Gibbs paradox of entropy discontinuity, even without using the dynamical part of the theory. Therefore, I would regard the solution of the Gibbs paradox as an important aspect of quantum theory.

Another remark about statistical order: From any aprioristic standpoint, it is a somewhat metaphysical assumption. However, the fact that the formal schema with matrices and unitary transformation perfectly fits the statistical events in the atomic sphere without much "extra fitting" is, for me at least, an indication that we have arrived at a very fundamental level of physical analysis. The simple quantal probability formalism cannot be applied to states of health, or states of laziness, or states of wealth, but only to states of atomic systems. For this reason, I believe that we are here very near to a fundamental level, without need for deeper levels.

VIGIER: I am afraid I don't think it is true that one could have derived wave mechanics just by an effort of the mind. If that were true, then over twenty-six years of experiments have been lost. The key question is: What are the qualitative aspects which distinguish quantum mechanics from classical mechanics? Granted, I agree with many of the arguments presented here. I recall a paper by Dirac which presented that argument of reversibility in relation to Heisenberg's matrix formulae.

However, I want to state emphatically that in quantum mechanics you just cannot shut waves out of the picture, for one simple reason: You need the boundary conditions to calculate the actual physical states. You cannot just reduce everything to an abstract formalism; you must put real physical boundary conditions on waves if you want to get the correct figures of actual known experiments, and no pure formalism that exists is going to explain that fundamental and simple fact.

Here is an example. If you want to calculate the Bohr orbits you must impose the condition, proposed by de Broglie, that you have an integral number of wavelengths along the electronic path, and so on. The problem of boundary conditions is a fundamental problem in wave mechanics.

Furthermore, I completely disagree with what Yourgrau said about Bohm's attempts. The real question raised by Bohm is not that one should try to throw out the formulae of quantum mechanics. His idea was that one could bring in sub-quantum fluctuations which would allow one to explain the statistics at the quantum mechanical level.

To continue my main argument: When we consider the Green functions and the Feynman graphs, we are meeting something which is completely alien to all the constructions of classical statistical theory. Now, a step toward an explanation of $pq - qp = h/2i\pi$. That is a fundamental qualitative property of quantum mechanics which has to be explained, and which cannot be brought into the usual frame of classical statistical mechanics. I do not believe one can possibly do it.

For example, there is recent paper by Norbert Wiener, written just before his death (it is in the *Annals* of the Institute Poincaré), where he shows, by modifying the measure in phase space, that one can eventually arrive at behaviors which resemble the statistical behavior of quantum mechanics. But this implies the introduction of qualitatively new ideas into statistical mechanics, if one wants to really explain facts. Thus, I don't believe that one can explain a thing unless one answers that question: What is the exact physical meaning of the boundary conditions which are imposed on the waves which we introduce into this picture? If we don't answer that question, we don't answer any experiments at the "plumber's" level.

LANDÉ: In the statistical interpretation, the boundary conditions arise automatically from the condition that the probabilities add up to unity (normalization to 1) in an infinite domain, and that they are univalent in an angular-periodic domain (around a path). And the ψ-function is a probability amplitude function. In certain special cases it has the form of a wave curve, which you can draw on paper. But it is not a real wave. Real waves have sines and cosines, whereas ψ is an exponential function, very essentially containing the imaginary i. Born clarified this in 1927. In my opinion, the arguments against the statistical interpretation are not convincing, and I hold to Born, at least at this stage of the game.

YOURGRAU: About the boundary conditions, I think Vigier raised a good point. Further, it should again be emphasized that Landé's derivation is a hindsight derivation. After the whole arduous development of quantum theory (in which he took part) one cannot simply announce: "Well, we could have actually done this by some sort of clever probabilistic approach." You can do this today, but it couldn't have been done thirty or forty years ago.

Chapter XIX

SOME PROBLEMS OF THE CONNECTION BETWEEN TECHNICAL DEVELOPMENT AND ECONOMIC HISTORY

György Ránki
Hungarian Academy of Sciences

Crane Brinton has made a sharp distinction between the speculations of prophets like Spengler and Toynbee, and the professional narrowness of scientific academic historians. I admit quite frankly that I feel myself rather closer to the professional narrow-minded academic historians than to the speculators, and I am afraid that when I try to speculate I am stepping on very marshy ground. To demonstrate my point, let me begin with a statement of Raymond Aron which appears in his *Introduction to the Philosophy of History* (p. 75):

> History, according to a classical formula, is the spontaneous memory of societies. The past which is of interest, is first that of the group: historical curiosity seems to be connected with the feeling each individual has of belonging to a whole which transcends him. This is certainly true for primitive forms of history, but in so far as individuals have become conscious of themselves, the historian is no longer limited to exalting by memory, to justifying by legend, or to consecrating by an ideal example, a collectivity or a power. He undertakes his inquiry as he approaches present problems following the variations of social reality and personal judgement.

It is certain and cannot be too emphatically stressed that the natural sciences cannot be arbitrarily transposed into history, as has at times occurred. Every science has its particular characteristics, a status appropriate to its object, and is thus independent of historical narrative. But on the other hand, humanity is bound up with nature, and so historians dealing with humanity have to keep an eye on technological factors which are at work directly or indirectly in society, due to their economic implications.

When we deal with history, we must keep in mind the fact that the modern world is intimately bound up with the profound transformation of the technical means of production. Although a technological conception of history is to be rejected, it must nevertheless be admitted that modern technology was one of the most important conditions which made possible the rise of the capitalist system. However much intellectual and spiritual factors may be emphasized as accounting for the growth of technical development, it is certain that the dynamics of modern economy is inseparable from modern technology.

Influenced by technological development, the economic and social structure of the nineteenth century reveals still more acutely the characteristics which took shape in the industrial revolution. We see here the tendency toward large-scale industry, organization and social regimentation of the workers, the decline in skilled labor, the rise of entrepreneurs who organize a strictly rationalized production to fit the market situation. With the expansion of modern industry there also occurs a transformation in social relationships.

An invention usually affects first the persons using it directly. But after the primary effects come drastic changes, first in the economic practice of production, and then in the economic organization as a whole. Derivative effects impinge on other social institutions such as the family and the government. Economic development consists essentially in the improvement of what Adam Smith called the productive force of labor, to which we may add, with Marx, the improvement of social organization.

Smith mentions three main causes of economic development connected with technology: the increase in dexterity and skill, the development of human specialization, and, most importantly, the invention of machines.

The relation between technique and economy has been studied particularly by Marx, especially in *Das Kapital* III, *The German Ideology,* and *The Poverty of Philosophy.* Technique appears in his works as the motive force and the foundation of the economy.

John Maynard Keynes has also shown in his *General Theory* that technical progress is an indispensable factor in the economy. In particular the importance of technical progress is central to the theory of investment.

Technical progress takes the form, first, of the creation, introduction, and employment of new and more nearly perfect tools and equipment involving a high degree of mechanization and automatization of the production process. Then comes the creation and use of new products and especially of new raw materials. Third, we have the development and introduction of a new productive technology, better use of productive equipment, and the profection of scientific methods of using labor. Last, we see the improvement of the organization of labor and an increase in the educational and technical qualifications of working people. The connection between technical progress and economic and social development is studied thoroughly in Georges Gurvitch's *Industrialisation et Technocratie* which appeared in Paris in 1949.

Economic history is a branch of general history, to be sure, but it has a methodological distinctiveness which is characterized by a marked quantitative interest. Further, it is (or it should be) more connected with technical development. The birth of economic history itself is, then, a result of modern economic technical development. According to the classical view, scientific economic history is dated from 1879, with the publication of Inama Sternegg's work, *Deutsche Wirtschaftsgeschichte*. I believe, however, that the date of its birth may be put earlier. Inevitably, the economic revolution of the eighteenth and nineteenth centuries raised historical problems and stirred the historical imagination. Such revolutionary (agrarian) changes aroused curiosity about the history of agriculture. The beginning of the nineteenth century brings the birth of the cotton industry; until the end of the eighteenth century 'textile industry' means linen and wool but not cotton. With the rise of the cotton industry came in 1835 *The History of Cotton Manufacture* by Baines. After only a few years it was followed by a monograph about another of the significant materials of the industrial revolution: iron. The book was entitled *A Comprehensive History of the Iron Trade*, by D. Scrivenor, published in 1841.

It would not be very difficult to prove how close is the connection between economic development and economic history, and how the topics of economic history have changed as new elements have been introduced into economic and technical life. Economic development has brought into being titanic forces of production, and if these forces are to be kept at work, there must be a particular kind of organization of production. Thus, the topic of economic history has always been two-sided: on the one hand are the new productive forces, and on the other the organization of production. Economic development introduced more than new notions into economic history. Let us take, for example, the notion of heavy industry and the development of the engineering sciences; we may find three absolutely different interpretations of the essential characteristics of heavy industry during the last seventy-five years.

Around the turn of the nineteenth century, metallurgy and the production of railway rolling stock and water craft dominated the writing of the history of technology; after the First World War, with the general use of electricity, mass production of combustion engines and machine tools became the focus of interest; while after the Second World War, in connection with complete automation, it was electrotechniques, the making of precision instruments, the manufacture of synthetic materials and the atomic power industry, which gained prominence.

Many people use the term 'industrial revolution' to refer only to a purely technical revolution. In so doing, they have lost sight of the special significance of that transformation in the structure of industry, and in the social relation of production which was the consequence of technical change at a certain crucial level. Marxists took into consideration not only the technical factors but also the reciprocal relationship between men and the production process. According to them the vital factor was the form of social organization. Similarly, non-Marxist historians — such as W. W. Rostow in *The Stages of Economic Growth* — stress the importance of social organization. Thus changes in technology during the last hundred and fifty years have left their mark on economic history in Europe and America.

The leading sectors during the industrial revolution in a given country are always dependent not only on the historical development and circumstances of that country, but also on the changing flow of technology. The pre-conditions for sustained industrialization generally require radical change in three non-industrial sectors. To build, for instance, social overhead capital means one thing if one deals with the early nineteenth-century technology, another in the late nineteenth, and yet a third in the twentieth century.

The technical revolution of communication and transport was preceded in Great Britain by the industrial revolution; in the majority of the Western European countries the two revolutions occurred at the same time. But in the countries of Eastern Europe the revolution of communication and transport, the building of modern railways and the utilization of water ways preceded the industrial revolution (as Ivan Berend and I have tried to show in a number of monographs).

The period since the First World War witnessed to an even greater extent the effects of technical progress. The specific features to which I am referring are the extension of the use of machines to new fields, which results in a great saving of human and animal power, and the most economical use of already existing machine production. Technical progress has probably never been so universal or so rapid as during the present era. One of the causes of the recent acceleration of technical progress and connected economic development is the dissemination of scientific methods, which are now systematically developed in research institutes for almost every industry.

So we can draw a conclusion. In the first period technical progress was still more empirical than scientific; it depended more on the response to particular and immediate problems of industrial practice than on the autonomous development of scientific knowledge. Technical development was, therefore, likely to take the form of slow modification of details rather than spectacular leaps to a new technique decisively superior to its predecessors from the start, as Charles Singer shows in *The History of Technology*.

In our time of organized scientific research, when the innovations are not regarded in a *laissez-aller* fashion, everything has changed. Research, education, and economy are not organized as entirely separate institutions with different specialized functions: they merge into each other. The distinction between basic and applied research is tending to disappear as economic progress becomes more scientific. Thus, economic evolution cannot be sharply distinguished from research and science.

The effects of modern technology in the political field are also decisive. For instance, modern warfare with the necessary organization by the state of all industrial and agricultural production, of mass propaganda, and the rationing of every necessity of life, brings about comprehensive organization and an enormous concentration of power in the state.

Attitudes toward modern technology vary widely. It is often assumed that modern technology makes human labor monotonous; but that is only one side of the picture.

Technical progress has in fact often reduced alienation from nature and produced alternatives to the monotony of urban life. Decadence, another characteristic feature of our contemporary cultural situation, is indeed influenced in part by technological development, but not wholly so. It is true only insofar as decadence occurs primarily in periods of great but unequally distributed wealth and of inadequate technical development.

DISCUSSION

BRECK: Generally, historians shy away from 'philosophy of history', and are reluctant to reflect over-much on the 'meaning' of events and movements, leaving such work to the philosophers and the poets. The facts themselves lead us, we seem to say. But I wonder whether we should not candidly admit our dependence on the time we live in and on the varying economic and technological forces which sweep us along. I don't want to call them determinative, as I imagine Ránki and I have quite different views of the 'causes' of things. But I do feel that we historians have underemphasized certain physical forces in our teaching and writing of history.

I wonder, then, if Ránki can give us some examples of this relationship between theory and physical reality with which the East Central European historians have been concerned.

RÁNKI: I quite agree with Breck's statement, and would only add that there are close connections between scientific progress, economic development, and economic history. Let's take two innovations: one, the computer—the other, the new Keynesian system. Both have contributed to economic development in the period since the Second World War, the latter with its attention to planning and the central direction of economic life, the former with its rapid calculation and impetus to programming and, again, planning.

Finally, a word about the new economic history. The computer and the new economic conceptions have made necessary a thorough-going transition in our technical methods, rendering the methods of the nineteenth century obsolete. Thus a work based only on partial data, primary indices, or micro-economic analysis is no longer adequate. Consequently, the task of economic historians has changed. Now we have to deal with general problems of input-output, capital accumulation, national income, rate of growth. In brief: Economic historiography has changed with the times from microeconomic analysis to macroeconomic synthesis.

The philosophy of history is an interpretation of the present. It is just as linked to a philosophical concept of existence which recognizes itself, as it is inseparable from the epoch which interprets. But philosophy of history is also inseparable from technique and its social consequences.

POPPER: The greater part of the paper seems to elaborate, though not to formulate, the thesis that economic history, and, therefore, the history of technology, is very important—indeed, so important that without it we cannot understand 'history' (whatever we may mean by this word—say, 'universal history' or the 'history of literature' or the 'history of science'), and least of all 'political history'. Since this thesis is nowhere explicitly formulated in Ránki's paper, I am not sure whether this is what he wants to convey; but if this is not his thesis, I do not know what his thesis is.

Now, assuming this is his thesis, then I should say that it is in a very important sense a thesis which is widely accepted by western historians. There is hardly a university in the West in which economic history is not taught. And the importance of economic history for political history is considered a truism. At the same time, it is only fair to say that Marxism has done a great deal to make this thesis a truism.

But it is recognized as a truism even by opponents of Marxism, such as myself, and especially by the many non-Marxist or anti-Marxist economists who represent economic thinking in the Western World. So far there seems to be agreement. But I disagree with a number of minor remarks which seem to me to reveal a certain Marxist dogmatic bias.

Ránki, in rightly saying that "The effects of modern technology in the political field are also decisive . . . ", says something that is widely accepted. But he fails to say what is equally true, that effects originating in the political field—that is to say, in the field of power politics — *are also decisive upon modern technology.* Like most Marxists, he overstresses the effect of the technological and economic domain on the other domains, and does not say much on the opposite effects. (Most Marxists, if thus challenged, readily admit that these other effects exist; yet they always try to play them down, and they habitually overlook that these effects are just as important, *and often more important,* than the classical Marxist influence of the economic "substructure" on the political and ideological "super-structure".)

Take, for example, atomic energy. It is more than doubtful whether we would have today atomic power plants, had not Hitler challenged the West to develop nuclear weapons. What came here first, and what came here second, is easily decidable historically; power politics led to the economically important developments, and not vice versa. This is a simple historical fact.

It is also a fact that the development of air transport (an economic develop-ment) was helped tremendously by the military developments of the First and Second World Wars.

Such examples may be multiplied. I suppose Ránki will be ready to admit this. But he did not stress the point. Like a good dogmatic Marxist, he merely stressed the opposite point.

As I said in the beginning, the opposite point also occurs in history, and it is often very important: Railways were not, originally, developed in England for strategic purposes. But their development on the continent of Europe and in Russia had, as we know, partly strategic (and thus power-political) aims. This, again, is not mentioned by Ránki.

As a total result, he gives us a wrong picture of history, and a one-sided picture of the way in which the various domains of society interact.

There are three points in Ránki's paper which I wish to mention specifically.

1. He says that the difference between fundamental research and applied science is dwindling. This seems to me quite untrue. I admit that fundamental re-search, although usually undertaken and conceived without any thought of a possible application (Einstein still believed in 1919 his $E = mc^2$ would not lead to the development of any application), afterwards turns out to be applicable. But such a case is an example of the influence of pure science on economics — not the other way around. There are many such examples.

2. Towards the end of his paper, Ránki introduces somewhat abruptly the category of "decadence". This is a term which has become fashionable on the continent of Europe through the influence of Nietzsche, and it has had a great impact on German racialist philosophy. In the Anglo-Saxon scientific world it is looked upon as a mistaken slogan, derived from a misunderstanding of Dar-winism.

3. In his last paragraph, Ránki speaks about "the philosophic concept of existence". This, again, is a concept which has become fashionable through the German racialist philosopher Martin Heidegger. Unfortunately, he has a few followers even in the Anglo-Saxon countries. Fortunately, they are *very* few, apparently fewer than in the Marxist countries.

MERCIER: Let me first maintain that there is no question of science or chapters of science being roughly copies of history or of periods or aspects of history. History is not a science, for it does not know (know how); it only tells (tells what, and nothing but 'what is in the past'). By the way, mathematics is not a science either, since it neither tells (anything) nor knows, but just shows how to tell, at best, and even how to foretell if one knows (e.g. to predict), when applied to science. We may say: it can foretell and therefore it is a power.

Let me then insist upon the fact that—using 'morals' to mean recognition of usages and customs—any attempt to reduce economy and society as moral data to science is due to be a failure, because science searches for the truth about systems, whereas morals look for the good or the benefit of groups and/or individuals, the truth and the good being two irreducible, so-called cardinal values, as has been explained by experts in axiology (e.g. Le Senne).

On these premises, the three causes of economic development indicated by Adam Smith should not be identified with necessary causes in the scientific sense. We certainly agree that the invention of machines may develop economy. But weapons are kinds of machines and when they are used in a war, they do not just develop economy, but destroy more values in state and/or private property than are won by manpower in producing weapons. Hence, the decisions taken as to what machines shall be used for, represent the main part of the causality involved in the so-called economic development. This illustrates the fact that economy is not a science (it differs from ecology), but a chapter of morals—as all names ending with -nomy should be interpreted (except astronomy, for historical reasons)—since 'nomy' means giving the rules apt to foster human relations.

I would then say: not that technical progress is an indispensable factor in economy—though I think that I agree with what Keynes intends to say—but that a parallel improvement of morals and science allows their harmonious *rencontre,* and that their *rencontre* is nothing else but technics. Therefore, technical progress implies progress of both science *and* morals.

Ránki would, I believe, consent to this view, since he says, when commenting upon what he calls "economic history", that its topics are twofold: On the one hand, there is the coming of new productive forces (which certainly implies new scientific discoveries); on the other hand, the organization of production (which clearly implies new insights in the moral rules of communities).

However, science and morals do not necessarily progress simultaneously and in harmony, as is exemplified by successive interpretations of the notion of heavy industry. Ránki uses the phrases "purely technical revolution" or "technical progress". He seems to mean 'purely scientific revolution', which is nothing else but the very rapid scientific (physical, chemical, and possibly other) progress. Then I completely agree with his opinion that the industrial revolution means more than that, for it includes moral (social, economic, and other) aspects. By the way, the rapidity in the development of scientific knowledge, for instance, is not too amazing, for—in rough approximation—increase dx of knowledge during time dt is proportional to a momentary state of knowledge x, i.e. $dx/dt = kx$, hence $x = x_0 e^{kt}$; in other words, the amount of scientific knowledge increases exponentially.

It would, however, be incorrect to take for granted that scientific development (called technical development) is the immediate cause of changes in economy; it is a mediate cause only, since it is borne by people who feel responsible for mankind and who see that without an economic readjustment, the *rencontre* between progressive science and a non-progressive economy would, in the long run, be fatal to the harmony of human life.

But if, on the one hand, it can happen in certain countries that a rapid scientific progress precedes the desired parallel progress in the field of morals – 'morals' meaning economy or even social welfare (the latter does not seem to retain much of Ránki's attention, though it plays an eminent role in the historical development of societies) – it can, on the other hand, happen that in some countries scientific progress is delayed with respect to moral (social, economic or other) changes. This second phenomenon equally produces disjointment, and moral changes thus act as mediate causes of later scientific revolutions. In a way, this has recently happened in the USSR, and possibly (I am not a historian and, therefore, not an appropriate judge) also in France after the Revolution, which may partly be interpreted as a readjustment of the social situation with respect to a considerable scientific advance, but which, so to say, went beyond the re-establishment of harmony in the *rencontre* between morals and science.

Such alternating readjustments having taken place, it is broadly recognized today that a continuous dynamic and mutual adjustment of scientific and moral progress is a condition for a happy life. This is, I feel, what Ránki tries to suggest when he criticizes in his conclusions certain contradictions in what he calls philosophies of modern technology, sceptical attitudes, alienation, and decadence. In spite of the fact that I do not admit that decadence and the like need (although they can) come about in periods of unequally distributed wealth, because there exist counter-examples, I agree with Ránki in stating that a philosophy of history has to consider socio-economic, i.e. moral and scientific, factors in their interactions.

NAESS: Ránki discusses the profound influence of technical development and has offered us some new speculations – as he calls his careful reasoning. Let me comment a little on speculation.

Why does a person feel himself "rather closer to the professional narrow-minded academic historians" than to the speculators, to Toynbee or Spengler? Maybe one reason is that he takes his professional life more seriously as part of himself than his daydreamings and imaginations.

If we do not inhibit our imaginative powers, they will furnish us with vivid conceptions of the historical process as a totality. Frankly, I appreciate such conceptions very highly as part of a personality. When Ránki does not feel close to Spengler, I guess it is not because he belittles the tremendous value of historical total views as part of personal development, but because of the ridiculous pretensions of Spengler – his posing as a scientist and as an oracle.

Technical progress has, as Ránki rightly says, given rise to some measures against our alienation from nature. But it also makes great and wild nature disappear. The mountain Dhaulagiri is still more than 26,000 feet high, but with hotels in its neighborhood, helicopters buzzing around, and walkie-talkies connecting climbers with rescue organizations, its greatness is reduced by, let us say, ninety per cent. The same holds good of polar bears – modern guns and tourist dollars have reduced them to big mice. If we compare the number of people engaged in thoughtless development of technics to the number engaged in trying to assess and control its effects, it is not difficult to understand that irreparable damage is done on an increasingly vast scale. We look forward to a philosophy of history assessing not the influence of "technical development as such", but of technics in conjunction with value-positions and traditions, making it practically possible for our societies to effectively counteract its destructive potentialities.

RÁNKI: There is no disagreement between Sir Karl and myself in regard to the effect of politics upon modern technology. As he points out, most Marxists, if challenged, will admit it; and in this case, I do not want to be an exception.

But let me make a few further remarks in reply to my commentators:

First of all, my topic *today* was economic history and technical development, and it is for this reason that I did not stress any political aspects in general or, more specifically, any Marxist viewpoints. The influence of the *Unterbau* on the ideological *Überbau* and their interaction is a reciprocal relationship — and sometimes this interplay is of considerable significance. I freely admit all this. But if one accepts this point of view, then Popper's insistence upon the strong connection between fundamental research and applied science appears to be of no consequence.

Sir Karl has raised the issue of decadence. Well, as we all know, things do not cease to exist, just because we deny their existence. One need only recall the penetrating remark of Chesterton that everything in fact prolongs its existence by denying that it exists!

I should also like to take issue with Sir Karl's alluding to Heidegger during a discussion of my paper. If Sir Karl calls anyone a follower of Heidegger with as little justification as he has in my case, then, I am afraid, his statistical account of the number of Heidegger's disciples is not very much to be trusted.

Finally, I do not intend to involve our informal discussion in the big question of East and West. But if there has been progress in the scholarly life of the East, its very manifestation would be the fact that generalizations about racialism, such as those made by Popper — a good dogmatic anti-Marxist — are being less and less accepted.

The views of Mercier and Naess differ in some minor aspects from mine; yet in the principal issues we have much in common, and I am very grateful to them for not misinterpreting my ideas.

Chapter XX

IS SCIENCE HUMAN?

Hermann Bondi
University of London

The title of this paper is provoked by the attitude of mind that calls certain subjects "the humanities" and contrasts them with "science." Now there is no quarrel I have with those subjects. They certainly are very human in their character, but this title of "humanities" suggests, by exclusion, as it were, that the sciences are not human. This seems to me to be utterly false. On the contrary, I would argue that science is the most human of all our endeavors. We are nothing if we are not social animals. It is an essential part of all our work that we must be able to communicate and to cooperate.

What hinders us in so many fields is our inborn pride, and inborn megalomania, which make us think that, barring a few unfortunate mistakes, by and large we are infallible. The great thing about science is that it works on the assumption that we are utterly and completely and constantly fallible. This is the very essence of Popper's teaching on the philosophy of science. We have our noses constantly pushed into the dirt by experiments disproving our most beautiful theories. In this way, by showing us all the time that we are wrong, science takes due account of our disabilities. Because we are thus constantly taught human fallibility, we eventually even admit that we ourselves might occasionally make mistakes. This knowledge enables others to work together with us and allows us to accept the common yardstick of experiment and observation for disproof. It is, in my opinion, through this recognition of our disabilities in the very structure and nature of sciences that we are enabled to work together irrespective of our backgrounds, of our religions, of our ideologies, and the like. The common factor in human thinking lies in this recognition that we may be wrong and that we can be shown to be wrong by experiment and observation.

At the same time, this attitude points to the driving force in science. It is always the *disproof* of a theory that gives us the big push forward.

Now, why are theories always being disproved? I think there is no question that it is entirely due to the advances of technology which in turn are based on the advances of science. I believe that at this meeting there has been too little mention of the causative force of developing technology in modern science. Why did we have the burst of activity — revealing to us the electron and the atom and the nucleus — around the turn of the century? Because good vacuum pumps then became available. Why do we have a great burst of activity in molecular biology now? Because we have reliable and safe X-ray apparatuses. It is always through the advancement of technology that we can widen our horizons, that we can increase the range of our experiences, and it is in this way, I submit, that the mechanism of disproof goes further and further. It is at this point that we see certain links between science and other fields.

I was particularly interested in Lakatos' talk, in which he refers in the field of pure mathematics — which is not a science — to the similarly vital function of refutation; this function evidently plays the same part in pure mathematics (and I think we may include logic) that experimental disproof (falsification) does in science. But here I must say that as a scientist I feel a little conceited; indeed, I suffer from a superiority complex. The point is that in pure mathematics and logic we rely for refutation on human imagination inventing these awful monsters about which we heard so much earlier in this meeting. And so, while technology is the driving force in science, human imagination is the driving force (through our ability to concoct refutations) in pure mathematics. Of course, I recognize the enormous role of imagination in science in the formation of theories, and obviously a great deal of imagination has to go into experiment, but we do not *solely* rely on our imagination to produce our disproofs. In fact, when we look at the history and the development of physical science in particular, then we must see that the concepts — to which we have been driven, painfully driven, by experiment — have been forced upon us against all our wishes and desires.

I think the life of the modern physicist was very well described by Lewis Carroll when he made the Red Queen say that she imagined three incredible things every day before breakfast. This is clearly the point. Our imagination is not our strong side. In science we have the enormous help of experiments making us create things we would never have dreamt of otherwise, simply because nothing else will work. And in pure mathematics and logic we have to rely entirely on *inventing* our own monsters and not on finding them in our own backyard when we dig a little deeper through experiments.

It seems to me that all this shows very much how right scholars of all academic disciplines are when they pay such very high respect to originality and imagination. It is a very scarce human ability. It is a great driving force. As it is certainly something in which we are very deficient, creativity has to be encouraged as much as possible.

I come away from this conference very much with the feeling that (because of this similarity of structure) scientists on the one hand, and pure mathematicians and logicians on the other, live in very similar worlds, where beautiful theories and beautiful definitions may be destroyed by miserable counter-examples somebody thinks up or discovers in his laboratory. I am inclined to think, however, that the connection between science and history is of a very different character.

It is interesting, incidentally, that although one might regard both logic and history as belonging to the humanities, science – in the respect in which I am going to discuss it – lies midway between these two subjects.

In experimental disproofs (falsifications) and in refutations in logic and pure mathematics we have indefinitely repeatable situations (events, data, facts). Once somebody has thought up a particular monster which can destroy some theory, any fool can repeat that disproof. It is not quite so simple in experimental science, but again, an experiment that has been performed once can then be repeated with far less difficulty.

In history we are in a field where repetition is difficult. I don't want to stick my neck out further than is strictly necessary, but I think it can be agreed that it is much more difficult. Incidentally, science is such a broad field that we also have domains in our respective disciplines where repetition is not easy, since some scientific fields are historical in character. This is the case, for example, in astronomy. If we ask a question about the origin of the solar system, then this is a strictly historical question. We want to find out what happened and we can only find out what happened from the bits and pieces that have been left over to our day. And this subject is made exceptionally difficult just because we don't know enough about its repeatability. How probable, how likely are the conditions that lead to the formulation of a solar system? How many solar systems are there in our galaxy? If we knew the answer to this question, it would help us in reconstructing this piece of history. Conversely, if we could understand the history of our solar system, we could say a lot more about how often such developments are likely to be repeated elsewhere.

When we look at geology, we again have a subject where repetition is not of the order of the day. We encounter, however, this kind of situation in its extreme form in cosmology. It is, of course, frequently said that we have got just *one* universe. It is interesting that Bertrand Russell, in his *History of Mechanics*, opposes "Mach's Principle" by saying that the universe was repeated as often as one would like to put down initial conditions. However, I think this is a viewpoint with which few will now agree. The universe is essentially unique, and to look for the way it works is, therefore, something at least as difficult as what the historian tries to do. It bears other great similarities to the work of the historian. After all, history isn't just what happened, it is what happened *and* was important, and the decision about what is important is, I believe, one of the most crucial decisions the historian has to make. It is only when he has made his choice of what is important that he can look at the question of causality in history. It depends on where you draw the line as to how far you can follow any causal links.

I think we have very much the same situation in cosmology. It is, again, a question of what it is you regard as significant. We may not be all that many decades away from a satisfactory theory of cosmology, telling us why, if we average over many clusters of galaxies, we get what we find. But if we determine the level of importance rather lower, if we ask, "Why do we have in this region eight clusters full of ellipticals while in another region there are clusters that have only sixty per cent of ellipticals?" — if we ask questions of this kind, then I think we will have to wait a lot longer for the answers. The difficulty of the subject depends on what you regard as important.

CONCLUSION

The Editors

Now that we have come to the end, we ask ourselves whether we have indeed built bridges among the disciplines of logic, physical reality, and history. As far as we are concerned, our horizons have expanded, and we shall return to our diverse academic duties with renewed enthusiasm for philosophic and scientific inquiry, as well as for the difficulty task of communicating our newest and best thinking to others. Our historic sights have been raised, too. The Greeks sent us the Pre-Socratics and the great Triad, even Sextus Empiricus. The German 'delegation' was rather large: Kant, Marx, Dilthey, Weber, Mannheim, Planck, Heisenberg, Born, and Schrödinger. The Danes, of course, presented Bohr. From North Africa, the ghost of Ibn Khaldun descended upon us. We welcomed French thinkers like Descartes, Pascal, Bayle, Simon, and Poincaré. Spinoza was also here, from the Netherlands; and the British Isles generously sent us Locke, Berkeley, Hume, Gibbon, and Rutherford. Two visitors got a somewhat mixed reception: Teilhard de Chardin and C. P. Snow. Let us, however, admit that without the latter's controversial assertions we might never have considered such a symposium at all.

We are bound to confess that we did not succeed in every respect initially intended. We are still far removed from *a synthèse raisonnée*, and often we merely accomplished *catalogues raisonnés*. None the less, we were taught a great lesson: The only permissible, proper intoxication for scholars is stimulating, challenging conversation.

Have we made any, ever so minute, ever so humble, progress? Well, we have noticed the emergence of new questions in our respective fields, where we were sure to have the answers. We also realized how difficult it is to concentrate on issues in the realm of ideas exclusively without digressing into personalities. Without claiming too much, one might say that a *rapprochement*, however vague and tentative, was the positive outcome of our manifold exchanges. The results seem to indicate that our efforts point clearly in this direction.

Being discerning students, we made a momentous discovery: There is always the danger of attempting to build bridges joining our disciplines out of sheer joy of doing something constructive! And thus we are resigned to the fact that there are many pseudo-bridges, quasi-bridges, so-called royal roads that we dare not enter, if we do not wish to sacrifice intellectual integrity and betray the tenets of our respective academic pursuits. Facile compromise and forced agreements can easily turn out to be worse than the conflicting viewpoints one wishes to subsume under a common denominator. Sound inquiry into dissident positions is already, in itself, a worthwhile enterprise. And so we finish without rigid conclusions any definite message.

Voltaire — so legend has it — once remarked that the secret of being tiresome is to say everything. In this respect we have definitely succeeded: we have not said everything. Is there any hope that we shall one day perceive what Baudelaire was able to see solely in Beauty: *L'infini dans le fini*?

The Editors

INDEX

action constant, 305
Acton, Lord, 62
aerodynamics, 307
Ajdukiewicz, K., 173, 181
Alexander, S., 117, 125
Allen, D. C., 214
Alpher, R. A., 207
Ambarzumian, V. A., 106
analysis, semantical, 284–6, 290
antineutrino, 114
Aquinas, St. Thomas, 124
Archimedes, 261
Aristotle, 100, 102–3, 146, 211, 238, 325
Aron, R., 71, 311
Artaud, Antonin, 123
atomic events, 150
atomic statements, 149, 154
atomistic universe, 164
Augustine, St., 39, 116, 124
Austin, J. L., 287–8
automation, 312, 314
axiom of choice, 189
Ayer, A. J., 89, 100

Bacon, Francis, 227, 241
Baines, E., 313
Bar-Hillel, Y., 147, 167
baryon, 107, 110, 112, 196–7
  number, 192
Bayes, T., 153, 159–60, 166
Bayle, Pierre, 209, 215–29, 325
Becquerel, A. H., 199
Bell, L., 109
Bellarmine, Robert, 210
Bentham, Jeremy, 182–3
Berend, I., 314
Bergmann, G., 91, 98
Bergson, Henri, 117, 125, 201
Berkeley, George, 10, 137, 325
Bernays, P., 93, 97–8
Bernstein, J., 109
Bible, 214–30
biconditionals, 284, 292
biology, theories of, 73, 146, 205–6

biosphere, 115–17, 120–2
Birge, R. T., 273–5
Birkhoff, G., 35
Black, M., 25
Blanchard, W., 69
Boethius, 116
Bohm, D. J., 140–1, 308, 310
Bohr, N., 11, 33, 37, 45–6, 132, 141, 143,
    200, 305, 308, 310, 325
  model of, 199
  orbits, 310
  selection rule, 305–6
Boltzmann, L., 31–2, 253, 261
  constant, 207–8
Bolzano, B., 20
Bondi, H., 72, 75, 126, 143–4, 199, 202,
    208, 247–51, 262, 265–76, 307,
    321–4
Bondiday, 273, 275
Born, M., 13, 15, 30, 397, 303–4, 307, 325
  commutation rule, 303, 305
boson, 108
Brandt , R., 71
Breck, A. D., 37, 72–3, 75, 126, 229, 249,
    315–6, 325–6
Brejnev, V. S., 106
Brill, D., 106
Brinton, C., 311
Brodski, A. M., 109–10
Brouwer, D., 19
Brown, N. O., 37
Brunschvicg, L., 56
Buddha, 12
Bugbee, H., 101

c, 203, 207–8
Cabibbo, N., 109
Calvin, John, 210
cardinal number, 287
Carnap, R., 7, 89, 94–5, 98, 123, 147–8,
    154–62, 277, 279–89
  Carnap's p^+, 154–5, 165
    p^*, 155, 158

Carnap (cont'd)
  Frege–Carnap theory of descriptions,
    279, 285–6
Carr, E. H., 235–6, 245
Cartesian tetrad, 113
Cartan, E., 194, 197
Casserley, J. V. L., 102, 115–27
causality, 115–9, 135, 201, 247, 250–1,
    324
  and determinism, 135–6
causation
  historical, 68–70
  in history, 231–51
Chagall, Marc, 246
Charron, P., 223
Chasson, R., 73
Chew, G. F., 52
Chillingworth, W., 212
Chomsky, N., 291
Christology, 125
Christosphere, 118, 121–2, 126
Church, A., 89, 98, 134, 179, 181
Cicero, 60, 225
Cohen, E. R., 273–5
Cohen, R. S., 76, 231–51, 262, 274–6, 295
Collingwood, R. G., 60, 62, 64, 66, 75
communication theory, 147
commutation rules, 31, 33
compenson, 109
complementarity principle, 33
Compton wavelength, 200
Comte, Auguste, 49, 237
connectives, propositional, 149, 150, 184
conservation laws, 192–3
constants
  in logic, 284–6, 292, 295
  in physics, 203–8, 273, 275
constituent, 149–50, 156–60, 163–4,
    168–9, 289
Copenhagen interpretation, 18, 33–5,
    55, 134
Copernicus, 53, 79, 210, 233
cosmological principle, 273
cosmology, 274–5, 323–4
  and history, 253–64
  and microphysics, 105–14
  models in, 79–88, 251
  and philosophy, 127, 253–64
coupling constant, gravitational, 273
CPT invariance, 108
creation, hypotheses of, 275
cybernetics, 56, 142

da Costa, U., 214

Daille, J., 211
Dancoff, S. M., 110
Darwin, C., 4, 63, 124, 317
deBroglie, L., 50, 111, 141, 191, 197,
    200–02, 310
decadence, 317, 320
De Cet, B., 110
decision theory, 169
Dedekind, J. W., 96
definitions, 51–61, 282, 286, 289, 290–5
delta function, 41, 304
Derain, A., 246
de Sales, St. Francis, 210
Descartes, 223, 325
descriptions, Russell's theory of, 277–95
Deser, S., 105
de Sitter
  model, 110
  space, 196
d'Espagnat, B., 138
determinism, 115–6, 135, 201, 244
development, technical, 311–20
Dicke, R. H., 263, 273
Dilthey, W., 325
Dirac, P. A. M., 34, 42, 106, 113, 143,
    196, 202, 204–8, 259, 273, 309
  Dirac–Jordan hypothesis, 106
disproof of a theory, 321–3
DNA, 56, 142–3, 145, 201
dogmatism, 53–4, 145–6, 228
Doppler effect, 256–7, 261
Dray, W., 61, 71
dualism, particle-wave, 138, 297, 306
Duane, W., 16, 306
  quantum rule, 306
Duhem, P., 132, 137, 143
DuMond, J. W. M., 273–5
du Perron, Cardinal, 210
Dürr, H. P., 111

ϵ, 173, 178–181
eclecticism, 141–2
Eddington, A. S., 125, 133, 137, 204
Edel, A., 245
Egils Saga, 66
Egorov, I., 194, 197
Egyged, L., 106
eigenvalues, 258, 298, 302–3
Einstein, A., 15, 41, 46, 49–50, 53–5, 105,
    112, 114, 132, 142–4, 192,
    198–208, 211–2, 253–68, 271–3,
    297–9, 308, 317
  Einstein-space, 194, 196
  gravitational radius, 200

electromagnetic theory, 199–200
elementary particles, 191–202
Eliot, T. S., 246
emergence, 5–10, 243
enlightenment, 229
entropy, 148, 203
Erasmus, 217
ergodics, theory of, 253
evaluation rules, 283–6, 289
evolution, 115–27, 246
existence (*see also* reality), 35, 89–103,
    173–190
existentialism, 102–3, 115–26, 288–9
expansion, in logic, 174–5, 190
experiment, thought, 257, 273
explanation, reductive, 242

falsification, 152, 161
Farrington, B., 241, 245
fascism, 241
Feigl, H., 7
Fermi, E., 111
fermion, 105
Feynman, R. P., 144, 272, 310
field concept, 144
field theory, unified, 105–14, 191
Findlay, F. N., 139
fine-structure constant, 203, 273
Finkelstein, D., 106, 111
Flato, M., 193, 197
Fock, V. A., 113, 141
    and Ivanenko, tetrads of, 113
Ford, K. W., 114
formalism, quantum, 298
formulas, atomic, 285
Fourier, J. B. J., 31, 305–6
Franck–Hertz experiment, 305
freedom, 74, 76, 117–21
free will, 116–7
Frege, G., 91, 100, 141, 277–9, 285–8,
    Frege–Carnap theory of descriptions,
        279, 285–8
    Frege–Strawson theory of descriptions,
        279, 285–6
Freud, Sigmund, 37, 77
Friedmann, A., 106–8
Frolov, B., 106, 109–10
Fubini, G., 194, 197
functor, 177, 179, 180–1
Furth, M., 277

Galilean metric, 106
    invariance, 304–5, 307
Galileo, 146, 204, 237, 267

Gallie, W. B., 66, 71
Gamow, G., 203–8, 260, 263–4, 274–5
Gandhi, M. K., 69–70
Gardiner, P., 71
Gargan, E. T., 71
Gassendi, Pierre, 227
Gell-Mann, M., 110, 192
    Gell-Mann–Okubo formula, 192
    Nishijima–Gell-Mann relation, 192
geology, 65–6, 77, 323
geometry, Riemannian, 44, 51
geosphere, 115, 117, 122
Gibbon, Edward, 219, 229, 325
Gibbs, J. W., 253, 261
    paradox, 309
Gilson, E., 126
Glanvill, J., 223, 227
Gold, T., 107
goldston, 108–9, 114
Goldstone, J., 108, 112–14
Gontery, J., 210
gravitation, 41, 45, 50, 105, 107, 113,
    195, 198, 274
gravitational constant, 106, 204–5, 207
gravitodynamics, 105, 112
gravitino, 114
graviton, 105, 107, 109, 112–14
Greek fathers, 124–5
Grenness, C. E., 141
Grotius, H., 212
Gumplowicz, L., 122
Gurvitch, G., 313

Hadamard, J. S., 307
hadrons, 110
Halévi, J., 213
Halmos, P. R., 93
Hamiltonian, 34, 52, 303
Hardouin, J., 212
Hasan, Y. F., 225–6, 248–9
Hayek, F. von, 238, 245
Hecataios, 68
Hegel, G. W. F., 14, 129, 133, 136–7,
    238, 243
Heidegger, M., 33, 35, 123, 126, 201,
    317, 320
Heisenberg, W., 10, 15–17, 33, 109–12,
    113–5, 125, 133, 137, 141, 207,
    304, 307, 309, 325
    matrix formulas, 309
Hempel, C. G., 168
Henderson, G. P., 92, 98
Herodotus, 68
Hertz, H., 46

Hervet, G., 210, 224
Hesiod, 68
Heyting, A., 19
Hilbert, D., 282
Hilbert space, 43, 51
Hilferding, R., 241, 245
Hill, E. L., 15, 30
Hintikka, J., 87–8, 101, 147–72, 226–7, 230, 286, 288, 292–3, 295
historians, non-Marxist, 314
historical materialism, 311–20
historicism, 210
historiography, 59–60
history
  autonomy of, 59–71
  as a behavioral science, 59–72
  causality in, 324
  causation in, 231–51
  economic, 311–20
  as evolution, 115–27
  experimentation in, 66–9, 72
  and existentialism, 33
  individualism in, 68–72
  laws of, 60–1, 201–2, 227–8, 238–9
  model building in, 79–88
  philosophical, 219
  philosophy of, 315–6, 319
  and physics, 188
  primitive forms of, 311
  providential view of, 209–30
  Sartre on, 201–2
  scepticism and the study of, 209–30
  and science, 1–17, 36–7, 56–8, 76, 79–88, 324
  as science, 59–77, 247–51, 316–20
  of science, 53–4, 316
  technological, 312
  theory of, 199
Hitler, Adolf, 36, 120, 317
Hogarth, C. A., 257
Hoyle, F., 263, 267
Hubble's constant, 106, 108
Hume, David, 53, 209–11, 215, 219, 229, 325
Husserl, E., 33, 35
Huxley, Aldous, 229
Huygens, Christian, 236
hydrodynamics, 307
hypercharge, 109, 192, 195, 202
hyper-photons, 114
hypotheses, law-like, 247–51

Ibn al-Athir, 226
Ibn Khaldun, 226, 325

idealism, 121, 130, 137, 141
imaginary number, in quantum theory, 303–4, 307
immortality, 145
indeterminacy principle, 115–6, 202
indeterminism, 10, 15–17, 31, 33, 36, 116
induction, 161–3, 165–6, 199
  axiom of, 189
information
  measure of, 150–55, 158, 160–3
  semantic, 147–72
  statistical theory of, 147–8
instrumentalism, 53, 222
interaction, electromagnetic, 199–200
interactions
  strong, 191–2, 194, 196, 198
  weak, 191, 196, 276
interdependence law, 301, 306
interference law, 303, 309
intuitionism, 139
invariance group
  Poincaré, 192
  de Sitter, 192
  strong, 192
irreversibility, 253, 261
Ishihara, S., 194, 197
Islamic civilization, philosophy, 225–8
isometry groups of motion, 193, 195–6
Ivanenko, D., 105–14
  and Fock, tetrads of, 113
  Ivanenko–Tamm theory, 113

Jaspers, K., 139, 228
Jeans, Sir James, 125
Jerome, St., 216
Jona-Lasinio, G., 111
Jordan, C., 194, 259
Jung, C. G., 37

Kaila, E., 123
Kant, Immanuel, 45, 53, 198, 325
Kantor, W., 207–8
Kaplan, D., 99, 185–7, 277–95
Kepler, Johannes, 13, 79–88
  laws, 247
Keyes, D., 33, 35, 126, 199, 201–2, 228, 249
Keynes, J. M., 312, 316
Kibble, T. W., 109, 113
Kierkegaard, Sören, 102, 129, 228
K-meson decay, 253
kinematics, classical, 309
knowledge, growth of, and physics, 46–51
Kobayashi, H., 195

Koish, F., 106
Koran, 225–6
Kotarbiński, T., 173
Kraft, J., 35
Kurdgelaidze, D. F., 108, 111
Kutta–Youkowski theorem, 307

Lagrangian, 42, 112–3
Lakatos, I., 9, 53, 130, 220–3, 322
lambda operator, 292
Lambert, D., 286, 288
Lamprecht, K., 68
Landé, A., 10, 16–7, 51–2, 133, 137, 141,
    297–310
language and metalanguage, 26–7
language
    logically imperfect, 285
    logically perfect, 279, 282–95
    as ontic commitment, 94–5
    ordinary, 279, 282, 286, 295
    second-order, 189
La Peyrère, I. (Pereira), 214–5, 228
Laplace, P. S. de, 240
    Laplace–Beltrami equation, 194–6
La Placette, 211–2
law, statistical or probabilistic, 298
Lazerowitz, M., 89, 93, 98
Leibniz, 113
Lejewski, C., 173–90, 293–4
length
    elementary, 207–8
    fundamental, 110, 196, 199, 200, 202
Lenin, V. I., 241
Le Senne, R., 318
lepton, 110, 112
Leśniewski, S., 177
Lessing, T., 246
Lewis, C. I., 141
Lichnerowicz, A., 193–4
Lie algebras, 193, 195
Liebel, H., 64, 71
linguistic analysis, 122–4
    and semantic information, 147–72
Locke, J., 53, 212, 221–2, 325
logic
    Aristotelian, 100, 103
    definitions, 30–1
    first-order predicate, 174
    inductive, 87
    intuitionist, 19–21
    metalanguage in, 26
    modal, 97
    Platonic, 178–81
    propositions, 177
    propositional, 149–50, 152–4, 157–8

logic (cont'd)
    quantification and ontology, 173–90
    realism and relativism in, 17–30
    second-order, 189
    semantic information, 147–72
    stoic, 146
    two truth-values, 177
logical reconstruction of physics, 231–2
London, Jack, 241, 245
long-range forces, 202
Lorentz, H. A.
    invariance, 51, 109, 113
    transformation, 34
    systems, 43
Louis XVI, 248
Lucretius, 240

McNaughton, R., 97
McVittie, G. C., 106
Mach, Ernst, 106, 323
Machiavelli, 214
macrophysics, 44–5
Maldonado, J., 210
Mannheim, K., 233, 245, 325
Marcel, Gabriel, 101
Marshak, R. E., 110–11
Marx, Karl, 77, 129, 233, 237–8, 248,
    312–3, 325
Marxism and economic history, 311–20
Marxists, 314, 317
materialism, dialectical, 130, 138, 141
Mates, B., 134
Maxwell, James Clerk, 17, 271–2
mean values, 302, 304
measurement in physics, 203–08
mechanics
    classical, 298, 303, 306, 309–10
    statistical, 310
Menasseh ben Israel, 213
Mendel, Gregor, 68
Mercier, A., 31, 36, 39–58, 74, 101,
    112–3, 188–90, 198–9, 201, 207,
    223, 260, 274–5, 307, 310,
    317–8
Mersenne, M., 223, 227
Merton, R. K., 233, 245
metahistory, 59–60, 311
metalanguage, 26–30
metaphysics, 122, 125, 127, 137, 139,
    222, 238, 245
metascience, 130
metatheory, 131
methodology, comparative, 250
microcanonical ensemble, 309
microphysics, 44, 105–14, 298

Mieli, A., 245
Mies van der Rohe, L., 246
Mill, John Stuart, 238–9
Millikan, R. A., 42
Milne, E. A., 274–5, 259
Minkowski, H., 113, 255
Mitter, H., 111
models, 140, 145, 159, 163, 168–9, 192,
    202–3, 227, 237, 251, 287
  cosmological, 107
  theoretical, in history, 79–88
Möller, C., 105, 111
momentum activity, quantized, 306
monism, 14, 141
Montaigne, 209, 223, 228
Moréri, L., 217
multiplet, 192, 196
music, framework for, 246
mutual interdependence, law of, 300
mysticism, 115–27
myth, 235

Nadel, G. H., 71
Naess, A., 54–5, 61, 69, 71, 73, 77,
    129–46, 223–5, 319–20
Nambu, Y., 111
names
  improper, 293
  proper, 279, 281–3, 293
nature (see also cosmology and physics),
    laws of, 201–2, 309
Naumov, A. I., 110–11
Ne'eman, Y., 107, 110
  thermodynamic time, arrow of, 107
Nef, J. U., 75, 146, 246, 250
Neumann, J. von, 34–5, 97
Neurath, O., 233–4
neutrino, 107, 109–10, 114
  antineutrino, 114
Newman, E. T., 269–71
Newton, Isaac, 17, 33, 41, 52, 79–88, 198,
    203–4, 221, 237, 264–75
Nicole, P., 211
Nietzsche, Friedrich, 12–13, 102, 125,
    249, 317
Nishijima, K., 192
  Nishijima–Gell-Mann relation, 192
nominalism, 91, 178–9, 183
noosphere, 115–8, 120, 122

observables, 299–300, 303, 305
observation-statement, 162
observer in quantum theory, 15
Ockham, Wm. of, 8, 9, 14, 36–7, 98, 287

Ockham's razor, 7–9, 14–5, 89–98
Okubo, S., 111, 192
  Gell-Mann–Okubo formula, 192
  mass formula for bosons, 195
Olbers paradox, 261–2, 307, 309
ontological commitment, 98–9, 103,
    184–5, 288
ontology, 173–90
operationalism, 130
operators
  modal, 293
  variable-binding, 292
Opticks, 87
order, general postulate of, 303
origin of life, 205
oscillator, harmonic, 305
Osiander, A., 53

paradoxes, 20, 23–4, 28, 39
paraphrasis, 182
Parmenides, 14
particle events, 307
particle 'picture', 297, 306
particulars, 97, 189
Pascal, 52, 145, 325
Paul, St., 124
Pauli, W., 188
Peirce, C. S., 10
Pellison, P., 211
Penrose, R., 269–71
Penzias, A. A., 107
Pereira, I., 214–5
  pre-Adamite theory of, 214
Perring, J. K., 109
"personality", notions of, 59–77
Petiau, J., 111
phase space, 310
philosophy
  and evolution, 125
  of history, 311–20
  hypotheses in, 169–73
  pluralism in, 129–46
  of science, 234
  of science and history, 129–46
photon, 105, 107–10, 114
  hyper-photon, 109, 114
physical systems, historicity of, 253–64
physical theories, historically developed,
    79–87
physics
  and history, 74, 238–44
  philosophy of, 141
  pluralism in, 129–46
  realism and subjectivism in, 15–17

physics (cont'd)
  regression in, 112
  statistical, 208, 253–61
  unification in, 112–14
Pietarinen, J., 159
Pirani, M. S. von, 114
Planck, M., 15–7, 31–2, 132, 134, 207,
    273, 308, 325
  constant, 32, 203, 208
  unit of length, 200
Plato, 14, 133, 173–81, 325
Platonism, 50, 133
Plekhanov, G., 244
pluralism, 5–30, 133–50, 140–1
  of forms, 124
  in history, 133–4, 144–5
  in logic, 139–40
  in mathematics, 139
  in philosophy and physics, 129–46
Poincaré, H., 132, 273, 325
Pomeranchuk, I., 113
Popkin, R. H., 53, 144, 209–30
Popper, K. R., 1–37, 48, 50, 52, 54–9, 71,
    98, 103, 132, 141, 151–4, 168–71,
    189–90, 199–200, 220–22, 238,
    241–5, 308, 312, 316–21
positivism, 86, 143–4
  logical, 127
postulates, non-quantal, 306
pramatter, 110
Pre-Socratics, 325
Principia Mathematica, 278, 281, 283,
    285, 293
probabilism, 53
probabilities, interference law of, 303
probability, 166–69
  amplitude, 303
  amplitude function, 310
  a priori, 154–5, 158, 160, 162–6
  calculus of, 148, 169–70
  conditional, 152
  function, 308
  logical, 150, 156, 166
  matrices, 298, 300, 309
  measure of, 149, 151, 162–70
  prior, 153
  posterior, 153, 158–60, 169
  problems, 147–72
problem shift, 3, 9, 60–62, 274–6
propositions, calculus of, 176
protothetic, 177, 183
Pyrrhonism (*see also* scepticism), 210,
    216–7, 220–9

Q-predicate, 156–9, 161–5
quantification, 90, 93
  existential, 90, 92, 100
  in logic, 98–103
  and ontology, 173–90
  theory, 90–92
  universal, 90, 100
quantifier
  existential, 90, 154, 292
  universal, 156
quantum mechanics, 15, 42–3, 114,
    191–204, 206, 297–310
  completeness of, 34, 130, 200
quantum numbers, 34
  'new', 191–2, 196–7
  'old', 191–2, 197
quantum rules, 297, 304–5
quantum theory, 15–7, 33–4, 42–6, 52–3,
    137, 191–201, 273
  formalism of, 298
quark, 110–12
quasi-theories, 114
Quine, W. V. O., 7–8, 20, 30–1, 35,
    89–103, 134, 139, 173, 178, 181,
    183–7, 189, 275, 287–9, 291–2,
    294–5
  Russell–Quine elimination, 295

Rabi, I. I., 269
Raczka, R., 194
Ranke, L. von, 64, 68
Ránki, G., 220, 246–7, 249–51, 311–20
rationalism, 130–1, 139–40
realism, 55, 133, 135, 139, 143
reality (*see also* existence) 1–37
  physical, 39–58, 133, 135, 137,
    140–1, 143, 145–6
reconstrual, grammatical, 286, 288
reduction, 5–35
Reformers, 210, 215
Reichenbach, H., 18
reification, 95–6
relativity
  general theory of, 34, 41, 44–5, 51–2
    55, 113, 193, 199, 208, 265–76
  special theory of, 31, 41, 55, 113
  theory, 253–64
Renaissance, 146, 209–10, 225–9
research *vs.* technology, 317
resonance, 105, 109, 197
reversibility, 253, 299
revolution
  industrial, 312, 314, 318
  Keynesian, 247

Ricci, lemma of, 259
Richert, H., 60, 68, 71
Riehl, A., 60, 68
Riemannian space, 44, 113, 193–4
Robinson, J. H., 60, 62, 64–5
Rodičev, V. I., 105, 111
Rosenfeld, L., 133, 135–7, 141
Rossi, P., 71
Rostow, W. W., 314
Rousseau, J. J., 244–5
Russell, B., 277–95, 323
    Russell's iota, 292
    Russell–Quine elimination, 295
    theory of descriptions, 90, 289–95
    theory of variables, 90–3, 97
Rutherford, Ernest, Lord, 143, 325
Ryle, G., 89

S-matrix, 51
Sagitov, M. V., 106
Sakata, S., 110–11
Sakharov, A. S., 106
Sakita, B., 110
Sakurai, J. J., 109
Salam, A., 108
Salpeter, E. E., 273
Sanches, F., 227–8
Sarasin, P. and F., 71
Sartre, Jean-Paul, 201–2
scepticism
    in historical analysis, 209–29
    philosophical, 23, 53–4
    and religious problems, 233
sceptics, 53–4
Schaer, J., 112
Schaffhauser, E., 112
Schild, A., 106
Schlick, M., 24–5
Schilpp, P. A., 168
Schönberg, Arnold, 246
scholastics, Latin, 124
Schönfinkel, M., 92–3, 98
Schopenhauer, 12–3
Schrödinger, E., 13–4, 303–4, 307
    operator rule, 303, 305
Schwartz, J., 18, 30
Schwarzschild, M., 106, 272
Schwinger, P., 105, 109
science, philosophy of, 199
sciences
    behavioral, 64–6
    decline of Greek, 237–8
    and explanation, 231–51
    measurement system in, 203–8

sciences (cont'd)
    pluralism in, 129–46
    reduction in, 5–10
    scepticism in, 209–29
scientific knowledge, synthesis of, 201
scope conventions, 293
Scott, K., 286, 292
Scrivenor, D., 313
second order, variables of, 183
selection rule, 305
semantics, 132, 134, 141, 176–7, 180–3,
        186, 189, 292, 294–5
sentences
    atomic, 284, 286
    component, 284
    compound, 280, 284
set theory, 97
Sextus Empiricus, 210, 223–4, 325
short-range forces, 202
Simmel, G., 71
Simon, R., 214–7, 220, 325
Singer, C., 315
Skaldagrimson, E., 66
Skinner, B. F., 61, 71
Smith, A., 312, 318
Snow, C. P., 325
Snyder, H., 106
Sokolov, A. V., 114
Sommerfeld, A., 33, 273
solar system, origin of, 323
solipsism, 221–2
spectral lines, intensity of, 308
Spengler, O., 59, 311, 319
spin, 192
    isobaric, 192, 195
spinor, 105, 107, 110–11
Spinoza, 215–8, 228, 236, 325
Startsev, A. A., 110
state
    macroscopic, 253
    microscopic, 253
statement
    content of, 151
    existential, 156
statistical mechanics, 310
Stebbings, S., 35
Sternegg, I., 313
Stevenson, C. L., 287–8
Stillingfleet, E., 222
Stravinsky, Igor, 246
Strawson, P. F., 139, 143, 278–9, 285–6,
        288
    Frege–Strawson theory of descriptions,
        279, 285–6

structure-descriptions, 155, 157–8
subjectivism, 15
sub-quantum
    fluctuations, 310
    parameters, 197
substitution-sense of quantification, 186,
    189–90
Sumner, W. G., 122
sun's energy, 205–6
symmetry group
    global, 192
    external, 193
    internal, 193
symmetry
    postulates of, 298, 308
    principles, 149, 309
system
    closed, 273
    logically perfect, 291

Tacitus, 212
Tamm, I. Y., 110–11
    Tamm–Ivanenko theory, 114
Tarski, A., 20–8, 30, 36, 93, 183, 283,
    287–8
tautology, 152
technology, 121
    and history, 311–20
    history of, 314
    productive, 313
Teilhard de Chardin, P., 117–22, 125–7,
    142, 249, 325
Teller, E., 205
tempon, 204
Tennessen, H., 59–78, 122–5, 145, 228–9,
    247, 250, 290–1
terms, singular, 294
tetrads of Fock and Ivanenko, 113
tetradic schema, 3
theories
    scientific, 1–3, 12
    statistical, 17
thermodynamics, 44, 47
    second law of, 309
Thiry, Y. R., 194
Thomson, J. J., 42
Thucydides, 68
Tillich, Paul, 228
time, in physical theory, 31, 50, 202
    direction of, 109, 113, 253–60
    measurement of, 274–6
    physical, 303, 305
topology, 207–8
Törnebohm, H., 79–88, 171, 248, 251

Toynbee, A., 59, 63, 311, 319
Traditionists, 226
transformation
    orthogonal, 302–3
    rules, 289, 292
    unitary, 302–3
transition value, 302–6
Treder, H. J., 33–4, 51–2, 73, 76–7, 87,
    105, 112–3, 142–3, 200–4, 208,
    253–64, 272, 309
truth, 94–5, 133, 151, 181, 187–8, 211,
    221, 238
    coherence theory of, 21
    conditions, 279
    correspondence theory of, 33–5
    function, 284
    logical, 152
    problems of, 20–30
    pragmatic theory of, 21
    value, 277–8, 283–4
    "value-gap" theory of, Frege's, 278
Tuchman, B., 249
two-place predicate, 96
two-way symmetry of probabilities,
    postulate of, 299, 308

uncertainty relations (see also
    indeterminacy principle), 125
unification of cosmology and
    microphysics, 105–14
unified theory, 105, 111–2, 114, 191, 274
unit, imaginary, 303
unitary SU(3) symmetry, 110
unitary transformation, 302–3
universals, 183, 189
    and particulars, 89–103
universe
    age of, 204, 206, 208
    expanding, 273
Ussher, J., 214
Utiyama, R., 51, 109, 113

validity in philosophy of science, 132–5,
    141
value, semantical, 283
variables, 284–6
    bound, 293
    hidden, 138
verification, 152
Vermeer, J., 236
Veron, F., 210
Vigier, J. P., 33, 55–8, 74, 76–7, 109, 141,
    145–6, 191–202, 268, 271–3, 309
Vladimirov, Y. S., 106

Vogt, J., 71
Voltaire, 326
Vranceanu, G., 194

Wang, H., 97
wave
  function, 303, 305, 307
  mechanics, 310
  'picture', 297
Weber, J., 114
Weber, M., 325
Wegener's continental-drift theory, 208
Weinberg, S., 108
Weizsäcker, C. F. von, 112, 141
Wells, H. G., 229
Weyl, H., 14–5, 110, 193–5, 259
  group, 193–5
Wheeler, J. A., 106, 111, 144, 272
Whitehead, A. N., 125, 287–8
Wiener, N., 142, 310
Wiener, P. P. 287–8
Wigner, E. P., 198, 201
Wilkins, J., 223, 227

Wilson, T., 106–7
Windelband, W., 64, 67–8, 71
Wittgenstein, L., 25, 30
Wöhler, F., 63, 146
Wright, F. L., 246

Yang, C. N., 111
Yang–Mills equations, 202
Yano, K., 194
Yourgrau, W., 32, 36, 51–2, 57, 73–4, 86,
  100, 103, 113–4, 125, 134, 139,
  171, 184–5, 197–8, 201–2,
  207–8, 229, 246, 250, 261, 263,
  271, 273, 275–6, 289, 308–10,
  325–6
Yukawa, H., 106
  field, 202

Zeldovič, Y. B., 106
Zeno, 211
Zermelo, E., set theory of, 97
Zweig, G., theory of, 110